UNITEXT for Physics

UNITEXT for Physics series publishes textbooks in physics and astronomy, characterized by a didactic style and comprehensiveness. The books are addressed to upper-undergraduate and graduate students, but also to scientists and researchers as important resources for their education, knowledge, and teaching.

Stefano Pierini

Oceanic and Atmospheric Fluid Dynamics

Stefano Pierini
Department of Science and Technology
Parthenope University of Naples
Naples, Italy

ISSN 2198-7882 ISSN 2198-7890 (electronic)
UNITEXT for Physics
ISBN 978-3-031-77993-0 ISBN 978-3-031-77991-6 (eBook)
https://doi.org/10.1007/978-3-031-77991-6

Translation from the Italian language edition: "Fluidodinamica dell'Oceano e dell'Atmosfera" by Stefano Pierini, © The Editor(s) (if applicable) and The Author(s), under exclusive license to Springer Nature Switzerland AG 2025. Published by Springer Nature Switzerland. All Rights Reserved.

The translation was done with the help of an artificial intelligence machine translation tool. A subsequent human revision was done primarily in terms of content.

This Springer imprint is published by the registered company Springer Nature Switzerland AG
The registered company address is: Gewerbestrasse 11, 6330 Cham, Switzerland

To my wife Antonella and my son Francesco

Preface

This text is based on my teaching experience, which consisted, since 1984, in teaching the courses of (i) "Physical Oceanography" and "Fluid Mechanics" at the Naval University Institute (IUN) of Naples, (ii) "Climatology and Meteorology" and "Mathematical Methods of Physics" at the University of L'Aquila and (iii) "Fluid Dynamics", "Meteorological, Oceanographic and Climate Modeling" and "Chaos and Climate" at the Parthenope University of Naples. This introductory text, of upper undergraduate level, aims to provide a basic preparation of fluid dynamics and its applications concerning the dynamics of the ocean and the atmosphere.

The most salient features of this text are the simplicity of the treatment and its compactness. As for simplicity, I have paid much attention to the connections between the various topics, not skimping on simple and intuitive explanations, so that the reader can acquire a unitary vision of the subject. I have often accompanied the necessary mathematical treatment with detailed explanations, so as to make the derivations easier to follow for the reader not particularly inclined to analytical details. I have paid particular attention to the quantitative aspects that connect the mathematical results to real contexts, estimating the main parameters in play and comparing them with the experimental data. Finally, I have taken particular care in preparing the 137 original figures and in choosing the 30 figures by other authors, aware that the graphic part contributes significantly to the understanding of the concepts.

As for the compactness of the text, this is manifested in the limited—although sufficiently broad—choice of topics analyzed, which I considered fundamental. I avoided considering a wider range of topics because this could be misleading for those who wish to obtain a solid preparation regarding the essential aspects of the sector. Nonetheless, I have made reference on several occasions to important topics not covered in detail but, nevertheless, of considerable importance, which I suggest to study in more specialized texts. For some of them (nonlinear waves, intrinsic variability, nonlinear dynamical systems, chaos, baroclinic and barotropic instability, climate tipping points, thermohaline circulation and AMOC), I present an introductory treatment, in some cases also with reference to my research activity (Sects. 11.4, 17.5, 20.5), and an essential bibliography.

This text requires knowledge of calculus and classical physics and is divided into two parts. Part I (Chaps. 1–11) introduces fluid dynamics in inertial reference frames, in which only the active forces come into play. Chaps. 1–8 include the main fundamental concepts of incompressible fluids and may, therefore, be of interest to anyone wishing to acquire a basic preparation in fluid dynamics, whatever the subsequent specialization one intends to follow. The subsequent Chaps. 9–11 deal with surface and internal gravity waves, of great importance in oceanography and meteorology. Part II (Chaps. 12–20) deals with topics of specific oceanic and meteorological interest for which the apparent Coriolis force plays a fundamental role. Barotropic geostrophic dynamics (Chaps. 12 and 13) is considered the starting point for subsequent generalizations (Fig. 13.3), which include baroclinic geostrophic flows and thermal wind (Chap. 14), ageostrophic flows in atmospheric and oceanic boundary layers (Chaps. 15 and 16), and quasigeostrophic flows (Chap. 19). Finally, Chap. 20 provides an introductory discussion of a series of phenomena of great interest, the in-depth analysis of which is deferred to more advanced studies. In addition, five appendices discuss aspects whose inclusion in the main text would have made reading more difficult and discontinuous.

In conclusion, simplicity and compactness make this text particularly suitable for a wide audience of upper undergraduate students. However, the book may also be of interest to M.Sc. and Ph.D. students, as well as to researchers who carry out their scientific activity in the broad context of climate physics and related sectors. Ultimately, this book aims to provide a solid physical-mathematical and conceptual basis of oceanic and atmospheric fluid dynamics that will then allow one to proceed to more in-depth and specific studies of ocean, atmospheric and climate physics.

Naples, Italy
October 2024

Stefano Pierini

Contents

Part I
Fluid Dynamics in Inertial Reference Frames

Chapter 1
Introduction

In this chapter we begin by discussing the nature of fluids and the differences between gases and liquids from a microscopic point of view. The continuum hypothesis is then introduced, which underpins the mathematical treatment of fluid dynamics. Galileo's principle of relativity and the Newtonian laws of dynamics are then recalled. Finally, the division of the text into two parts concerning fluid dynamics in inertial and rotating reference frames is discussed.

1.1 Fluids

Matter presents itself in three states of aggregation: solid, liquid and gaseous. Although some similar characteristics can be identified between the solid and liquid states (such as density, the degree of molecular interaction, etc.), one difference above all others distinguishes solids on one hand and liquids and gases on the other: while the action of forces, even intense ones, can only deform a solid, producing at most small deformations, fractures or breaks, liquids and gases share the characteristic of being a *fluid*, which allows small forces to produce large redistributions of mass within them.

In other words, a solid body has a well-defined shape and each elementary mass that composes it has its own position. A fluid mass, on the other hand, has a shape determined by the solid that contains it (in liquids a liquid–gas separation surface can form) and the elementary masses that compose it do not have their own position, being able to move easily. The purpose of this text is to analyse the physical laws underlying these movements and to introduce a wide variety of phenomena that characterise their dynamics—or *fluid dynamics*—with particular reference to the movements that occur in the oceans and in the atmosphere.

Naturally, liquids and gases present great differences in physical nature, which will now be considered, but these often do not translate into substantial differences

S. Pierini, *Oceanic and Atmospheric Fluid Dynamics*, UNITEXT for Physics,
https://doi.org/10.1007/978-3-031-77991-6_1

from the fluid dynamics point of view, so it is often possible to speak generically of *fluid dynamics* (or-equivalently- fluid mechanics), even though in some contexts it is preferable to use more specific terms, such as *hydrodynamics, aerodynamics, magnetohydrodynamics, gas dynamics*, etc. Considering the applications that will be discussed, in this text we will always refer to water—or seawater—as a liquid and to air as a gas.

The first important difference between water (in the liquid state) and air concerns their *density* $\rho = dm/dv$ defined as the ratio between the mass dm of a fluid element and the volume dv that contains it. At sea level, at a temperature of 4 °C for pure water, we have $\rho_{water} = 1000$ kg m^{-3} while for air $\rho_{air} \cong 1.29$ kg m^{-3} (in Sect. 3. 1 we will see how this density decreases exponentially with altitude). From a fluid dynamics point of view, this difference simply means that the same force will impart a much greater acceleration on a mass of air rather than on one of water.

Another important difference, more relevant in fluid dynamics, concerns the *compressibility coefficient* K, defined as the ratio between the percentage increase in density $\delta\rho/\rho$ and the increase in pressure δ that produces it. Under the same conditions indicated above, K in water is about *five* orders of magnitude smaller than in air, so much so that it is possible to state, albeit in a non rigorous way, that water, but more generally any substance in the liquid state, is *incompressible.* On the contrary, air behaves with excellent approximation as a *perfect gas*, for which its equation of state linking pressure, density and temperature is well known. Another significant difference concerns the *dynamic molecular viscosity coefficient* (see Sect. 2.4), whose value for air is two orders of magnitude smaller than that of water.

Why do water and air show such large differences in density, compressibility and viscosity? A comprehensive answer to this question must be based on the molecular structure of the fluid and on the nature of the interaction forces between the molecules, but this goes well beyond the scope of this text. Here we will resort to a very simplified explanation, but sufficient for the purpose of the present discussion (for further information, reference can be made, for example, to Landau and Lifshitz 1980).

For non-ionised molecules of simple shape capable of establishing a chemical bond, for distances r of the order of 10^{-8} cm an intense attractive force manifests itself that leads to the formation of a new molecule. If the chemical bond is not possible the force is repulsive ($F > 0$), decreasing very rapidly with distance (Fig. 1.1). For distances greater than an *equilibrium* value r_0 a weak attractive force arises ($F < 0$) which tends to zero very quickly. For simple molecules, we have $r_0 \sim 3 - 4 \times 10^{-8}$ cm.

Naturally, the forces at play are of a quantum nature; it can be stated in a non-rigorous but intuitive way that, for distances comparable to the size of the molecules ($r < r_0$) the outermost spatial distribution of electrons (the molecular orbital), of negative charge, produces the electrical repulsion between two molecules. On the other hand, for distances greater than r_0, the *polarisation* that one molecule induces on the other leads to a weak electrical attraction, which tends to zero very quickly with their distance.

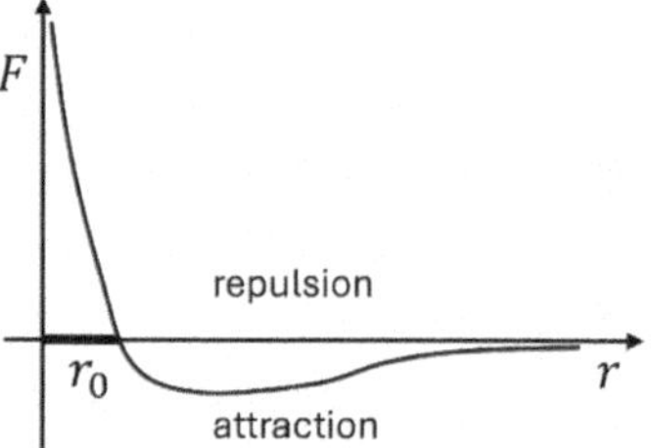

Fig. 1.1 Schematic representation of the interaction force between two molecules unable to form a chemical bond

Now consider the air. The number of molecules contained in a m^3 of perfect gas at 0 °C at sea level is given by the *Loschmidt number* $n_L \cong 2.69 \times 10^{25}$. Imagining the molecules arranged on a regular cubic lattice, each line of 1 *m* in length would contain $n_c = \sqrt[3]{n_L} \cong 3 \times 10^8$ molecules, therefore, the *average distance* between two molecules at sea level turns out to be $\overline{r} = 1/n_c \cong 3.33 \times 10^{-7}$ cm, or, $\overline{r} \sim 10\, r_0$. This implies that in a perfect gas there is no interaction between molecules except very episodically, when a molecule intercepts the cross section of another, producing scattering. It is estimated that, at sea level, between one collision and another, a molecule travels—at a constant speed along a straight line—a distance $\Delta r \cong 7 \times 10^{-6}$ cm (the so-called *mean free path* of the *kinetic theory of gases*); therefore, $\Delta r \sim 200\, r_0$. To these properties, add the obvious one for which the kinetic energy of a molecule is far greater than its potential energy. In light of all this, it is easy to understand why a perfect gas (and therefore air) is compressible.

Conversely, in a liquid, $\overline{r} \sim r_0$; therefore, each molecule strongly interacts with those nearby, remaining at an average distance that is just that allowed by the repulsive intermolecular force. The molecules organise themselves into groups, locally giving rise to partially ordered structures that change continuously. In light of this, it is easy to understand why a liquid is almost incompressible: the impulse transferred to the molecules by an increase in pressure cannot produce a significant variation in the average distance between molecules, and therefore of the fluid volume, due to the repulsive forces that oppose this process. Moreover, the difference of about a factor 10 between the average molecular distance $\overline{r}$ in the air at sea level and that in the water also explains the difference of about a factor 10^3 in the density of the two fluids. Finally, a similar argument can explain the large difference between the value of the dynamic molecular viscosity coefficient of water and that of air.

1.2 The Fluid as a Continuous System

Considering the enormous number of molecules contained even in the smallest conceivable macroscopic volume in practical situations (Sect. 1.1), a mass of fluid can be considered as a *continuous system*, that is, describable by a set of adjacent macroscopic volumes, called *fluid elements*, each of volume *dv* and mass *dm* (a similar description is also adopted for a solid within the theory of elasticity, as used,

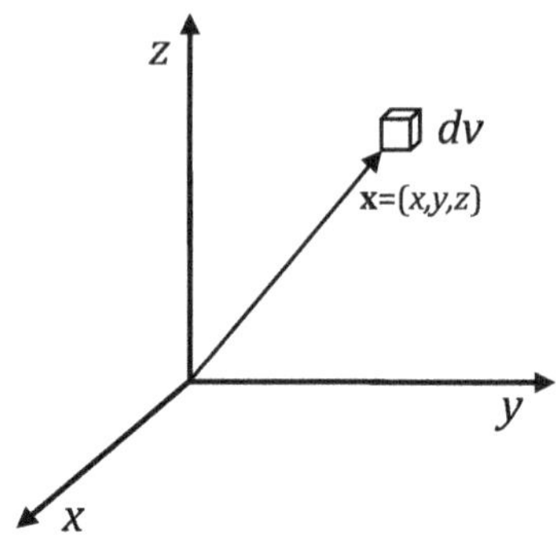

Fig. 1.2 Small but macroscopic element of fluid schematically represented by a cube. The choice of this geometric figure is arbitrary and -like any other shape- can change over time, but it will prove very useful in the treatment. A system of orthogonal Cartesian coordinates identifies the position of the fluid element

for example, in the study of the geophysics of the solid Earth). This is equivalent to introducing the so-called *continuum hypothesis*, according to which the characteristics of the fluid vary continuously from point to point despite the discontinuity of the microscopic structure of matter.

The fluid elements must also have linear dimensions small compared to the length scale on which a noticeable variation of the macroscopic parameters in play is recorded. It will thus be possible to apply differential and integral calculus to them, an essential requirement to be able to formulate the physical laws that govern fluid dynamics with the language of mathematics.

To fix ideas, it is convenient to think of initially cubic volumes (Fig. 1.2); despite this geometric figure being arbitrary and -like any other initial shape- can change over time, this choice will often prove very useful in the treatment.

In fluid dynamics, the state of a fluid is described using the so-called *Eulerian representation*. In a given reference frame (see the next section), a coordinate system is defined (for example, an orthogonal Cartesian coordinate system); at time t, each fluid element is therefore identified by the position vector $\mathbf{x} = (x, y, z)$ (Fig. 1.2). Consequently, every variable related to the fluid element that at time t is located at the point $\mathbf{x}$ will be expressed as a function of the three spatial coordinates and time t. This, for example, applies to the velocity vector $\mathbf{u}(x, t) = (u, v, w)$, which represents the mean velocity of the molecules contained in the volume, applies to the density $\rho(\mathbf{x}, t)$ (already defined) as well as to all thermodynamic variables such as the pressure $p(\mathbf{x}, t)$, the temperature $T(\mathbf{x}, t)$, etc. Indeed, the macroscopic volume dv, no matter how small, fully represents a thermodynamic system.

1.3 Fluid Dynamics in Inertial and Rotating Reference Frames

The fluid dynamics discussed in this text falls within the scope of classical mechanics and must, therefore, comply with the laws of Newtonian dynamics. There are also fluid dynamics applications based on the laws of quantum mechanics (such as the phenomenon of superfluidity) or on the theory of special and general relativity (such as problems of relativistic plasma and, more generally, of high-energy physics and relativistic astrophysics, see Appendix E for further information). These fascinating aspects are beyond the scope of this text.

Specifying the reference frame is obviously an essential prerequisite for correctly formulating the laws of dynamics, both for a system of material points and for a continuous system such as a fluid. The concept of *inertial reference frame* (IF) will now be introduced, discussing its implications in Newtonian dynamics (see Appendix E for its extension to the Einsteinian special theory of relativity).

The main contribution of Galileo Galilei (1564–1642), which marked, perhaps like no other scientific discovery, the birth of modern science, consisted in recognizing that, unlike the Aristotelian conception (in which the effects of friction were ignored), the uniform motion of a body is not produced by a force applied to it but is, rather, the natural dynamic manifestation in the absence of forces, provided that one places oneself in a reference frame called *inertial*. In an IF, the *principle of inertia* holds, according to which a body (which here is represented for convenience as a material point) not subject to forces, remains in its state of rest or of uniform rectilinear motion $\mathbf{u} = const$. On the other hand, if S is an IF, every other frame S' that translates with respect to S in uniform rectilinear motion with velocity $\mathbf{U} = const$ is also inertial. The passage from one IF to another is called *Galilean transformation*, for which, if $\mathbf{U} = (U, 0{,}0)$ one has $x' = x - Ut$; moreover, the law of composition of velocities $u' = u - U$ holds. The law of inertia is also known as Newton's first law.

Newton's second law states that, in an IF, the force $\mathbf{F}$ applied to the point mass m produces an acceleration $\mathbf{a} = \mathbf{F}/m$. It is of fundamental importance to note that this law is invariant with respect to a Galilean transformation ($\mathbf{a}' = \mathbf{a}$). This expresses the *Galilean principle of relativity*, according to which the laws of mechanics manifest themselves in the same way in every IF; in other words, no mechanical experiment can allow one to distinguish one IF from another (in Appendix E we will see how this principle extends to the case of velocities close to that of light).

The Earth is not an IF, in fact, with respect to an IF the Earth rotates around its own axis with angular velocity $|\boldsymbol{\Omega}_e| \cong 7.292 \times 10^{-5}$ rad^{-1}s^{-1} (Foucault's pendulum provided an elegant experimental proof of this in 1851). A reference frame of this type is called *rotating*.

Newton's second law can be extended to non-inertial reference frames as long as the balance of forces includes so-called *apparent* forces that account for the acceleration that the body undergoes in the absence of active forces; for this reason, such forces are proportional to the mass of the body itself. It is therefore clear that in the context of fluid dynamics these forces fall into those *of volume* (Sect. 2.1) like the force of gravity, while differing significantly from them. In applications related to oceanic and atmospheric fluid dynamics, the apparent force to consider is the *Coriolis force*.

In light of these considerations, this text is divided into two parts. Part I introduces the basic concepts of fluid dynamics and covers a series of phenomena (also of direct meteorological and oceanographic interest, such as surface and internal gravity waves) whose spatio-temporal scales are so small that the Coriolis force is completely irrelevant. Part II is instead dedicated to a wide class of fluid dynamic phenomena for which, on the other hand, the Coriolis force plays an essential role. It should be noted that while Part I is completely independent of Part II, many concepts introduced in Part I are preparatory to understanding those covered in Part II.

Bibliography

Landau, L.D., Lifshitz, E.M.: Statistical Physics. Pergamon Press, Oxford (1980)

Further Recommended Reading

Batchelor, G.K.: An Introduction to Fluid Dynamics. Cambridge University Press, Cambridge (1967)
Kundu, P.K., Cohen, I.M., Dowling, D.R.: Fluid Mechanics. Elsevier, Amsterdam (2012)
Landau, L.D., Lifshitz, E.M.: Fluid Mechanics. Pergamon Press, Oxford (1987)
Tokaty, G.A.: A History and Philosophy of Fluid Mechanics. Dover Publications, New York (1971)
White, F.M.: Fluid Mechanics. McGraw-Hill, New York (2011)

Chapter 2
Forces in Fluid Dynamics

In fluid dynamics, there are two types of forces: volume forces and surface forces. The former have a perfect counterpart in the Newtonian dynamics of the material point, while the latter are specific to the fluid as a continuous system. This chapter analyses the two categories of forces.

2.1 Volume Forces, Gravity

The *volume forces* (also known as *long-range forces*) act on the fluid element of volume dv. Let $d\mathcal{F}_v$ be the volume force acting on it. Since, as we will see, the equations of fluid dynamics are expressed in terms of quantities per unit mass (or volume), it is convenient to introduce the *volume forces per unit mass*:

$$\mathbf{F}(\mathbf{x}, t) = \frac{d\mathcal{F}_v}{dm}$$

and *per unit volume* $\rho\mathbf{F}$. In the cases treated in this text, the volume forces are reduced to the (conservative) force of gravity and the (non-conservative) apparent force of Coriolis (Sect. 12.1), both proportional to the mass dm.

The weight to which the small volume is subject is $d\mathcal{F}_g = dm\ \mathbf{g}$. Assuming that x and y are the horizontal coordinates (that is, lying on the geoid) and z is the vertical one oriented upwards (Fig. 1.2), we will have $\mathbf{g} = (0, 0, -g)$ where g is the acceleration due to gravity (at sea level $g \cong 9.81 ms^{-2}$). Therefore, the gravitational force per unit mass will be:

$$\mathbf{F}^{(g)} = \mathbf{g} = -\nabla V_g, \tag{2.1}$$

where $V_g = gz$ is the gravitational potential energy.

S. Pierini, *Oceanic and Atmospheric Fluid Dynamics*, UNITEXT for Physics,
https://doi.org/10.1007/978-3-031-77991-6_2

2.2 Surface Forces: The Pressure

- **Surface forces**

The *surface forces*, also known as *short-range forces* (the *pressure* and the *molecular viscosity*) act on a surface that separates two fluid elements or a fluid element from a solid body.

In the first case (fluid-fluid), in a perfect gas (like air) the force derives from the transfer of momentum associated with the passage of molecules from one side to the other of the surface. In a liquid, in addition to a similar effect—much more intense than in a gas—there is also an important opposing contribution from the intermolecular forces exerted across the surface, which balances *almost* exactly the transfer of momentum. In the second case (fluid-solid body), in a gas the force derives from the transfer of impulse associated with the *collisions* of the fluid molecules with the solid body; in a liquid, molecular interactions between liquid and solid also come into play.

The surface force $d\mathcal{F}_s$ applied to an elementary surface of area ds is proportional to the area (as is natural if one considers the physical origin of the force) and depends on the orientation of ds, which is identified by the unit vector $\mathbf{n}$ normal to the surface itself (Fig. 2.1) Hence we can write:

$$d\mathcal{F}_s = \mathbf{S}(\mathbf{x}, t, \mathbf{n})ds.$$

The vector $\mathbf{S}$, referred to as *stress*, expresses the surface force *per unit area* (Fig. 2.1). By convention, $\mathbf{S}$ is the stress exerted by the fluid towards which $\mathbf{n}$ points on the fluid located on the opposite side.

Based on this convention, an obvious consequence of the third principle of Newtonian dynamics is the following relationship:

$$\mathbf{S}(\mathbf{x}, t, -\mathbf{n}) = -\mathbf{S}(\mathbf{x}, t, \mathbf{n}). \tag{2.2}$$

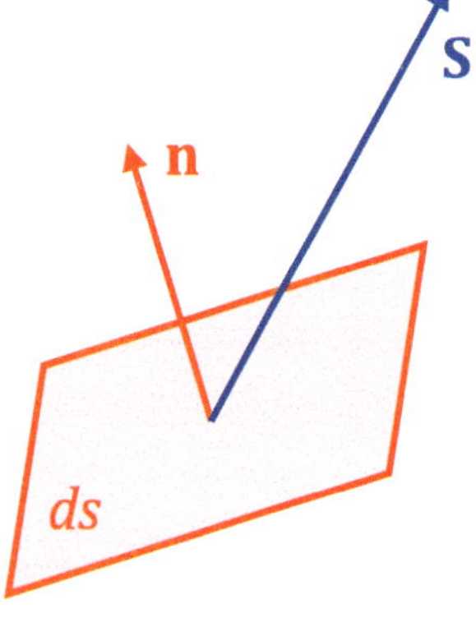

Fig. 2.1 The stress **S** exerted on an elementary surface of area *ds* perpendicular to the unit vector **n**

In Appendix B (Eq. B.4) it is demonstrated that the stress can be expressed as follows,

$$S_i(\mathbf{x}, t, \mathbf{n}) = \sum_{j=1}^{3} \sigma_{ij}(\mathbf{x}, t) n_j \tag{2.3}$$

where σ is called *stress tensor*. Essentially, $\mathbf{S}$ depends linearly on the components of $\mathbf{n}$ and the coefficients of this linear combination are given by the matrix elements σ_{ij}.

From Eq. (2.3) it can be deduced that the generic σ_{ij} represents the i-th component of the stress applied to a surface normal to the direction x_j. The three diagonal elements (σ_{11}, σ_{22} and σ_{33}) are called *normal stresses*, as each of them gives the normal component of the stress applied to a surface lying on one of the three coordinate planes. For example, in Fig. 2.2 the normal stress σ_{33} represents the component along z of the stress applied to a surface lying on the x-y plane. The six non-diagonal components of σ are instead called *shear stresses*.

It is important to note that the tensorial character of σ ensures the necessary invariance of Eq. (2.3) under rotation of the coordinate axes. Let $\mathbf{A}$ be a *unitary transformation* ($\mathbf{A}\mathbf{A}^T = \mathbf{I}$, where $\mathbf{I}$ is the unit matrix) that rotates the orthogonal Cartesian axes by passing from the coordinates $\mathbf{x}$ to the new coordinates $\mathbf{x}'$. Under such rotation, a scalar quantity (such as, for example, the density ρ, the temperature T, etc. and, as we will see, the pressure) remains unchanged, while the components of the generic vector $\mathbf{V}$ transform as $\mathbf{V}' = \mathbf{A}\mathbf{V}$. Similarly, the elements of a tensor (such as for example σ, but we will encounter other tensors in the subsequent discussion) transform as $\sigma' = \mathbf{A}\sigma\mathbf{A}^T$. It is immediate to verify that this ensures the formal invariance of Eq. (2.3) under rotation of the axes.

In general, the nine matrix elements σ_{ij} represent unknowns of the fluid dynamics problem. However, based on considerations regarding the moment of the forces acting on a control volume, it can be demonstrated (Appendix B, Eq. B6) that the following fundamental relation holds:

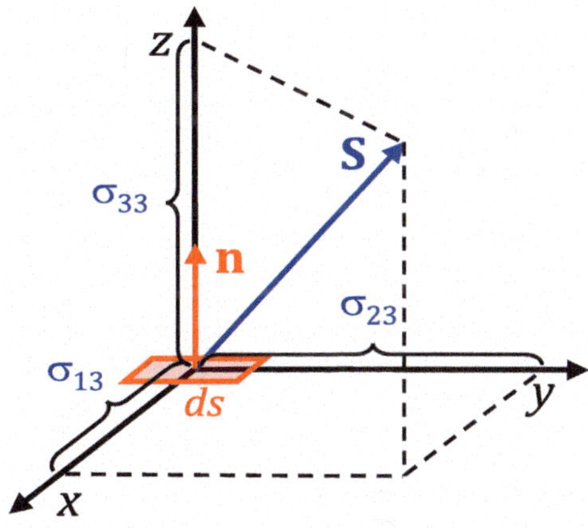

Fig. 2.2 Example of stress for $\mathbf{n} = (0, 0, 1)$

$$\sigma_{ij} = \sigma_{ji}. \tag{2.4}$$

Therefore, σ is a symmetric tensor. Two important properties follow from this. The first is that the independent unknown elements are not 9 but only 6. Furthermore, since every symmetric matrix is diagonalisable under an appropriate unitary transformation, at every point and at any time it will be possible to rotate the axes so that $\boldsymbol{\sigma}'$ locally assumes a diagonal form; those are called the *principal axes* of the tensor.

We have already mentioned that surface forces include pressure forces and viscous forces; the rest of this paragraph will consider pressure, both in a fluid at rest and in a moving fluid, while viscosity will be dealt with in Sect. 2.4.

- **Pressure**

We start by considering a fluid at rest; in this case the stress is always perpendicular to the surface on which it is applied, and is called *static pressure*. In accordance with this general property, the *Pascal's principle* (Blaise Pascal, 1623–1662) states that a change in pressure applied to a confined incompressible fluid is transmitted to all its points with equal intensity and always perpendicular to any surface. In particular, it is immediate to verify experimentally that the outflow of fluid from orifices made on the wall of a container following an increase in pressure occurs everywhere perpendicular to the wall itself. This principle underlies a series of hydraulic devices (siphon, press and hydraulic brake, jack etc.) among which the hydraulic press, which allows to lift a large weight by applying a small force on a limited portion of fluid.

The character of the stress in a fluid at rest is subject to a theoretical deduction based on the symmetry of σ, and therefore, on its diagonalisability, as discussed in Appendix B. In such a case the stress (here synonymous with static pressure) takes the form (ref. Eq. B10–11)

$$\mathbf{S}_p = -p\mathbf{n}, \tag{2.5}$$

(where $p > 0$ is a scalar) to which corresponds the tensor

$$\sigma_{ij}^{(p)} = -p\delta_{ij}, \tag{2.6}$$

where the symbol *Kronecker delta* δ_{ij} ($\delta_{ij} = 1$ if $i = j$; $\delta_{ij} = 0$ if $i \neq j$) represents the matrix elements of $\mathbf{I}$ (in compact formalism, $\sigma^{(p)} = -p\mathbf{I}$). Figure 2.3 shows the pressure applied to the surface ds with normal $\mathbf{n}$. Equation (2.5) shows that the static pressure represents an *isotropic stress*, from which it follows that a small element of fluid of spherical shape is subject to uniform compression (see Fig. 2.2), compatible with the absence of motion.

It is clear that in a fluid at rest all coordinate systems are principal axes: in fact, $\sigma^{(p)\prime} = -p\mathbf{AIA}^T = -p\mathbf{I} = \sigma^{(p)}$. Furthermore, the scalar character of p derives from the relation

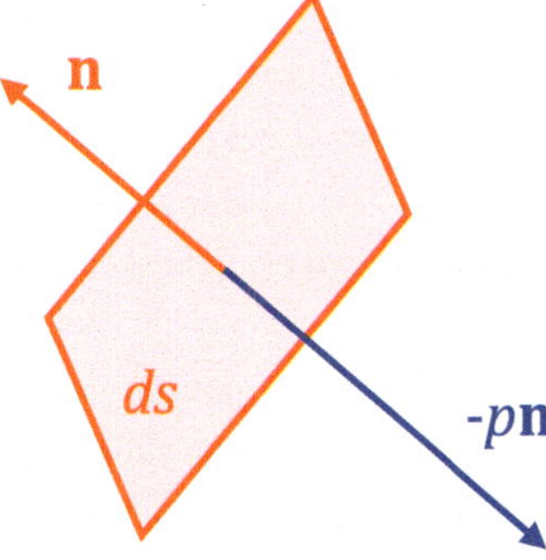

Fig. 2.3 The pressure exerted on an elementary surface

$$p = -\frac{1}{3} tr[\sigma] \tag{2.7}$$

implied by Eq. (2.6); in fact, the trace of a tensor is invariant under rotation of the coordinate axes. The discussion on pressure in a static fluid concludes by observing that, in the absence of motion in the fluid, it is entirely natural to assume that the molecular dynamics and any interaction between molecules (in the case of a liquid) across a generic surface *ds* are isotropic and that, therefore, the corresponding surface force is perpendicular to *ds* regardless of its orientation.

In the case of a moving fluid, the tangential stresses are not null and the normal component of the stress acting on a surface will depend on the orientation of the same. It is however convenient to introduce a mechanical definition of pressure that shares with the static pressure the scalar and isotropic character of the variable and gives a measure of the local uniform compression of the generic fluid element. This is achieved by introducing the *pressure in a moving fluid* using the same formal definition of pressure (Eq. 2.7) valid in the static case (Appendix B).

In a moving fluid, the stress tensor can therefore be written as

$$\sigma_{ij} = -p\delta_{ij} + d_{ij} \tag{2.8}$$

(in compact formalism, $\sigma = -p\mathbf{I} + \mathbf{d}$), where the first term represents, as seen, an isotropic stress, while $\mathbf{d}$ represents the *anisotropic contribution to the stress tensor*, including both tangential stress and normal stress with a null sum (since $tr[\mathbf{d}] = 0$ by the very definition of p). In Sect. 2.4 it will be shown how $\mathbf{d}$ can, for a wide class of fluids, be expressed through an appropriate combination of velocity gradients.

In conclusion, it is worth remembering that in the *international system of units* (SI) pressure is measured in Newton m^{-2} and in the *cgs* system in dyn cm^{-2}. 1 Newton m^{-2} is called *Pascal* (Pa); 1 Pa $= 10\,\mathrm{dyn\,cm}^{-2}$. A unit of measure widely used in meteorological and oceanographic applications is the bar: 1 bar $= 10^5$ Pa. Its usefulness lies in the fact that the mean *hydrostatic pressure* of the atmosphere at sea level (1 Atm = 760 mm Hg, first determined by Evangelista Torricelli in 1644, see Fig. 3.4) is almost exactly equal to 1 bar: 1 Atm = 1.01325 bar. In meteorology, it is convenient to use the *mbar*: 1 mbar = 100 Pa = 1 hectoPa (1 hPa). In oceanography, the *dbar* is widely used (see Sect. 3.1).

2.3 Pressure Gradient Force

The governing equations of fluid dynamics that need to be determined will refer to the generic small volume dv. Therefore, regarding surface forces, it is necessary to derive the *volume resultant* of these, that is, the vector sum of all the elementary surface forces applied to the facets that delimit the small volume dv itself: in fact, this is the force that imparts an acceleration to the fluid element and that will therefore appear in the dynamic equations.

We will now move on to deriving the resultant of the pressure forces acting on the volume, known as *pressure gradient force*, following two different procedures. First, an elementary method will be used that requires only knowledge of the concept of derivative and allows for an intuitive understanding of the essence of the pressure gradient force. Subsequently, a less intuitive but very powerful and general mathematical procedure typical of field theories (such as electromagnetism, where it is adopted to obtain Maxwell's equations from the laws in integral form; see Appendix A, Application 2) will be used, as will be done several other times in the rest of the text.

The first procedure refers to an elementary cube aligned with the coordinate axes subject to a pressure field that varies only along the x-direction (Fig. 2.4).

The pressure forces acting on the two faces perpendicular to y cancel out and the same applies to the two facets perpendicular to z. Using Eq. (2.5), the resultant $d\mathcal{F}_1^{(gp)}$ of the pressure forces acting on the two facets perpendicular to x turns out to be:

$$
\begin{aligned}
d\mathcal{F}_1^{(gp)} &= \left[S_{p1}(x) + S_{p1}(x+dx)\right]ds \\
&= \left[-p(x) \bullet n_1(x) - p(x+dx) \bullet n_1(x+dx)\right]ds =
\end{aligned}
$$

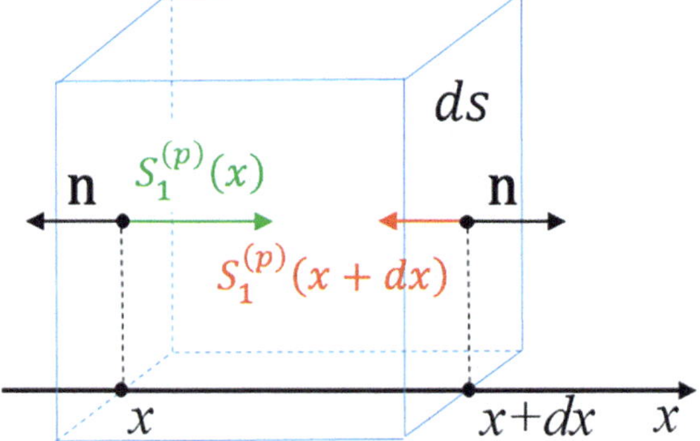

Fig. 2.4 Elementary volume subject to a pressure field that varies along the x-direction

$$= \big[p(x) - p(x+dx)\big]ds = \left[p(x) - p(x) - \frac{\partial p}{\partial x}dx\right]ds$$

$$= -\frac{\partial p}{\partial x}dv = -\frac{1}{\rho}\frac{\partial p}{\partial x}dm.$$

Therefore, the pressure gradient force per unit mass is given by:

$$F_1^{(gp)} = \frac{d\mathcal{F}_1^{(gp)}}{dm} = -\frac{1}{\rho}\frac{\partial p}{\partial x}.$$

If the pressure varies in every direction, the following general expression for the pressure gradient force per unit mass will be obtained:

$$\mathbf{F}^{(gp)} = -\frac{1}{\rho}\nabla p, \tag{2.9}$$

or per unit volume $\rho\mathbf{F}^{(gp)} = -\nabla p$ (the linear differential operator *nabla* is defined as $\nabla = (\partial/\partial x, \partial/\partial y, \partial/\partial z)$ and, if applied to a scalar φ, $\nabla\varphi$, provides the vector *gradient of* φ, also indicated as *grad* φ). $\mathbf{F}^{(gp)}$ originates from surface forces but has the nature of a volume force and as such, it will be included in the balance of forces applied to the generic elementary volume.

The second procedure refers to a finite volume V independent of time (which from now on will be called *control volume*) bounded by the closed surface ∂V (Fig. 2.5).

This procedure is described here in detail also in view of subsequent applications, for which we will proceed in a more synthetic way. Let $\rho\mathbf{F}^{(gp)}$ be the pressure gradient force per unit volume, now again assumed as unknown; the resultant of all the pressure forces applied to V will therefore be given by:

$$\mathcal{F}^{(gp)} = \iiint\limits_V \rho\mathbf{F}^{(gp)}dv. \tag{2.10}$$

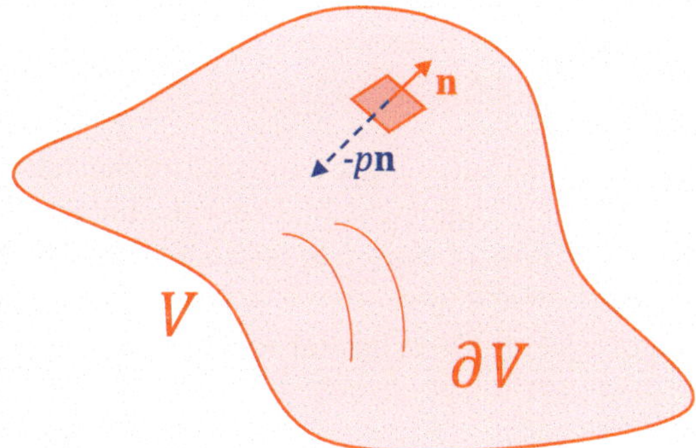

Fig. 2.5 Generic control volume V

The same quantity must be equal to the total pressure force exerted at the boundary ∂V by the surrounding fluid. This surface integral can, in turn, be transformed into a volume integral by resorting to Gauss's theorem (Appendix A, Eq. A1):

$$\mathcal{F}^{(gp)} = -\oiint_{\partial V} p\mathbf{n}ds = -\iiint_V \nabla p dv. \tag{2.11}$$

By equating the two volume integrals in Eqs. (2.10–2.11), we obtain:

$$\iiint_V \left[\rho \mathbf{F}^{(gp)} + \nabla p\right] dv = 0.$$

Generally, the vanishing of an integral does not imply that the integrand function is null at every point of the integration domain. However, in this case, the integral must be null for any choice of V: this implies that the integrand function must be identically null, from which follows the expression $\mathbf{F}^{(gp)} = -\nabla p/\rho$ of Eq. (2.9) already obtained by the first procedure.

2.4 Surface Forces: Viscosity in Newtonian Fluids

We will now see how the anisotropic component $\mathbf{d}$ of the stress tensor that appears in Eq. (2.8) can be expressed in terms of an appropriate combination of the gradients of $\mathbf{u}$ for a wide class of fluids, and certainly, for the air and water of our interest.

Imagine conducting a laboratory experiment with the setup indicated in Fig. 2.6. A layer of liquid (for example water) of thickness h contained in a rectangular channel is subject to the horizontal traction of a solid slab in contact with the surface of the fluid, forced to move in purely horizontal motion by an appropriate mechanical device. The speed of the fluid layer in contact with the slab will move at its same speed; the same will happen for the fluid in contact with the bottom, which is at rest (in Sect. 5.4 it will be seen that this is the natural condition of the fluid in contact with a solid wall). Inside, the motion is transferred, from one layer of fluid to the one below, due to the shear stresses associated with molecular viscosity.

Let d_{13} be the shear stress with a single component along $x_1 = x$, applied to the upper solid surface perpendicular to $x_3 = z$. The result of such an experiment shows that, after a transient phase during which the fluid is accelerated, a *steady flow* (i.e., independent of time) is established with a linear profile of horizontal velocity (known as Couette flow, cf. Sect. 5.9) linked to the stress by the following formula valid for incompressible fluids:

$$d_{13} = \mu \frac{\partial u_1}{\partial x_3}, \tag{2.12}$$

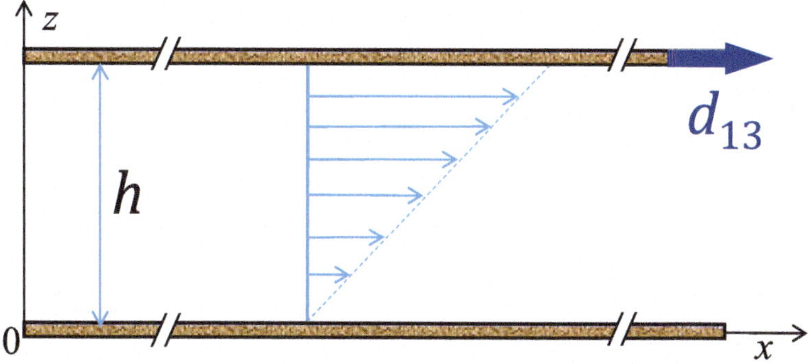

Fig. 2.6 Schematic illustration of the laboratory experiment revealing the constitutive relationship for Newtonian fluids

where μ is a parameter that depends on the fluid (and, weakly, on p and T) known as *dynamic molecular viscosity coefficient*. As an example, its values for air and water in standard situations are reported:

$$\mu_{air} \cong 1.8 \times 10^{-5} \ \mathrm{N\,s\,m^{-2}}, \ \mu_{water} \cong 1.0 \times 10^{-3} \ \mathrm{N\,s\,m^{-2}}.$$

This experimental relationship obtained by Isaac Newton is valid for a wide class of fluids called *Newtonian fluids* (the science that deals with fluids that do not respect this law is called *rheology*).

One might be tempted to generalise Eq. (2.12) for a generic pair of indices i, j as follows:

$$d_{ij} = \mu \frac{\partial u_i}{\partial x_j},$$

however, this tensor is not symmetric and this contradicts one of the fundamental properties of the stress tensor. The correct relationship, called *constitutive relation for a Newtonian fluid* is (for incompressible fluids) the following:

$$d_{ij} = \mu \left(\frac{\partial u_i}{\partial x_j} + \frac{\partial u_j}{\partial x_i} \right). \tag{2.13}$$

The tensor in brackets is obviously symmetric and is perfectly compatible with the experimental result described above (in fact $\partial u_3/\partial x_1 = 0$). This relationship can be written in the form

$$d_{ij} = 2\mu e_{ij} \tag{2.14}$$

where

$$e_{ij} = \frac{1}{2}\left(\frac{\partial u_i}{\partial x_j} + \frac{\partial u_j}{\partial x_i}\right) \tag{2.15}$$

is called the *rate-of-strain* tensor as it appears in the description of the deformation of a fluid element (see Appendix C for a detailed discussion of this tensor in relation to its role in the deformation of a fluid element without volume change). It has already been emphasised that Eqs. (2.13–2.14) strictly apply to incompressible Newtonian fluids; the effects of compressibility involve an additional term that is however omitted here, as it is beyond the scope of the current discussion.

Just as was done for pressure forces, for viscous forces we will need to determine the respective volume resultant applied to the generic small volume dv: for this derivation, refer to Sect. 5.3.

Recommended Readings

Batchelor, G.K.: An Introduction to Fluid Dynamics. Cambridge University Press, Cambridge (1967)

Çengel, Y.A., Cimbala, J.M.: Fluid Mechanics, Fundamentals and Applications. McGraw-Hill, Boston (2006)

Kundu, P.K., Cohen, I.M., Dowling, D.R.: Fluid Mechanics. Elsevier, Amsterdam (2012)

Landau, L.D., Lifshitz, E.M.: Fluid Mechanics. Pergamon Press, Oxford (1987)

Tritton, D.J.: Physical Fluid Dynamics. Van Nostrand Reinhold, New York (1977)

White, F.M.: Fluid Mechanics. McGraw-Hill, New York (2011)

Chapter 3
Elements of Fluid Statics

This chapter analyses some fundamental aspects of static fluids. Hydrostatic pressure is studied along with some of its important properties. Finally, Archimedean thrust is analysed as applied not only to a solid body immersed in a fluid, but also to a mass of fluid with a density that is different from that of the environment.

3.1 Mechanical Equilibrium, Hydrostatic Pressure

A fluid subjected to the force of gravity can be in equilibrium (that is, in a state of absence of motion) only if the balance of forces vanishes. Therefore, by cancelling the sum of the force of gravity (Eq. 2.1) and the pressure gradient force (Eq. 2.9) we have:

$$\mathbf{F}^{(g)} + \mathbf{F}^{(gp)} = \mathbf{g} - \frac{1}{\rho}\nabla \mathrm{p} = 0.$$

Remembering that $\mathbf{g} = (0, 0, -g)$, we obtain:

$$\frac{\partial p}{\partial z} = -g\rho. \tag{3.1}$$

This equation expresses the so-called *hydrostatic balance* and the $p(z)$ that satisfies it is called *hydrostatic pressure*. This law explains the variation of pressure with respect to z in the atmosphere and in the ocean.

As already observed in Sect. 1.1, in the atmosphere the equation of state of perfect gas applies with great accuracy:

$$p = \rho R_d T_v, \tag{3.2}$$

S. Pierini, *Oceanic and Atmospheric Fluid Dynamics*, UNITEXT for Physics,
https://doi.org/10.1007/978-3-031-77991-6_3

where $R_d = R^*/M_d$ is the gas constant for dry air ($R^* = 8.314$ J K^{-1}mol^{-1} is the *universal gas constant* and $M_d = 0.029$ Kg mol^{-1} is the molar mass of dry air) and T_v is the virtual temperature in degrees Kelvin, which includes the contribution due to the presence of water vapour. Substituting T_v with T (their difference is negligible in this context) and combining the hydrostatic balance (Eq. 3.1) with the equation of state (Eq. 3.2) we get:

$$\frac{\partial p}{\partial z} = -\frac{p}{H} \tag{3.3}$$

where the scale height H is given by:

$$H = \frac{R^* T}{g M_d}. \tag{3.4}$$

From the average temperature profile in the troposphere and stratosphere (Fig. 3.1) an average value $\overline{T} \cong 255$ K is estimated, which can be adopted for an ideal approximately isothermal atmosphere (the variation of T around $\overline{T}$ is not negligible but is still only about ~ 30%, so the isothermal approximation is acceptable in this approximate calculation).

Under this condition, from Eq. (3.4) we get $H \cong 7.5$ km. With this constant H the hydrostatic balance can be immediately integrated, providing an exponential decrease of p with altitude:

$$p(z) = p_0 e^{-\frac{z}{H}} \tag{3.5}$$

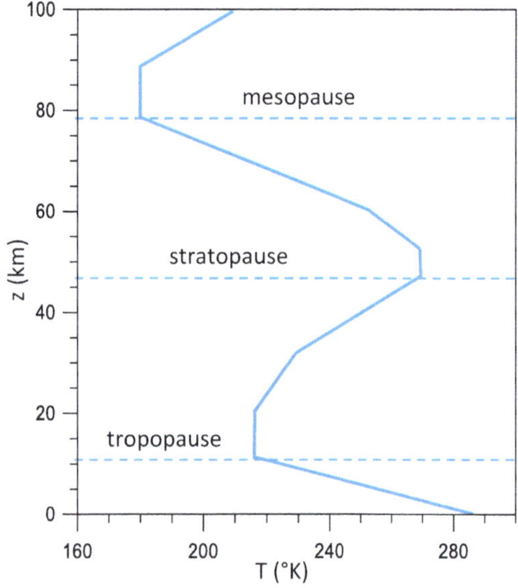

Fig. 3.1 Average T profile in the atmosphere

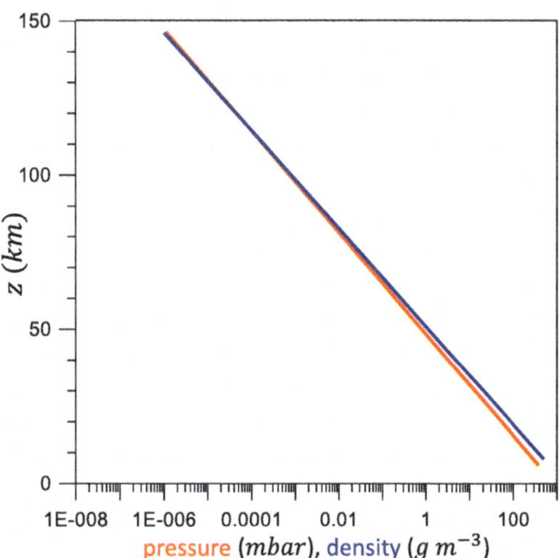

Fig. 3.2 Average profiles of p and ρ in the atmosphere obtained from experimental data

where p_0 is the pressure at sea level. From the equation of state for an isothermal atmosphere it then results that $\rho \propto p$, so the same dependence on altitude will also apply to density:

$$\rho(z) = \rho_0 e^{-\frac{z}{H}} \tag{3.6}$$

where ρ_0 is the density at sea level. The profiles (in logarithmic scale) of p and ρ as a function of z obtained from experimental data (Fig. 3.2) confirm this trend, with a scale height $H \cong 7$ km; the theoretical value determined above is, therefore, in significant agreement with the experimental one. It is interesting to note how the simple hydrostatic balance can, together with the equation of state of perfect gases, account for such a significant property of the Earth's atmosphere.

In the ocean, the situation is quite different, as a liquid is virtually incompressible (Sect. 1.1), with a consequent completely different state law. If we assume to know $\rho(z)$ from experimental data, by integrating Eq. (3.1) we get:

$$p(0) - p(-h) = -g \int_{-h}^{0} \rho(z) dz.$$

For a homogeneous ocean ($\rho = \overline{\rho} = const$) we will have:

$$p(-h) = p(0) + \overline{\rho} g h. \tag{3.7}$$

At the depth $z = -h$, to the surface pressure $p(0)$, equal to the atmospheric one at sea level (1 *Atm*), the *sea pressure* $\overline{\rho} gh$ is added. This result, in general, is also known as *Stevino's law* (Simon Stevin, 1548–1620).

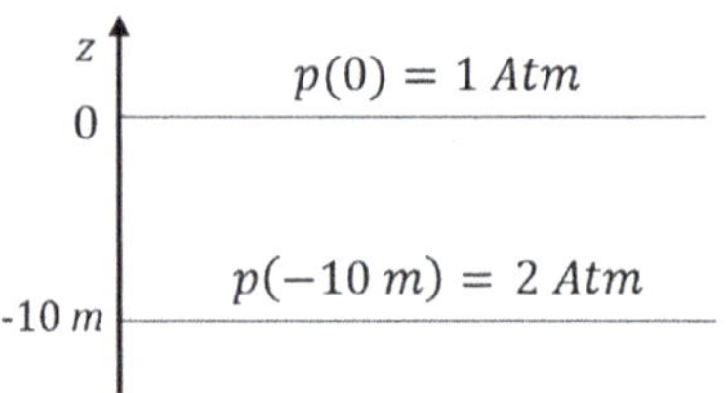

Fig. 3.3 Hydrostatic pressure in the ocean

Assuming a typical value of sea water density $\overline{\rho} = 1028$ kg m^{-3} and a depth of 10 m, we get $\overline{\rho}gh = 1.00847$ bar $= 0.995 \times 1$ Atm (see Sect. 2.2): therefore, 10 m of sea water produce a hydrostatic pressure almost exactly equal to 1 Atm (Fig. 3.3). For this reason, in oceanography, the *dbar* is widely used; in fact, this unit of measurement almost exactly corresponds to the hydrostatic pressure produced by one metre of sea water. On the other hand, if we consider a column of fresh water ($\overline{\rho} = 1000$ kg m^{-3} at 4 °C at sea level), the pressure of 1 Atm corresponds to a column of height $\Delta z = 101325/9810 \cong 10.33$ m.

We now describe an experiment of great interest devised in 1644 by Evangelista Torricelli (1608–1647), a student of Galileo Galilei, in which the hydrostatic pressure of a liquid is used to measure the hydrostatic pressure of the atmosphere (this simple apparatus can be seen as the first barometer in the history of fluid dynamics).

In Fig. 3.4 a liquid of density ρ is contained both in a wide container and in a tube closed at one end (in blue); initially the tube is also closed at the opposite end by a stopper (Fig. 3.4a). Subsequently, the tube is turned over, immersed in the container and the stopper is removed (Fig. 3.4b). Part of the liquid will flow into the container creating a vacuum in the upper part of the tube (in reality, vapour will form in the interstice, but its pressure will be negligible for the purposes of the experiment). The flow will cease when the difference in level Δz between the free surface of the liquid in the container and that in the tube is such that $\rho g \Delta z$ equals the atmospheric pressure exerted on the liquid–air separation surface in the container: only in this way will there be no horizontal pressure gradients at a generic depth below the tube, which would otherwise produce a horizontal flow tending to rebalance the hydrostatic pressure.

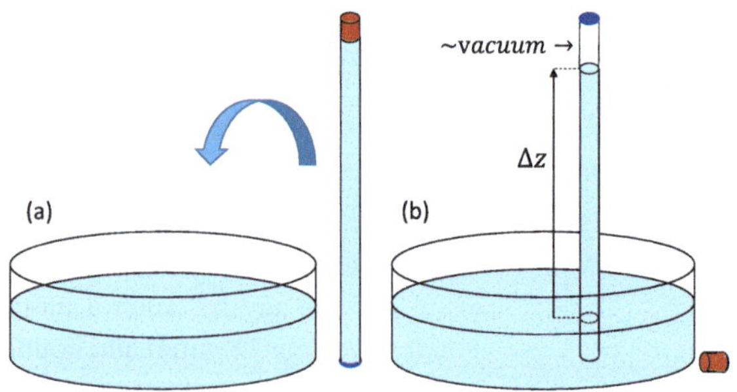

Fig. 3.4 Measurement of atmospheric pressure through the experiment of Evangelista Torricelli

Torricelli used mercury, the only metal to be in a liquid state at room temperature and with a very high density: $\rho = 13579$ kg m^{-3}. This allowed him to estimate the atmospheric pressure (also called 1 Torr in honour of the Italian scientist) with a very short tube, resulting in $\Delta z = 760$ mm. It is interesting to note that Galileo Galilei noticed that the water contained in a well and pumped from above could only emerge as long as the vertical distance between the water level in the well and that of the nozzle on the surface was less than $\sim$ 10 m. This essentially implied the same experiment as Torricelli's, but carried out with water (it has already been seen that, for fresh water, 1 Atm corresponds to $\Delta z = 10.33$ m).

It is worth noting that Torricelli's experiment helped overcome the Aristotelian-scholastic belief that a vacuum could not exist in nature (*horror vacui*). In fact, following the Galilean method, his experiment showed how the lowering of the mercury level in the tube left a vacuum in the upper gap. Subsequently, Blaise Pascal allowed the complete overcoming of the horror vacui with a similar experiment carried out on the top of Puy de Dôme in 1648.

3.2 The Hydrostatic Paradox, the Communicating Vessels

This paragraph delves into the property of hydrostatic pressure to depend only on the vertical extension of the overlying fluid, being otherwise independent of its spatial distribution.

Let's start by considering a famous experiment carried out by Blaise Pascal in 1646 (known as *Pascal's barrel*) exemplified in Fig. 3.5. Consider a closed barrel full of water and connected with a tube overlaid on it. In Fig. 3.5a the tube has a rather large diameter and can contain a volume of water Q for a very limited vertical extension: for example, imagine $Q = 1$ litre (equal to the mass of 1 kg) and $\Delta z = 0.1$ m. Consequently, the pressure inside the barrel will increase by the hydrostatic pressure of the water column, $\Delta p \cong 0.01$ Atm: this will not cause any structural alteration of the container. Now imagine (Fig. 3.5b) that the same volume of water Q is contained in a tube with a diameter 10 times smaller, therefore with a vertical extension 100 times greater ($\Delta z = 10$ m): in this case the pressure inside the barrel will increase by $\Delta p \cong 1$ Atm. The pressure gradient that will be established between the liquid inside the barrel and the air outside will be increased by a full 1 Atm on a small wall thickness, and this can cause the fracture of the wall and the lateral outflow of water, if not the complete destruction of the structure. It is natural that the small tube must however be more resistant than the barrel, otherwise it will be the one to explode before it happens to the barrel; in fact, in the original experiment Pascal used a metal tube (not even too thin!).

It is interesting to note that the same result would be obtained if the small tube, instead of being arranged vertically, were inclined in any way (Fig. 3.5c). This fact illustrates the so-called *hydrostatic paradox*, which consists in the apparent contradiction between the dependence of hydrostatic pressure on the vertical extension of the fluid and the total independence from the spatial distribution of the same. To

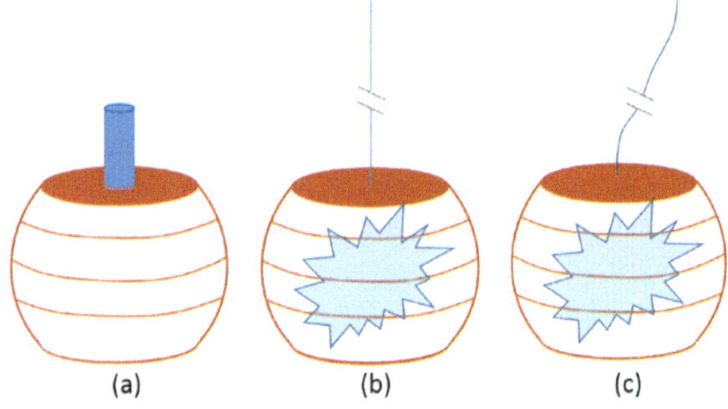

Fig. 3.5 Schematic representation of Pascal's barrel

understand how the paradox can be resolved, we now move on to describe another classic experiment in fluid statics, also due to Pascal.

In Fig. 3.6 three different containers are schematised with the bottom part identical with a base of surface S, while from a certain height upwards their shape differs. In all cases the level of the liquid is equal to h. In the first case (Fig. 3.6a) at each point of the base is superimposed a column of liquid of thickness h to which corresponds a hydrostatic pressure ρgh: consequently, it is entirely intuitive that the total force of hydrostatic pressure applied to the base is $F = \rho ghS$. However, exactly the same pressure force will be exerted on the base also in the other two cases, despite the weight of the liquid in the container being respectively greater (Fig. 3.6b) and less (Fig. 3.6c). This can be experimentally verified by measuring with a suitable scale the force F applied to S.

In the case of Fig. 3.6b, the paradox is explained as follows. At a point on the inclined wall, the red arrow—perpendicular to it—indicates the force that the wall exerts on the fluid as a reaction to the pressure force exerted by the fluid directed outward. The horizontal component (green arrow) is balanced by an equal and opposite force exerted by the opposite side of the container, while the vertical component (blue arrow) exactly balances the hydrostatic pressure exerted at that point by the

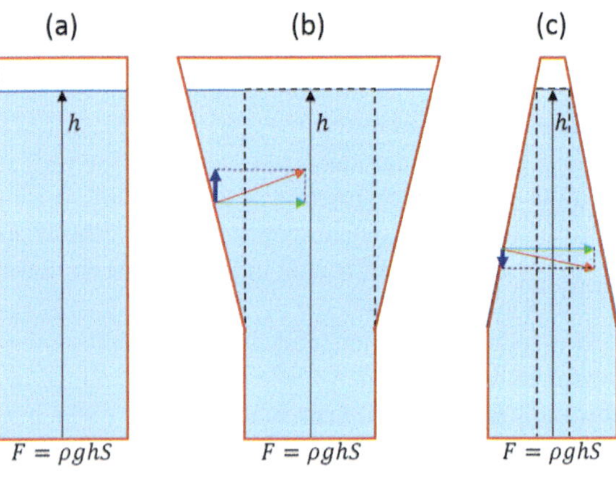

Fig. 3.6 Schematic representation of the hydrostatic paradox and its resolution

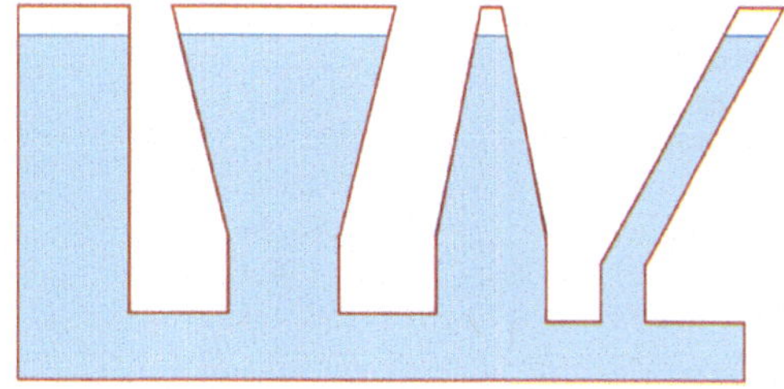

Fig. 3.7 The principle of communicating vessels

fluid above it. Consequently, the pressure force exerted on the base is given solely by the weight of the column of liquid delimited by the dashed lines. In the case of Fig. 3.6c, in the base portion delimited by the dashed lines, the hydrostatic pressure is due to the weight of the column of liquid above it. Outside, however, the hydrostatic pressure of the local column of fluid is added to the reaction of the wall (blue arrow) corresponding to that point, which exactly compensates for the weight of the missing liquid in the vertical: along that vertical, therefore, the pressure at the base will be equal to $\rho g h$. A geometrically more complex, but perfectly analogous explanation solves the hydrostatic paradox for containers of arbitrary shape, like the small tube represented in Fig. 3.5c.

An immediate consequence of the resolution of the hydrostatic paradox is the so-called *principle of communicating vessels* (also known as the *Stevin's second law*). It is clear that, in light of what has just been discussed, only if the level of the different communicating vessels is the same (Fig. 3.7) will there be an absence of horizontal pressure gradients in the duct that connects the vessels, and therefore the static nature of the liquid.

3.3 Archimedes' Principle, Reduced Gravity

Consider a body (solid or fluid indifferently) of constant density ρ' contained in a control volume V immersed in a fluid of ambient density ρ (Fig. 3.8). If $\rho' = \rho$, the fluid is static if the force of gravity acting on V, $\mathcal{F}_g = -\rho g V$, is exactly balanced by the resultant of the pressure forces $\mathcal{F}_p$ acting on ∂V; in this case $\mathcal{F}_p = \rho g V$. Assuming that the fluid is (initially) in equilibrium, $\mathcal{F}_p$ is the same even if $\rho' \neq \rho$. Therefore, the upward force $\mathcal{F}_p = \rho g V$, called *Archimedean thrust* (from the great ancient mathematician Archimedes, c. 287-212 BC, of the then Greek Sicilian colony of Syracuse), is equal to the weight of the ambient fluid displaced by volume V. Consequently, the resultant $\mathcal{F}_{tot}$ of the two forces acting on V is:

$$\mathcal{F}_{tot} = (\rho - \rho')gV. \tag{3.8}$$

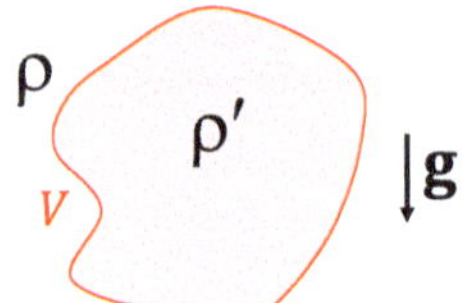

Fig. 3.8 Schematic representation of Archimedes' principle

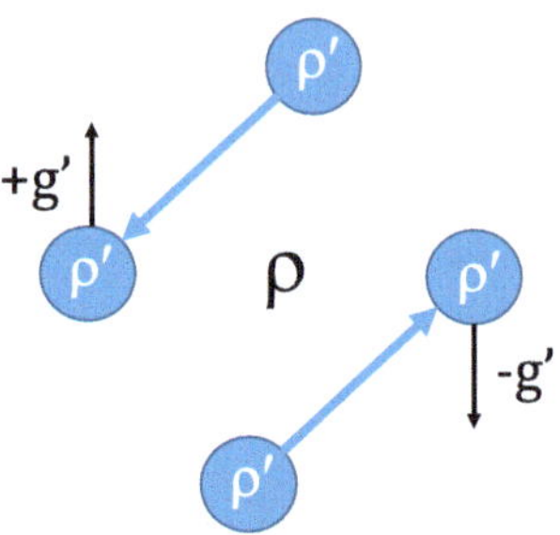

Fig. 3.9 Schematic description of reduced gravity

This relation expresses *Archimedes' principle*. If the density of the body is greater (less) than that of the fluid, there is a resultant directed downwards (upwards), while if the two densities are equal there is perfect buoyancy.

In the atmosphere and in the oceans, a stable stratification requires (for negligible effects of adiabatic temperature variations) a decrease in ρ with z (upwards). In such a situation, imagine (Fig. 3.9) a fluid element of density ρ' that moves by a small $\Delta z > 0$ from its equilibrium position due to vertical motions under the assumption that its density does not vary due to this small displacement, thus finding itself in an environment of density $\rho < \rho'$.

For Eq. (3.8), such an element will be subjected to a force per unit of mass directed downwards equal to

$$F_3 = -g',$$

where g',

$$g' = g\frac{(\rho' - \rho)}{\rho} \tag{3.9}$$

is called *reduced gravity* (for small displacements, $g' \ll g$). Obviously, if the initial displacement is downwards (see the same figure), the resulting force is upwards and always equal to g' in magnitude. In conclusion, in a stable stratification (that is, with ρ increasing with depth or decreasing with altitude) the vertical displacement of a fluid mass results in a *restoring force* that tends to re-establish its original position. It is therefore natural that g' appears in the analysis of internal gravity waves (Chap. 11), which originate precisely from this restoring force.

Recommended Readings

Batchelor, G.K.: An Introduction to Fluid Dynamics. Cambridge University Press, Cambridge (1967)

Çengel, Y.A., Cimbala, J.M.: Fluid Mechanics, Fundamentals and Applications. McGraw-Hill, Boston (2006)

Kundu, P.K., Cohen, I.M., Dowling, D.R.: Fluid Mechanics. Elsevier, Amsterdam (2012)

Landau, L.D., Lifshitz, E.M.: Fluid Mechanics. Pergamon Press, Oxford (1987)

Tritton, D.J.: Physical Fluid Dynamics. Van Nostrand Reinhold, New York (1977)

White, F.M.: Fluid Mechanics. McGraw-Hill, New York (2011)

Wallace, J.M., Hobbs, P.V.: Atmospheric Science, An Introductory Survey. Elsevier, Amsterdam (2006)

Chapter 4
Elements of Fluid Kinematics

This chapter will cover the main kinematic aspects related to fluid dynamics. The acceleration of a fluid element, its rotation, the necessity of mass conservation are all geometric aspects that are independent of the forces acting on the fluid, and therefore deserve a separate discussion.

4.1 Lagrangian Derivative

The Eulerian representation of fluid dynamics requires that an appropriate derivative be defined that takes into account the displacement of the fluid element in the time interval dt. Suppose that the element is initially at point $\mathbf{x}$ at time t: what is the derivative of the generic quantity $A(\mathbf{x}, t)$ with respect to time following the fluid element? The partial derivative of A with respect to time, $\partial A/\partial t$, does not provide the correct answer, as in the interval of time dt the fluid element has moved from $\mathbf{x}$ to $\mathbf{x} + d\mathbf{x}$, with $d\mathbf{x} = \mathbf{u}dt$ (Fig. 4.1).

The correct answer to the previous question is given by the so-called *Lagrangian* (or *substantial* or *material*) derivative (from the Italian mathematician and physicist Giuseppe Luigi Lagrangia, later naturalized French, 1763–1813),

$$\frac{dA}{dt} = \lim_{\Delta t \to 0} \frac{A(\mathbf{x} + \Delta\mathbf{x}, t + \Delta t) - A(\mathbf{x}, t)}{\Delta t}, \tag{4.1}$$

where $\mathbf{x}$ is a function of time and $\dot{\mathbf{x}} = \mathbf{u} = (u, v, w)$ is the velocity of the fluid element. That said, based on the expression of the total derivative of a function of several variables and the calculus chain rule, we obtain:

$$\frac{dA}{dt} = \frac{\partial A}{\partial t} + \frac{\partial A}{\partial x}\frac{dx}{dt} + \frac{\partial A}{\partial y}\frac{dy}{dt} + \frac{\partial A}{\partial z}\frac{dz}{dt} = \frac{\partial A}{\partial t} + \mathbf{u} \cdot \nabla A \tag{4.2}$$

S. Pierini, *Oceanic and Atmospheric Fluid Dynamics*, UNITEXT for Physics,
https://doi.org/10.1007/978-3-031-77991-6_4

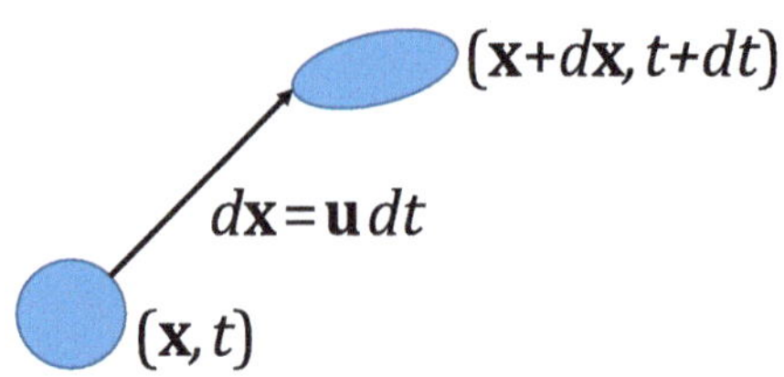

Fig. 4.1 Displacement of a fluid element

where with $\boldsymbol{a} \cdot \boldsymbol{b} = a_1 b_1 + a_2 b_2 + a_3 b_3$ we represent the scalar product between the two vectors **a** and **b**. The terms $\mathbf{u} \cdot \nabla A$ are called *advective* as they represent the contribution to the derivative due to the displacement of the fluid element in the infinitesimal time dt. If $dA/dt = 0$, A is *constant following the fluid element*; in such a case we will say that A is *materially conserved*.

The expression just obtained applies to all the dependent variables of the fluid dynamics problem, and obviously also to each component of the velocity vector **u**, in which case we obtain the *acceleration of the fluid element* (a fundamental quantity in the dynamic context, as will be seen in the next chapter):

$$\frac{d\mathbf{u}}{dt} = \frac{\partial \mathbf{u}}{\partial t} + (\mathbf{u} \cdot \nabla)\mathbf{u} \qquad (4.3)$$

with $(\mathbf{u} \cdot \nabla) = u\partial/\partial x + v\partial/\partial y + w\partial/\partial z$.

4.2 Continuity Equation

Reflect on the fact that **u** and ρ must satisfy an appropriate condition so that the total mass of the fluid is conserved. Consider a generic control volume V in which the mass $M(t)$ (Fig. 4.2) is contained; this can obviously depend on time, as the net mass of fluid crossing the boundary of V (∂V) can well be different from zero and depend on time.

In order for the total mass of the fluid to be conserved, it must be required that the derivative of the mass contained in V,

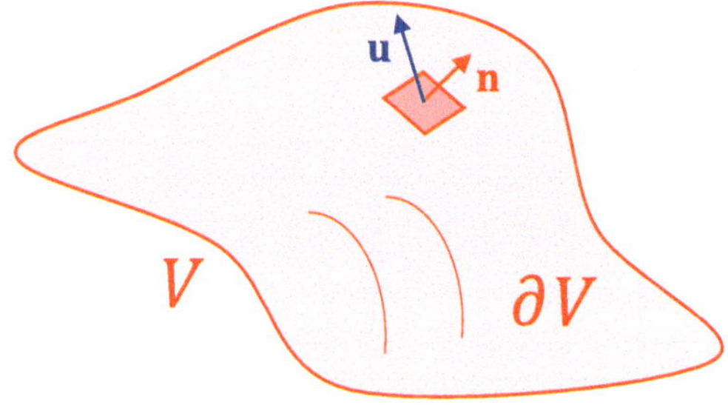

Fig. 4.2 Generic control volume V containing the fluid mass $M(t)$

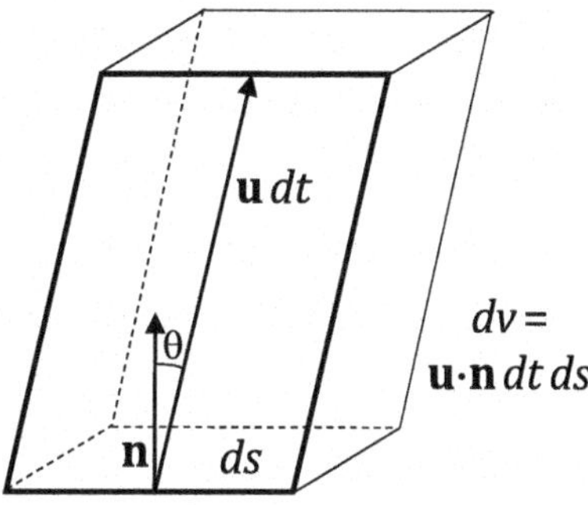

Fig. 4.3 Illustration of the volume dv that crosses the elementary surface ds in time dt

$$\frac{dM}{dt} = \frac{d}{dt} \iiint_V \rho dv \tag{4.4}$$

be equal and opposite to the net mass that crosses ∂V in time dt, called *mass flux* Φ:

$$\Phi = \oiint_{\partial V} \rho \mathbf{u} \cdot \mathbf{n} ds \tag{4.5}$$

Indeed, the quantity $\mathbf{u} \cdot \mathbf{n} ds dt$ (where $\mathbf{n}$ is the normal to the surface element ds) represents the volume dv invaded by the fluid that crosses ds in time dt (Fig. 4.3); therefore, $\rho \mathbf{u} \cdot \mathbf{n} ds$ in Eq. (4.5) represents the mass that crosses ds in the unit of time.

Imposing $dM/dt = -\Phi$ we get:

$$\iiint_V \frac{\partial \rho}{\partial t} dv = - \oiint_{\partial V} \rho \mathbf{u} \cdot \mathbf{n} ds \tag{4.6}$$

where the time derivative in Eq. (4.4) has been brought inside the integral and applied as a partial derivative to the integrand function due to the independence of the integration domain V from time. Now, similarly to what was done in Sect. 2.3, applying Gauss's theorem (Appendix A, Eq. A1) to the surface integral and finally bringing the resulting integral to the first member we will have:

$$\iiint_V \left[\frac{\partial \rho}{\partial t} + \nabla \cdot (\rho \mathbf{u}) \right] dv = 0. \tag{4.7}$$

It is recalled that the scalar product of *nabla* with a vector $\mathbf{A}$, $\nabla \cdot \mathbf{A} = \partial A_1/\partial x + \partial A_2/\partial y + \partial A_3/\partial z$, provides the (scalar) *divergence of* $\mathbf{A}$, also indicated as *div*$\mathbf{A}$. Since Eq. (4.7) must hold for every V, we will finally have:

$$\frac{\partial \rho}{\partial t} + \nabla \cdot (\rho \mathbf{u}) = 0. \tag{4.8}$$

This differential equation, known as the *continuity equation*, must always be verified in a fluid in any context.

Taking into account the identity $\nabla \cdot (\rho \mathbf{u}) = \mathbf{u} \cdot \nabla \rho + \rho \nabla \cdot \mathbf{u}$ and Eqs. (4.2) and (4.8) can be rewritten as follows:

$$\frac{d\rho}{dt} + \rho \nabla \cdot \mathbf{u} = 0. \tag{4.9}$$

The continuity equation in this form is subject to a simple interpretation. If we pass from the density ρ to the *specific volume* $\alpha = dv/dm = 1/\rho$ we get:

$$\frac{1}{\alpha}\frac{d\alpha}{dt} = \nabla \cdot \mathbf{u}.$$

The invariance of the mass dm thus leads to the following form of the continuity equation:

$$\frac{d(dv)}{dv} = \nabla \cdot \mathbf{u} dt. \tag{4.10}$$

The first member represents the relative variation of the volume that the fluid element undergoes in time dt, and this gives a very interesting and intuitive physical meaning to the divergence of the velocity field $\nabla \cdot \mathbf{u}$. At a point where $\nabla \cdot \mathbf{u} > 0$ the volume increases over time, thus, it expands as a consequence of the divergence of the flow lines (in fact the term "divergence" used for the operator $\nabla \cdot$ applied to a vector field originates precisely from its fluid dynamic meaning, as indeed happens for the operator "curl", as will be evident in Sect. 4.4). Conversely, where $\nabla \cdot \mathbf{u} < 0$ there is a contraction due to the convergence of the flow lines.

Of particular interest is the case in which it is possible to assume that the density of the generic fluid element is materially conserved, that is, following its trajectory:

$$\frac{d\rho}{dt} = 0. \tag{4.11}$$

In this case, it is said that the fluid is *incompressible* and the continuity equation is reduced to:

$$\nabla \cdot \mathbf{u} = 0 \tag{4.12}$$

or:

$$\frac{\partial u}{\partial x} + \frac{\partial v}{\partial y} + \frac{\partial w}{\partial z} = 0.$$

In this case, the flow is *solenoidal*, as is the case for the vorticity $\boldsymbol{\omega}$ defined in Sect. 4.4 and, in electromagnetism, for the magnetic induction vector $\mathbf{B}$ (or $\mathbf{H}$ in a vacuum, cf. Eq. A4 in Appendix A). Strictly speaking, in fluid physics the term "incompressible" implies that the fluid does not change its density as the pressure varies (a case which is never strictly verified, not even for liquids). Therefore, here we adopt an extensive interpretation of the term that also excludes non-isentropic processes, for which there can be density variations due to heat and mass exchanges mainly of a turbulent nature (see Chap. 6).

The incompressibility assumption expressed by Eq. (4.12) is certainly valid for liquids (except, of course, if one wants to study sound propagation), but Eq. (4.11) will have to be relaxed if one wants to take into account slow mixing processes that lead to significant changes in ρ, as typically happens in ocean circulation. For example, in oceanographic modelling Eq. (4.12) is always adopted but Eq. (4.11) is often replaced by suitable advection–diffusion equations for temperature and salinity (which determine ρ, together with pressure, through the state law of seawater) describing non-isentropic processes, as a result of which substantial changes in water masses, and therefore in their density, occur over long periods of time. Therefore, the most correct way to read the two Eqs. (4.11–4.12) in relation to Eq. (4.9) is as follows:

$$\left|\frac{d\rho}{dt}\right| \ll \rho\left(\left|\frac{\partial u}{\partial x}\right| + \left|\frac{\partial v}{\partial y}\right| + \left|\frac{\partial w}{\partial z}\right|\right). \tag{4.13}$$

This relationship expresses the fact that in an incompressible fluid (as defined here), density variations are negligible in the mass balance.

In this regard, it is important to note that even for a gas like air, Eq. (4.13) is verified, provided that the fluid velocity does not exceed 20–30% of the speed of sound c_s (but the approximation is considered good even for subsonic flows up to 80% of c_s, as will be seen in Chap. 8). Therefore, Eq. (4.12) can also be adopted for many aerodynamic and meteorological phenomena, but certainly not when considering convective motions, for which there are significant vertical displacements and, therefore, considerable effects of expansion and contraction of air masses.

In conclusion, it should be emphasised that $d\rho/dt = 0$ does not imply at all that ρ is constant, neither spatially nor temporally, but only that the density of a fluid element is conserved following its trajectory. Consider, for example, an incompressible ocean initially described by a certain density field $\rho_0(z)$: a complex motion field that would be established would produce a distribution of ρ generally dependent on both $\mathbf{x}$ and t: $\rho(\mathbf{x}, t)$. On the other hand, it is obvious that, if initially $\rho = const$, incompressibility would imply $\rho = const$ at all times. In this case ρ ceases to be an unknown and the fluid is said to be *homogeneous and incompressible*. This case will be the subject of various in-depth studies in the rest of the discussion.

4.3 Two-Dimensional and Incompressible Flows

We now introduce a case of great interest that will find application in oceanic and atmospheric dynamics (such as geostrophy, treated in Sect. 12.4) but also in external flows of aerodynamic and naval interest (Sects. 8.4–8.6).

A flow described by the velocity field of the type

$$\mathbf{u} = \left[u(x, y, t), v(x, y, t), 0\right] \tag{4.14}$$

is said to be two-dimensional. The components of the motion are only two (in Eq. (4.14) they are the horizontal ones, but they could also be those along the x and z axes) and do not depend on the third coordinate. In such a situation, the fluid motion can be thought of as composed of vertical columns that move without changing their orientation and thickness. It is also assumed that the hypothesis of incompressibility holds:

$$\frac{\partial u}{\partial x} + \frac{\partial v}{\partial y} = 0. \tag{4.15}$$

In this case, there will exist a scalar function $\psi(x, y, t)$, called the *streamfunction*, such that $\mathbf{u} = (u, v)$ can be expressed as follows:

$$\mathbf{u} = \left(-\frac{\partial \psi}{\partial y}, \frac{\partial \psi}{\partial x}\right) \tag{4.16}$$

The existence of ψ is guaranteed by the theorem that states that, for every solenoidal vector field $\mathbf{u}$, there exists a vector potential $\mathbf{A}$ such that $\mathbf{u} = \nabla \times \mathbf{A}$ (see the next Sect. 4.4 for the definition of the operator $\nabla \times$); for example, think of the magnetic vector potential implied by Eq. (A4). Naturally, the determination of ψ will require the use of appropriate dynamic differential equations; Eq. (4.16) only highlights the kinematic side of the problem. It is immediate -and obvious from what has been said- to verify that the continuity equation is automatically satisfied.

The streamfunction has a useful kinematic meaning: the isolines $\psi = const$ coincide with the streamlines, i.e., with the lines along which $\mathbf{u}$ is everywhere tangent. To verify this, it is sufficient to recognise that $\nabla\psi$ identifies the direction along which ψ records the maximum variation and, as such, is perpendicular to the local tangent to the line $\psi = const$. Moreover, given that

$$\nabla\psi \cdot \mathbf{u} = -\frac{\partial \psi}{\partial x}\frac{\partial \psi}{\partial y} + \frac{\partial \psi}{\partial y}\frac{\partial \psi}{\partial x} = 0,$$

$\mathbf{u}$ is in turn tangent to the isoline (Fig. 4.4), which is therefore a streamline. In Sect. 12.4 it will be seen that the isolines of ψ represent the isobars along which geostrophic winds and currents flow.

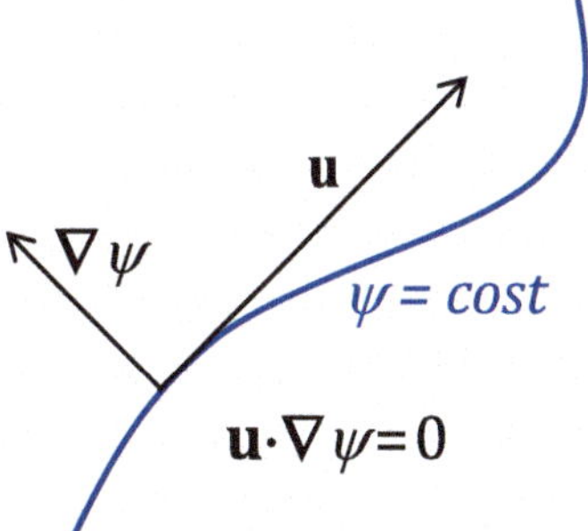

Fig. 4.4 The isolines $\psi = const$ are streamlines

4.4 Vorticity

The vorticity of a fluid element is a concept of great importance in fluid dynamics and is particularly so in meteorological and oceanographic applications. The *vorticity* $\boldsymbol{\omega}(\mathbf{x}, t)$ is defined as twice the angular velocity Ω of the fluid element (Fig. 4.5):

$$\boldsymbol{\omega}(\mathbf{x}, t) = 2\Omega(\mathbf{x}, t). \tag{4.17}$$

Ω is perpendicular to the plane of rotation and its modulus is

$$\Omega = \frac{d\alpha}{dt}$$

where $d\alpha$ is the angle swept by the rotation of the fluid element on the plane normal to the rotation in time dt.

It is obvious that vorticity is linked to the velocity field, therefore, the question arises: what is the relationship that links $\boldsymbol{\omega}$ to $\mathbf{u}$? In Appendix C (Eq. C10) the following fundamental relationship is demonstrated:

$$\boldsymbol{\omega} = \nabla \times \mathbf{u} \tag{4.18}$$

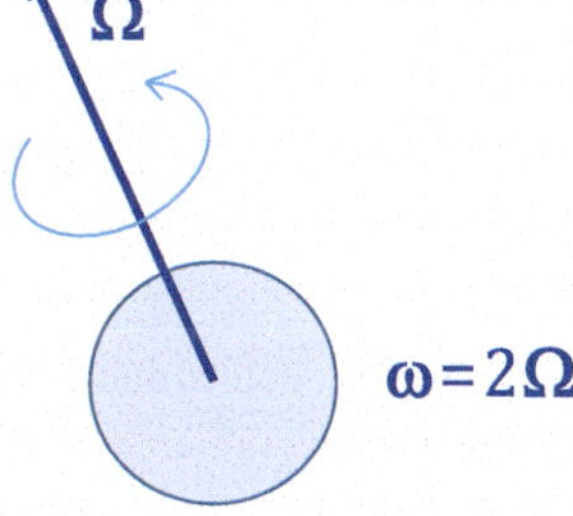

Fig. 4.5 Schematic representation of the vorticity of a fluid element

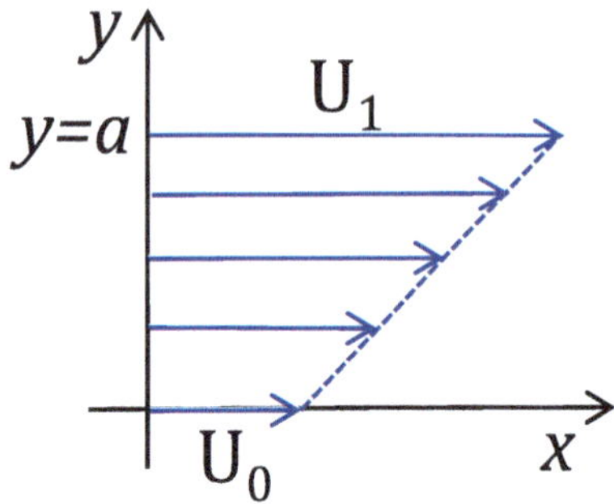

Fig. 4.6 Example of linear shear flow

where $\nabla \times \mathbf{u}$ is the *curl of* **u**, also indicated as *curl***u** (or *rot***u** in other languages). Therefore, $\boldsymbol{\omega}$ is a pseudovector whose components are given by:

$$\boldsymbol{\omega} = \begin{vmatrix} \mathbf{i} & \mathbf{j} & \mathbf{k} \\ \partial_x & \partial_y & \partial_z \\ u & v & w \end{vmatrix} = \left(w_y - v_z, u_z - w_x, v_x - u_y\right) \tag{4.19}$$

(for simplicity of notation the partial derivative here is indicated by a subscript; this useful convention will sometimes also be used in the following). In a two-dimensional motion like the one described in Sect. 4.3, $\boldsymbol{\omega} = (0, 0, \zeta)$, with $\zeta = v_x - u_y$.

As an example, consider a horizontal linear *shear flow* (Fig. 4.6):

$$\mathbf{u} = \left[U_0 + \left(\frac{U_1 - U_0}{a}\right)y, 0, 0\right] \tag{4.20}$$

The corresponding vorticity is the constant vertical vector

$$\boldsymbol{\omega} = \left(0, 0, -u_y\right) = \left(0, 0, \frac{U_0 - U_1}{a}\right)$$

and the angular velocity of the generic fluid element is:

$$\frac{d\alpha}{dt} = \frac{U_0 - U_1}{2a}$$

Figure 4.7 shows the rotation as seen from the reference frame of the generic fluid element. The displacement of the fluid element is, in this case, also accompanied by a change in its shape, as discussed in Appendix C.

It is very useful, especially in meteorological applications, to decompose the vorticity of a two-dimensional flow into two components, as follows (naturally, this decomposition can be extended to the three-dimensional case). Consider the natural

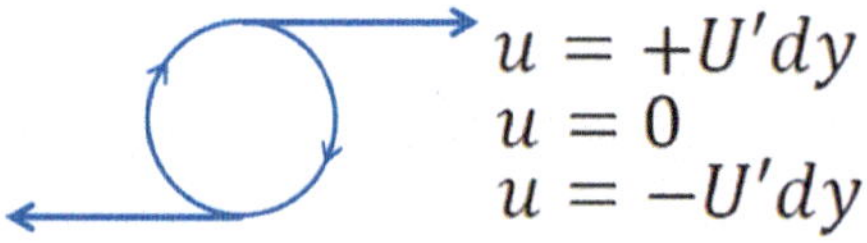

$u = +U'dy$
$u = 0$
$u = -U'dy$

Fig. 4.7 Initially circular fluid element with the indication of the velocities in the reference frame of the element itself

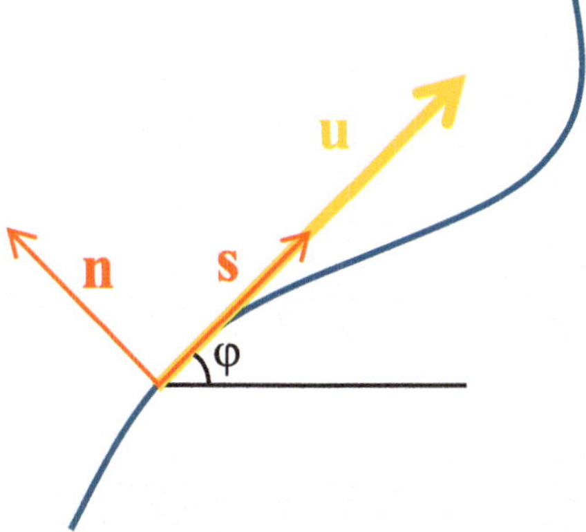

Fig. 4.8 The natural coordinate system

coordinate system defined in Fig. 4.8: **s** and **n** are the unit vectors respectively parallel and normal to **u**. Therefore, these directions are not fixed but depend on **u**.

It can be shown that ζ is the sum of the *shear vorticity* ζ_s due to the variation of **u** in the **n** direction and the *curvature vorticity* ζ_c associated with the curvature of the flow lines:

$$\zeta = \zeta_s + \zeta_c = -\frac{\partial |\mathbf{u}|}{\partial \mathbf{n}} + |\mathbf{u}| \frac{\partial \varphi}{\partial \mathbf{s}} \tag{4.21}$$

where the angle φ is defined in the figure and $\partial\varphi/\partial\mathbf{s} = R^{-1}$, where R is the local radius of curvature of the flow line (in the shear flow of Eq. (4.20), $\zeta = \zeta_s$). Note that, when ζ_c is positive the flow is counterclockwise, and vice versa.

In a radially symmetric vortex in which the fluid moves with *rigid body rotation* ($|\mathbf{u}| \propto r$) the two components are equal: $\zeta_s = \zeta_c = \zeta/2$ (Fig. 4.9a). In a vortex in which $|\mathbf{u}| \propto r^{-1}$ the two components are instead equal in magnitude but opposite, $\zeta_s = -\zeta_c$, so the motion is *irrotational* ($\boldsymbol{\omega} = 0$, $\zeta = 0$, Fig. 4.9b).

Examples of such a vortex, called *free vortex*, are provided by the circulation associated with the outflow of a liquid through a hole and, in environmental contexts, by tornadoes, waterspouts, etc. In all cases there is a centre of low pressure, with the resulting pressure gradient force balanced by the centrifugal force (*cyclostrophic balance*). In real cases the singularity present in the centre of symmetry of the theoretical vortex is replaced by a strongly rotational core. It is worth anticipating that the

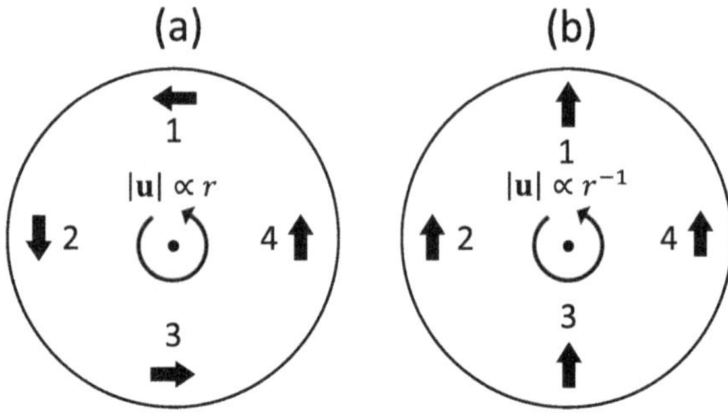

Fig. 4.9 Rotation of a fluid element highlighted in 4 successive instants in a vortex with rigid body rotation (**a**) and in an irrotational vortex (**b**)

irrotational vortex will be used in Sect. 8.5 to construct more complex irrotational flows.

4.5 Irrotational and Incompressible Flows

The dynamic conditions under which flows can be considered irrotational will be discussed in Chap. 8. Here it can be stated that $\boldsymbol{\omega} = 0$ is a necessary and sufficient condition for $\mathbf{u}$ to be written as follows:

$$\mathbf{u} = \nabla\varphi, \tag{4.22}$$

where $\varphi(\mathbf{x}, t)$ is an appropriate scalar function called *velocity potential* (φ is a monodromic function if irrotationality is verified in a simply connected domain). On the one hand, a velocity field expressed by Eq. (4.22) is irrotational, as the well-known identity $\nabla \times \nabla\varphi \equiv 0$ (*curl grad* $\varphi \equiv 0$) holds. On the other hand, the reverse can also be demonstrated: if $\nabla \times \mathbf{u} = 0$ an appropriate scalar φ exists such that Eq. (4.22) holds. Flows of this type are called *potential*; therefore, the adjectives irrotational and potential are synonymous.

It is important to note that an irrotational velocity field is analogous to a *conservative force field* $\mathcal{F}$ (like gravity or Coulomb's electrostatic force), for which $\nabla \times \mathcal{F} = 0$; in this case, a *potential energy* V can be introduced such that $\mathcal{F} = -\nabla V$, from which the law of conservation of total mechanical energy descends. This justifies the designation of "velocity potential" for φ, although this function does not have the dimensions of an energy and has a completely different physical meaning.

If in addition to irrotationality, incompressibility ($\nabla \cdot \mathbf{u} = 0$) also holds, from Eq. (4.22) the so-called Laplace equation follows:

$$\nabla^2\varphi = 0 \tag{4.23}$$

$$\nabla^2 \equiv \nabla \cdot \nabla = \frac{\partial^2}{\partial x^2} + \frac{\partial^2}{\partial y^2} + \frac{\partial^2}{\partial z^2}$$

∇^2 indicates the *Laplacian* linear differential operator (from the French mathematician and physicist Pierre Simon Laplace, 1749–1827). Therefore, the potential flows in an incompressible fluid are *harmonic* (i.e., they satisfy Eq. (4.23)).

The possibility of expressing the three components of $\mathbf{u}$ in terms of a single scalar function solution of a much-studied equation of mathematical physics like Eq. (4.23), allows for a wide class of analytical solutions in relatively simple but significant cases. In Sect. 8.5, some relevant solutions to this problem in terms of two-dimensional external flows of aerodynamic and naval interest will be derived, while in Chap. 9 the problem of linear surface gravity waves will be solved in this same mathematical context.

What do two such different fluid dynamics phenomena as a solid body moving in a fluid and gravity waves have in common? Apart from the dynamic aspects that we are not yet able to discuss, there is a kinematic aspect that is common to all irrotational and incompressible flows: the presence of a part of the boundary of the domain in motion (the contour of the body in the first example, the air-sea interface in the second).

It is easy to understand the reason. First of all, it is noted that $\nabla \cdot (\varphi \mathbf{u}) \equiv \varphi \nabla \cdot \mathbf{u} + \mathbf{u} \cdot \nabla \varphi$, but the first term is zero for incompressibility, therefore:

$$\nabla \cdot (\varphi \mathbf{u}) = |\mathbf{u}|^2 > 0.$$

Now, integrate $\nabla(\varphi \mathbf{u})$ over the entire domain Ω in which the fluid is contained; applying Gauss's theorem we get:

$$\iiint_{\Omega} \nabla \cdot (\varphi \mathbf{u}) dv = \oint\!\!\oint_{\partial\Omega} \varphi \mathbf{u} \cdot \mathbf{n} ds > 0 \tag{4.24}$$

We will soon see what the boundary conditions to impose on the contour of a fluid are, but it already seems obvious that $\mathbf{u} \cdot \mathbf{n}$ must cancel out along the contour at rest: however, in the case we are considering this cannot happen along the entire contour $\partial\Omega$, otherwise the surface integral in Eq. (4.24) could not be greater than zero. Consequently, in a potential flow there must necessarily be a portion of the contour in motion: it is precisely there that the forces will intervene and the possible time dependence of the flow will appear. This resolves the apparent paradox of a Laplace equation devoid of forcing terms and time derivative.

4.6 Circulation, Vorticity Tubes

The discussion of vorticity concludes by introducing the concepts of circulation and vorticity tube. The vorticity $\boldsymbol{\omega}$ is a vector field that provides a point measure of the rotation of the generic fluid element; an integral measure of the same property can be obtained by calculating the flux (Sect. 4.2) of $\boldsymbol{\omega}$ through a generic surface Γ:

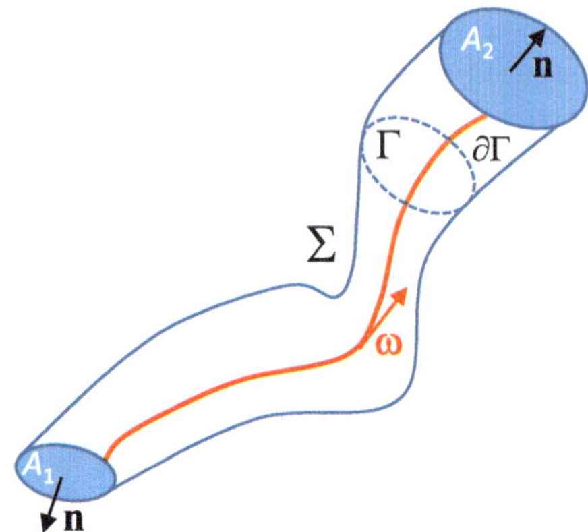

Fig. 4.10 Example of a vortex tube

$$C = \iint_{\Gamma} \boldsymbol{\omega} \cdot \mathbf{n} ds. \tag{4.25}$$

For Stokes' theorem (Eq. A2 in Appendix A) C is equal to the curvilinear integral of $\mathbf{u}$ extended to the closed curve $\partial\Gamma$ that delimits Γ, called *circulation*:

$$C = \oint_{\partial\Gamma} \mathbf{u} \cdot d\boldsymbol{\ell}. \tag{4.26}$$

A *vortex tube* is defined as a closed tubular surface such that, along its lateral wall, $\boldsymbol{\omega}$ is everywhere tangent. Consider a vortex tube Σ (Fig. 4.10) delimited by two sections, A_1 and A_2; the red line, along which $\boldsymbol{\omega}$ is tangent, is called *vortex line*. The following theorem holds:

- The flux of $\boldsymbol{\omega}$ through a surface that cuts the vortex tube Σ (like A_1, A_2 and Γ in the figure) is independent of the surface itself; equivalently, the circulation C is the same along every closed curve that wraps around the vortex tube (like $\partial\Gamma$ in the figure).

To demonstrate this property, it is first recognised that the flux of $\boldsymbol{\omega}$ on the entire surface $\partial\Sigma$ is null,

$$\oiint_{\partial\Sigma} \boldsymbol{\omega} \cdot \mathbf{n} ds = \iiint_{\Sigma} \nabla \cdot \boldsymbol{\omega} dv = 0, \tag{4.27}$$

since $\boldsymbol{\omega}$ is a solenoidal vector field (the divergence of the curl of a vector field is identically zero, as already observed). Moreover, since the flux of $\boldsymbol{\omega}$ on the lateral surface is null because $\boldsymbol{\omega} \cdot \mathbf{n} = 0$, the flux on the entire boundary $\partial\Sigma$ is reduced to the sum of the flux through the two sections A_1 and A_2,

$$\oiint_{\partial\Sigma} \boldsymbol{\omega} \cdot \mathbf{n} ds = \iint_{A_1} \boldsymbol{\omega} \cdot \mathbf{n} ds + \iint_{A_2} \boldsymbol{\omega} \cdot \mathbf{n} ds. \tag{4.28}$$

Combining Eqs. (4.27–4.28) proves the theorem (note that the flux on A_1 is opposite to that on A_2 because by convention $\mathbf{n}$ points towards the outside of a closed surface).

A vortex tube can be seen as a bundle of vortex lines and is characterised by its *intensity* $\mathcal{k}$, defined as the flux of $\boldsymbol{\omega}$ through a generic surface Γ that cuts the tube (Fig. 4.10) or, equivalently, as the circulation C along the generic closed curve that wraps around the tube $\partial\Gamma$:

$$\mathcal{k} = \iint_{\Gamma} \boldsymbol{\omega} \cdot \mathbf{n} \, ds = \oint_{\partial\Gamma} \mathbf{u} \cdot d\boldsymbol{\ell}. \tag{4.29}$$

Thanks to the theorem just demonstrated, $\mathcal{k}$ uniquely characterises the vortex tube. It is emphasised that this is a purely kinematic property.

Here it is worth anticipating a fundamental dynamic property: for Kelvin's theorem demonstrated in Appendix D (see Sect. 8.3), for barotropic motions ($\rho = \rho(p)$) and in the absence of viscous effects, the circulation C is materially conserved; that is, the circulation along any *material curve* (i.e., made up of fluid elements) remains constant during the evolution of the curve itself. From this property it follows that, under the same conditions, *every vortex tube is a material volume.* In other words, a vortex tube moves and deforms over time (with constant $\mathcal{k}$) but always contains the same mass of fluid. Moreover, by contracting a vortex tube, it reduces to a vortex line (or *vortex filament*, for example, the red line in Fig. 4.10); therefore, under the same conditions, *every vortex filament is a material line.*

An important kinematic property of vortex tubes and filaments is linked to the solenoidal nature of $\boldsymbol{\omega}$ (shared with the magnetic induction field $\mathbf{B}$): the envelope lines of a solenoidal vector field must close in on themselves. In fact, since the flow of $\boldsymbol{\omega}$ through any closed surface must cancel out, every line of the field that exits that surface must re-enter it. Therefore, the vortex tubes must have a ring-like character or, alternatively, must end on the boundary that delimits the fluid or extend to infinity in an unconfined fluid.

A relevant example of ring-like vortex tubes immersed in an environment with substantially irrotational motion is given, for example, by the thin, very resistant vortical ring structures produced by the impulsive emission of a gas different from the ambient one, known as *vortex rings*. For example, consider the *smoke rings* emitted by a smoker or those sometimes detectable above a volcanic cone, as happens for an effusive volcano like Etna.

Recommended Readings

Batchelor, G.K.: An Introduction to Fluid Dynamics. Cambridge University Press, Cambridge (1967)
Çengel, Y.A., Cimbala, J.M.: Fluid Mechanics, Fundamentals and Applications. McGraw-Hill, Boston (2006)
Kundu, P.K., Cohen, I.M., Dowling, D.R.: Fluid Mechanics. Elsevier, Amsterdam (2012)
Landau, L.D., Lifshitz, E.M.: Fluid Mechanics. Pergamon Press, Oxford (1987)
Shapiro, A.H.: Illustrated Experiments in Fluid Mechanics. The MIT Press, Cambridge, Massachusetts (1972)
Tritton, D.J.: Physical Fluid Dynamics. Van Nostrand Reinhold, New York (1977)
White, F.M.: Fluid Mechanics. McGraw-Hill, New York (2011)

Chapter 5
The Equations of Fluid Dynamics

In this chapter, the governing equations of fluid dynamics for inertial reference frames are derived. As for surface forces, only the effects of pressure appear in the so-called Euler equation, while in the so-called Navier–Stokes equation, the effects of molecular viscosity are also considered. The use of these equations allows the derivation of Bernoulli's theorem, the dynamics of vorticity, and various energy balances.

5.1 Derivation of the Governing Equations

The starting point is Newton's second law of dynamics for a material point of mass m in an inertial reference frame (the case of a rotating reference frame, essential for the atmosphere and oceans, will be dealt with in Part II of the text), expressed by the vector ordinary differential equation:

$$\frac{d(m\mathbf{u})}{dt} = \mathcal{F} \tag{5.1}$$

where m is the (inertial, equal to the gravitational) mass of the material point, $\mathbf{u}$ is its velocity, $m\mathbf{u}$ is the corresponding momentum, and $\mathcal{F}$ is the sum of all active forces applied to the body. The task now is to transpose this equation into the context of a continuous system. The idea is that an elementary volume of fluid of mass dm corresponds to a material point of mass m, but, as has already been seen, when dealing with fields it is convenient to start from integral expressions referring to finite volumes, then passing to the infinitesimal.

Consider then a generic finite *material volume* dependent on time $\tau(t)$ (Fig. 5.1). This volume differs substantially from the time-independent control volume V evoked several times previously because now τ, as the word itself says, contains

S. Pierini, *Oceanic and Atmospheric Fluid Dynamics*, UNITEXT for Physics,
https://doi.org/10.1007/978-3-031-77991-6_5

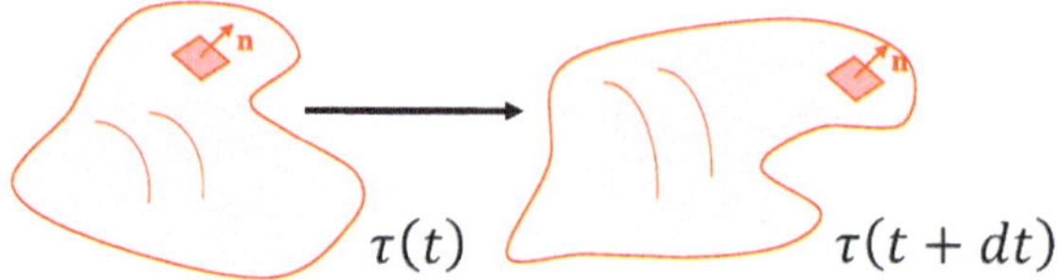

Fig. 5.1 Schematic representation of a material volume $\tau(t)$ and its evolution

matter, that is, a portion of fluid that moves and deforms over time following the velocity field $\mathbf{u}$. The figure schematically shows a possible evolution of τ.

The fluid dynamic analogue of Eq. (5.1) will therefore be:

$$\frac{d}{dt}\iiint\limits_{\tau(t)} \rho\mathbf{u}d\tau = \mathcal{F} \tag{5.2}$$

where $d\tau$ is the infinitesimal volume, itself dependent on time, as it is transported by the velocity field, and $\mathcal{F}$ is the result of all volume and surface forces acting on τ. How to transport the temporal derivative inside the volume integral, which is now a material integral? It will not be possible to proceed as in the derivation of the continuity equation (see Eq. 4.6), because now the limits of integration depend on time. The temporal derivative can be brought inside the integral, but it will have to act as a Lagrangian derivative not on the integrand function but on the entire object of the integral, including the differential:

$$\frac{d}{dt}\iiint\limits_{\tau(t)} \rho\mathbf{u}d\tau = \iiint\limits_{\tau(t)} \frac{d}{dt}(\rho\mathbf{u}d\tau) = \iiint\limits_{\tau(t)} \frac{d\mathbf{u}}{dt}\rho d\tau + \iiint\limits_{\tau(t)} \mathbf{u}\frac{d}{dt}(\rho d\tau).$$

The Lagrangian derivative of $\rho d\tau = dm$ (last integral) is zero as the mass dm is invariant; from Eq. (5.2) we therefore get:

$$\iiint\limits_{\tau(t)} \rho\frac{d\mathbf{u}}{dt}d\tau = \mathcal{F}. \tag{5.3}$$

Ultimately, the first member is nothing more than the sum of the products of mass and acceleration of all the elements that make up the material volume τ.

We now move on to determining the expression of the forces. $\mathcal{F}$ will be the sum of all volume and surface forces acting on τ:

$$\mathcal{F} = \iiint\limits_{\tau(t)} \mathbf{F}\rho d\tau + \oiint\limits_{\partial\tau(t)} \mathbf{S}ds. \tag{5.4}$$

In conclusion, the (vector) equation of fluid dynamics in integral form is given by

$$\iiint_{\tau(t)} \left(\rho \frac{d\mathbf{u}}{dt} - \rho\mathbf{g}\right) d\tau - \oiint_{\partial\tau(t)} \mathbf{S} ds = 0, \tag{5.5}$$

where **F** has been replaced by the force of gravity per unit of mass **g**. The only other volume force considered in this text will be the Coriolis force for fluids in rotating reference frames (Part 2).

5.2 Euler's Equations

Euler's equations (in honour of the great Swiss mathematician and physicist Leonhard Euler, 1707–1783, who first formulated them) refer to a so-called *perfect fluid*, for which shear stresses (the viscosity) are negligible. In this case **S** is reduced to pressure: $\mathbf{S} = -p\mathbf{n}$ It has already been seen (Eq. 2.11) how Gauss's formula allows to transform the surface integral into a volume one, so Eq. (5.5) provides:

$$\iiint_{\tau(t)} \left(\rho \frac{d\mathbf{u}}{dt} - \rho\mathbf{g} + \nabla p\right) d\tau = 0$$

Given the arbitrariness of τ, the Euler (vector) partial differential equation is obtained:

$$\frac{d\mathbf{u}}{dt} = \mathbf{g} - \frac{1}{\rho}\nabla p \tag{5.6}$$

where the acceleration is given by Eq. (4.3).

As will be seen in Chap. 8, the hypothesis of a perfect fluid is valid for motions at high Reynolds numbers outside the boundary layers; this is a far from rare case in nature, therefore, Euler's equations find a wide field of application. In the same chapter, relevant examples of these flows, of an irrotational nature, will be presented.

5.3 Navier–Stokes Equations

For real fluids, in which molecular viscosity is also active, it is convenient to abandon the compact vector notation and rewrite Eq. (5.5) in terms of the three vector components:

$$\iiint_{\tau(t)} \left(\rho \frac{du_i}{dt} - \rho g_i\right) d\tau - \oiint_{\partial\tau(t)} \sum_j (-p\delta_{ij} + d_{ij}) n_j ds = 0$$

where S_i has been replaced by its expression, Eq. (2.3), in terms of the stress tensor $\sigma_{ij} = -p\delta_{ij} + d_{ij}$ (Eq. 2.8). Applying once again Gauss's theorem to the surface integral, we obtain:

$$\iiint_{\tau(t)} \left(\rho \frac{du_i}{dt} - \rho g_i + \frac{\partial p}{\partial x_i} - \sum_j \frac{\partial d_{ij}}{\partial x_j} \right) d\tau = 0$$

(where, of course, $x_1 = x, x_2 = y, x_3 = z$), from which it follows:

$$\frac{du_i}{dt} = g_i - \frac{1}{\rho} \frac{\partial p}{\partial x_i} + \frac{1}{\rho} \sum_j \frac{\partial d_{ij}}{\partial x_j}. \tag{5.7}$$

This equation differs from that of Euler (Eq. 5.6) for the presence of the divergence of the tensor **d** in the balance of forces, which represents the volume resultant of the viscous forces. Substituting the expression given by Eq. (2.13) (which is remembered to be valid for incompressible fluids) we obtain:

$$\frac{1}{\rho} \sum_j \frac{\partial d_{ij}}{\partial x_j} = \frac{\mu}{\rho} \sum_j \frac{\partial^2 u_i}{\partial x_j^2} + \frac{\mu}{\rho} \frac{\partial}{\partial x_i} \sum_j \frac{\partial u_j}{\partial x_j} = \frac{\mu}{\rho} \nabla^2 u_i.$$

The second summation, in which the order of derivation has been swapped (thanks to Schwarz's theorem), is null due to the incompressibility of the fluid $\left(\sum_j \partial u_j / \partial x_j = \nabla \cdot \mathbf{u} = 0 \right)$.

In conclusion, the dynamics of a real and incompressible fluid is governed by the (vector) partial differential equation known as Navier–Stokes:

$$\frac{d\mathbf{u}}{dt} = \mathbf{g} - \frac{1}{\rho} \nabla p + \nu \nabla^2 \mathbf{u}. \tag{5.8}$$

The parameter $\nu = \mu/\rho$ is called the *coefficient of kinematic molecular viscosity* and weighs the viscous effects on the acceleration of the fluid element. The fact that the viscous terms are represented by the Laplacian of the velocity shows that molecular viscosity has the effect of diffusing velocity (or momentum). Indeed, it is well known from the diffusion theory based on Fick's laws (from the German physician and physiologist Adolf Eugen Fick, 1829–1901), that the diffusion of any quantity A produced by the gradients of the same manifests itself through the presence of second spatial derivatives in the evolution equation of A (see Eq. (8.2) and the related discussion).

As examples of ν, its values for air and water in standard situations are reported:

$$\nu_{air} \cong 1.50 \times 10^{-5}\ \mathrm{m^2 s^{-1}},\ \nu_{water} \cong 1.01 \times 10^{-6}\ \mathrm{m^2 s^{-1}}$$

It is interesting to note that, while $\mu_{air} < \mu_{water}$ (see Sect. 2.4), we have $\nu_{air} > \nu_{water}$: the greater density of water means that inertia prevails over the even greater dynamic molecular viscosity. Another example is given by mercury: $\mu_{Hg} \cong 1.5 \times 10^{-3} N\ s\ m^{-2}$ is comparable to that of water but $\nu_{Hg} \cong 1.16 \times 10^{-7} m^2 s^{-1}$ is ~ 10 times less than that of water due to its very high density (see Sect. 3.1).

For completeness, the three Eq. (5.8) are reported, making all the terms explicit (for simplicity, the compact notation for partial derivatives is adopted):

$$u_t + uu_x + vu_y + wu_z = -\frac{1}{\rho}p_x + \nu\left(u_{xx} + u_{yy} + u_{zz}\right)$$

$$v_t + uv_x + vv_y + wv_z = -\frac{1}{\rho}p_y + \nu\left(v_{xx} + v_{yy} + v_{zz}\right)$$

$$w_t + uw_x + vw_y + ww_z = -g - \frac{1}{\rho}p_z + \nu\left(w_{xx} + w_{yy} + w_{zz}\right)$$

A very particular role is played by the advective terms $(\mathbf{u} \cdot \nabla)\mathbf{u}$. These are *nonlinear terms* as, in this case, they are the product of two dependent variables, however derived. Their presence has a profound effect on the nature of the solutions to the differential problem. If these are null or negligible (in some relevant cases this is actually true, such as for a wide class of gravity waves, see Chap. 9) the *principle of superposition* applies, with consequent simple integrability of the equations, at least for simple geometries. Their presence instead removes this property, represents a big problem for the analytical solution of the equations (but not for their numerical solution) and characterises the solutions in a very peculiar way. We will see in Sect. 6. 2 how the nonlinear terms are at the basis of the existence of turbulence in fluids.

5.4 Initial and Boundary Conditions

In order to be solved, the governing equations must be accompanied by initial and boundary conditions. The former are also required in a system of ordinary differential equations (like those of the mechanics of material point) while the latter only concern partial differential equations, like those of fluid dynamics.

The number of initial conditions depends on the order of the time derivative. In Newton's equations, the second derivative $\ddot{\mathbf{x}}$ requires the imposition of two initial conditions: one for position and one for velocity. In fluid dynamics, the Eulerian representation sees $\mathbf{u}$ as the kinematic variable, with a consequent first derivative with respect to time in the dynamic equations. As we have seen, also for the continuity equation there is a first derivative, and so it is in general for all other prognostic equations (not discussed in this text). Consequently, only one initial condition will have to be prescribed for each of the dependent variables, but naturally for each point of the domain:

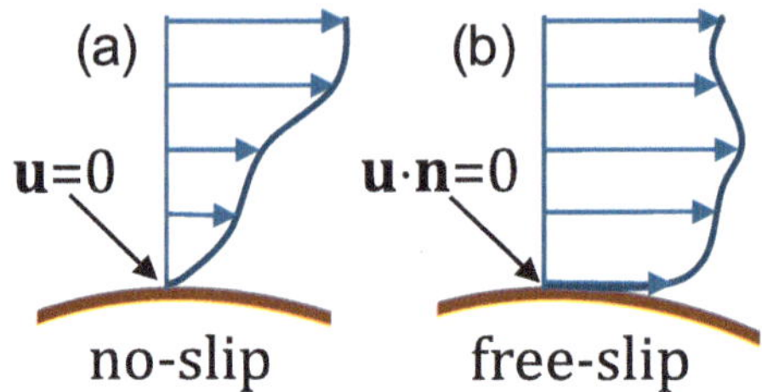

Fig. 5.2 Schematic representation of the no-slip and free-slip boundary conditions

$$\mathbf{u}(\mathbf{x},0)=\mathbf{u}_0(\mathbf{x});\ \ \rho(\mathbf{x},0)=\rho_0(\mathbf{x});\ \text{etc.} \tag{5.9}$$

The boundary conditions define the domain within which the fluid is contained (but in a more general context they may also prescribe flows at the boundaries, etc.). Let Ω be the domain containing the fluid. For real fluids, the velocity field must satisfy the following condition (called *no-slip*) along the boundary of the domain:

$$\mathbf{u}|_{\partial\Omega}=\mathbf{U}|_{\partial\Omega} \tag{5.10}$$

where $\mathbf{U}$ is the speed with which the boundary of Ω is in motion (think of the motion of a body within a fluid). If the boundary is at rest, $\mathbf{u}|_{\partial\Omega}=0$ (Fig. 5.2a). For a perfect fluid (that is, in the absence of viscous effects) the conditions to be imposed (called *free-slip*) will be:

$$\mathbf{u}\cdot\mathbf{n}|_{\partial\Omega}=\mathbf{U}\cdot\mathbf{n}|_{\partial\Omega} \tag{5.11}$$

In this case, the fluid can slide along the boundary but, obviously, it cannot penetrate it, therefore, the normal component will still have to cancel out (Fig. 5.2b). As we will see, these conditions also apply to a real fluid at high Reynolds numbers outside the boundary layer.

5.5 Complete Set of Equations for an Incompressible Fluid

In summary, the complete differential problem under the assumption of incompressible fluid is as follows:

$$\begin{cases} \frac{\partial\mathbf{u}}{\partial t}+(\mathbf{u}\cdot\nabla)\mathbf{u}=\mathbf{g}-\frac{1}{\rho}\nabla p+\nu\nabla^2\mathbf{u}\\ \frac{d\rho}{dt}=0\\ \nabla\cdot\mathbf{u}=0\end{cases} \tag{5.12}$$

There are 5 partial differential equations for 5 unknowns ($u,\ v,\ w,\ p,\ \rho$), therefore the problem accompanied by the initial and no-slip boundary conditions is well posed. In the more restrictive case of an incompressible and homogeneous fluid, ρ is a constant, therefore, the problem is reduced to 4 unknowns for 4 equations (the equation $d\rho/dt = 0$ becomes trivial).

It is worth noting that this is not the most general set of equations. As already mentioned in Sect. 4.2, even under conditions of incompressibility, the problem of describing the modification of air and water masses due to turbulent mixing (Chap. 6) arises. As for the atmosphere, incompressibility is not valid for small scale convective motions but can be considered satisfied in many cases of meteorological dynamics and subsonic aerodynamics (see Sect. 8.4).

5.6 Energy Flux, Viscous Dissipation

In this paragraph, two fundamental aspects concerning the energy balance in a fluid will be discussed: the energy flux and the viscous dissipation of mechanical energy. The analysis will concern the particularly relevant case of a fluid with constant density (the more general case of a compressible fluid in the presence of heat exchanges leads to more complex relationships, the treatment of which is beyond the scope of this text).

Consider an incompressible and homogeneous liquid ($\rho = const$) contained in a domain Ω. In this context, the *modified pressure* $p = p_{total} - p_h$ can be introduced, where p_h is the hydrostatic pressure. With this position, the force of gravity does not appear in the force balance as its effect is summarised in p_h, while p is due solely to the motion of the fluid. This simplification does not involve any approximation, but can only be made for fluids with constant density if, moreover, the boundary conditions do not include the force of gravity but only the velocity of the fluid (for example, the modified pressure cannot be introduced for surface gravity waves, see Eq. 9.10). With this simplification, it is convenient to rewrite the Navier–Stokes equation (Eq. 5.7) as follows:

$$\rho \frac{\partial u_i}{\partial t} + \rho[(\mathbf{u} \cdot \nabla)\mathbf{u}]_i = \sum_j \frac{\partial \sigma_{ij}}{\partial x_j} \tag{5.13}$$

where it is recalled that $\sigma_{ij} = -p\delta_{ij} + d_{ij}$. Taking into account the identity

$$(\mathbf{u} \cdot \nabla)\mathbf{u} = \frac{1}{2}\nabla|\mathbf{u}|^2 - \mathbf{u} \times \boldsymbol{\omega} \tag{5.14}$$

and defining the kinetic energy per unit volume

$$E_K = \frac{1}{2}\rho|\mathbf{u}|^2,$$

Eq. (5.13) is rewritten as

$$\rho \frac{\partial u_i}{\partial t} + \frac{\partial E_K}{\partial x_i} - \rho(\mathbf{u} \times \boldsymbol{\omega})_i = \sum_j \frac{\partial \sigma_{ij}}{\partial x_j}. \tag{5.15}$$

To move from a formulation of dynamics in terms of momentum to one in terms of energy, it is necessary to calculate the work done by surface forces per unit time: this is obtained by multiplying both members scalarly by $\mathbf{u}$ $\left(\sum_i u_i \cdot\right)$. In this way, the third term on the left-hand side is cancelled out and we obtain:

$$\sum_i \rho u_i \frac{\partial u_i}{\partial t} + \sum_i u_i \frac{\partial E_K}{\partial x_i} = \sum_{ij} u_i \frac{\partial \sigma_{ij}}{\partial x_j}$$

Noting that the first summation is equal to $\partial E_K/\partial t$, applying in the second summation $\partial/\partial x_i$ to $u_i E_K$, considering then the incompressibility of the fluid and, finally, decomposing the last summation we obtain:

$$\frac{\partial E_K}{\partial t} = -\sum_i \frac{\partial (u_i E_K)}{\partial x_i} + \sum_{ij} \frac{\partial \left(u_i \sigma_{ij}\right)}{\partial x_j} - \sum_{ij} \sigma_{ij} \frac{\partial u_i}{\partial x_j}.$$

Combining the gradient of the first two summations on the right-hand side (in the second, the two mute indices i and j are swapped and it is considered that σ_{ij} is symmetric) and considering the incompressibility of the fluid in the third summation $\left(p \sum_i \delta_{ij} \partial u_i/\partial x_i = 0\right)$, we obtain:

$$\frac{\partial E_K}{\partial t} = -\sum_i \frac{\partial}{\partial x_i}\left(u_i E_K - \sum_j u_j \sigma_{ij}\right) - \sum_{ij} d_{ij} \frac{\partial u_i}{\partial x_j}.$$

Furthermore, it is convenient to decompose σ_{ij} into its two components in the summation in brackets, finally obtaining:

$$\frac{\partial E_K}{\partial t} = -\sum_i \frac{\partial}{\partial x_i}\left[(E_K + p)u_i - \sum_j u_j d_{ij}\right] - \sum_{ij} d_{ij} \frac{\partial u_i}{\partial x_j}. \tag{5.16}$$

The first term in square brackets, $\mathbf{F}_E^{(rev)} = (E_K + p)\mathbf{u}$, represents the density of *reversible energy flux*; indeed, for a perfect fluid $\left(d_{ij} = 0\right)$ Eq. (5.16) takes the classic form

$$\frac{\partial E_K}{\partial t} + \nabla \cdot \mathbf{F}_E^{(rev)} = 0 \tag{5.17}$$

which expresses the conservation of a quantity in relation to the corresponding flux (see the analogous continuity equation, Eq. (4.8), which expresses the conservation of the fluid's mass). In this case $\mathbf{F}_E^{(rev)}$ is the energy flux associated with the simple mass transport in the moving fluid.

In real fluids the second term in square brackets (the vector **ud**) represents the density of *irreversible energy flux* transferred due to molecular viscosity; in this case, however, the last summation on the right-hand side of Eq. (5.16) is also present, the meaning of which will now be revealed by integrating the equation over a generic control volume V and using Gauss's theorem:

$$\frac{d}{dt}\iiint_V E_K dv = -\oint\!\!\!\oint_{\partial V}\sum_i\left[(E_K+p)u_i - \sum_j u_j d_{ij}\right]n_i ds - \iiint_V \Phi dv \tag{5.18}$$

where

$$\Phi = \sum_{ij} d_{ij}\frac{\partial u_i}{\partial x_j}. \tag{5.19}$$

In Eq. (5.18) the term on the left-hand side is the time derivative of the fluid's kinetic energy contained in V while the surface integral represents the energy flux (reversible and irreversible) that crosses the closed surface ∂V that delimits V.

The meaning of the last term is better highlighted if one integrates not over a generic volume V but over the entire volume Ω that contains the fluid. If it is imagined that $\partial\Omega$ is constituted by a solid wall, **u** is null on $\partial\Omega$, in which case Eq. (5.18) is reduced to:

$$\frac{d}{dt}\iiint_\Omega E_K dv = -\iiint_\Omega \Phi dv. \tag{5.20}$$

Therefore, Φ is nothing more than the *viscous dissipation rate* per unit volume or, in other words, the rate at which molecular viscosity irreversibly converts the fluid's mechanical energy into heat. Its expression is calculated from Eq. (5.19) by substituting **d** with its expression given by Eq. (2.14):

$$\Phi \equiv \sum_{i,j} d_{ij}\frac{\partial u_i}{\partial x_j} = 2\mu\sum_{i,j} e_{ij}\frac{\partial u_i}{\partial x_j} = 2\mu\sum_{i,j} e_{ij}e_{ij} = \frac{\mu}{2}\sum_{i,j}\left(\frac{\partial u_i}{\partial x_j}+\frac{\partial u_j}{\partial x_i}\right)^2$$

(the substitution of $\partial u_i/\partial x_j$ with e_{ij} is made possible by the symmetry of the latter). Therefore, in conclusion:

$$\Phi = \frac{\mu}{2}\sum_{i,j}\left(\frac{\partial u_i}{\partial x_j}+\frac{\partial u_j}{\partial x_i}\right)^2. \tag{5.21}$$

Three aspects should be noted. First of all, since $\Phi > 0$, from Eq. (5.20) it follows that the total kinetic energy of the fluid, also defined as positive, must decrease over time until all the mechanical energy initially contained in the system has been converted into heat (it is assumed that there are no energy inputs from the outside). Furthermore, Eq. (5.21) shows that the viscous dissipation rate is strongly dependent on the velocity gradients, being the sum of all possible symmetric combinations of gradients, squared; this implies that, the smaller the scales of motion, the more efficient is the dissipation (see the implications concerning the energy cascade in fully developed turbulence discussed in Sect. 6.1). Finally, since Φ is directly traceable to the symmetric rate-of-strain tensor e_{ij}, viscous dissipation turns out to be closely connected to the deformation of the generic fluid element, as discussed in Appendix C.

5.7 Bernoulli's Theorem and Its Applications

In physics, conservation laws play a very important role. Here we limit ourselves to mentioning the law of conservation of total energy (kinetic plus potential) of a material point (or a system of material points) subjected to a conservative force field (see Sect. 4.5 and Appendix A, Application 1). In addition to the formal elegance of the law, it is worth highlighting the possibility of using it to solve some relevant mechanical problems in an extremely simple way without having to resort to the solution of the differential equations of Newton's second law.

The theorem (or law, or principle) of Bernoulli (from the French mathematician and physicist Daniel Bernoulli, 1700–1782) constitutes the natural generalisation to a fluid, of the above-mentioned conservation law. Here the theorem will be formulated in its classic version, but more general versions are available for non-stationary flows and for non-perfect and compressible fluids. Only the extension for non-stationary flows will be considered in this text on the occasion of the study of gravity waves (Sect. 9.1).

The following conditions are assumed:

- stationary flow ($\partial \mathbf{u}/\partial t = 0$)
- perfect fluid ($\nu \nabla^2 \mathbf{u}$ negligible)
- homogeneous and incompressible fluid ($\rho = const$).

(the validity of the perfect fluid condition and its implications will be clarified later in Chap. 8). In this case, Eq. (5.8) is reduced to:

$$(\mathbf{u} \cdot \nabla)\mathbf{u} = \mathbf{g} - \frac{1}{\rho}\nabla p. \tag{5.22}$$

Taking into account the identity Eq. (5.14) and remembering that $\mathbf{g} = -\nabla(gz)$ (Eq. 2.1) we get:

$$\nabla H = \mathbf{u} \times \boldsymbol{\omega} \tag{5.23}$$

where (for simplicity $V = |\mathbf{u}|$)

$$H = \frac{1}{2}V^2 + gz + \frac{p}{\rho}.$$

Equation (5.23) can be interpreted by answering the following question: how does H vary along the streamline? Let $\mathbf{s}$ be the unit vector parallel to $\mathbf{u}$ (Fig. 4.8): this variation is given by the directional derivative of H along $\mathbf{s}$ which, for Eq. (5.23) provides:

$$\frac{\partial H}{\partial \mathbf{s}} = \nabla H \cdot \mathbf{s} = (\mathbf{u} \times \boldsymbol{\omega}) \cdot \mathbf{s} = 0.$$

Therefore, under the conditions listed above, Bernoulli's theorem is formulated as follows:

$$\frac{1}{2}V^2 + gz + \frac{p}{\rho} = \textit{constant along the streamline} \tag{5.24}$$

or, H is constant along each streamline (which for stationarity coincides with the trajectory of the fluid element), but its value may differ from line to line. The first two terms are the kinetic and potential energy (per unit mass) of the fluid element; the presence of the term containing the pressure therefore extends the law of conservation of mechanical energy of the dynamics of the material point to the fluid dynamics context (this was already implicit in the expression of the energy flow $\mathbf{F}_E^{(rev)}$ in a perfect fluid obtained in Sect. 5.6). Finally, it is very relevant to note that for irrotational flows (Sect. 4.5) Eq. (5.23) implies:

$$\frac{1}{2}V^2 + gz + \frac{p}{\rho} = \textit{constant throughout the fluid} \tag{5.25}$$

In other words, H has the same value along all streamlines. This case finds application in a wide range of external flows, such as, for example, those obtained in Chap. 8.

In many cases of practical interest, the variations of the gz term in H are negligible compared to those of the other two, for example for substantially horizontal flows or, to put it better, for small vertical variations. In this case, multiplying by ρ gives:

$$\frac{1}{2}\rho V^2 + p_s = p_{tot} \tag{5.26}$$

Bernoulli's law is now written in terms of pressure: $p_d = \rho V^2/2$ is called *dynamic pressure*, as it is associated with flow speed. The p_s component is the *static pressure*:

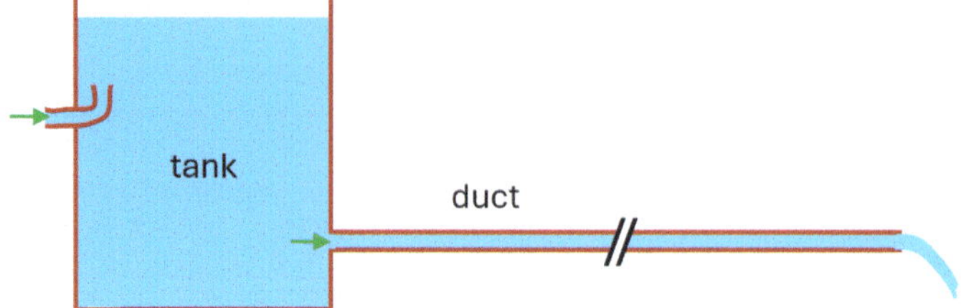

Fig. 5.3 A simple experimental apparatus that can produce the flow of a liquid through a duct

this depends on the random motion of the molecules, it is the thermodynamic pressure of the fluid and is the one that would be measured if one moved at the same speed as the fluid. Finally, their sum is obviously called *total pressure*, p_{tot}, but it is also called *stagnation pressure*, in fact, it is equal to the pressure that is in a stagnation point, where by definition $V = 0$ (see the analysis of the Pitot tube in this same paragraph).

We now move on to discuss three significant applications of Bernoulli's theorem for which the above conditions hold with good approximation. In particular, it is emphasised that the viscous effects will be considered totally negligible, but naturally a small effect on the flow is always present and will be highlighted in the discussion. In the following Sect. 5.9, on the contrary, the flow in a duct in which the viscous effects play a fundamental role will be analysed in detail.

In some subsequent applications, it will be assumed that a liquid flows through a duct, which can have different shapes and sizes. The Fig. 5.3 schematically shows the simplest way this can be achieved (in reality such an apparatus will have to be equipped with some technical adjustments, which are however ignored in the figure). Alternatively, a pump system can be used, etc.

- **Application 1: The Venturi tube**

Equation (5.26) is of immediate interpretation: the dynamic pressure can only increase at the expense of the static pressure, and vice versa, since their sum must remain constant. Therefore, in a horizontal duct with variable section S, since the velocity of the fluid (assumed incompressible) will have to be inversely proportional to S, the static pressure will be lower where the section is smaller.

To illustrate this property, in Fig. 5.4 a simple apparatus known as the Venturi tube is shown.

Imagine a horizontal duct of variable section containing air (white colour, density ρ_a) connected to a curved tube containing, for example, water (blue colour, density ρ_w). In the absence of motion (Fig. 5.4a) the pressure in the duct is the atmospheric one and the water level in the tube must be the same in both branches A and B due to the principle of communicating vessels (Sect. 3.2). In the presence of flow in the duct (Fig. 5.4b) the static pressure in the wide section, p_A, will be greater than that in the narrow section, p_B: as a result the water level in branch A of the tube will have to lower and consequently in branch B the water will be sucked in by the flow (the so-called *Venturi effect*, at the basis of many practical applications). The established

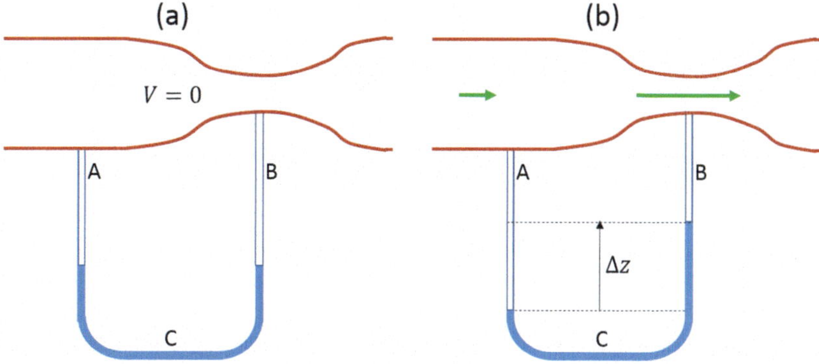

Fig. 5.4 The Venturi tube (dimensions not to scale)

difference in level Δz must satisfy the balance $p_A + \rho_a g \Delta z = p_B + \rho_w g \Delta z$ so that the horizontal pressure gradient at level C is zero, as required by the stationarity of the flow.

- **Application 2: The Pitot tube**

The Pitot tube allows the measurement of p_{tot} in Eq. (5.26) and, therefore, indirectly estimates the fluid speed. This device is therefore used as a speedometer for aircrafts, cars, boats, etc. Figure 5.5 illustrates the principle on which this technique is based.

Imagine a duct with a variable horizontal section (Fig. 5.5a) and a constant vertical section (Fig. 5.5b) in which a liquid (for example water) flows from left to right. Both at point A and, further downwind, at point B (both along the central -straight- flow

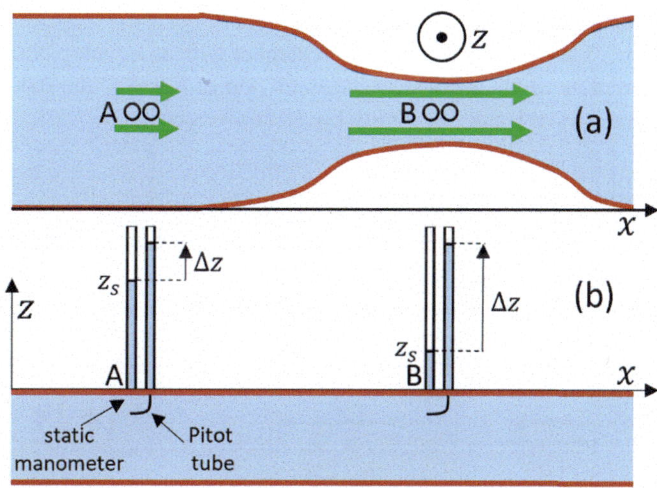

Fig. 5.5 The principle of the Pitot tube (dimensions not to scale)

line) a pair of vertical tubes are located (they are open at the top and are, therefore, in contact with the air, where $p = p_{atm}$). The first tube, called *static manometer* (or also piezometric tube) is also open at the bottom, therefore it is in contact with the water, which can rise inside it. The second tube, called the *Pitot tube*, is instead connected with a curved small tube open at the lower end and aligned with the fluid velocity, so as to intercept a flow line; obviously also in this tube the water can rise inside it. It is important that the horizontal part of the small tube is far enough from the solid wall so as to intercept the flow outside the boundary layer.

The pressure at the base of the static manometer corresponds to $p_s(x_A)$; in fact, the pressure through the boundary layer (Sect. 8.1) present between the solid wall and the non-viscous flow is practically uniform. With $p_s > p_{atm}$ the water will have to rise by a certain z_s with respect to the base of the tube until $p_{atm} + \rho_w g z_s = p_s$, thus stabilising the level. Therefore, z_s provides a direct measure of the local static pressure.

On the other hand, the Pitot tube intercepts a flow line that, at the mouth of the small tube (stagnation point) sees the cancellation of V. Therefore, the p_{tot} of Eq. (5.26) will manifest locally in a level $z_{tot} = z_s + \Delta z$ such that $\rho_w g \Delta z = \rho V^2/2$. Therefore, the combined measurement of the level of the static manometer and that of the Pitot tube allows the determination of the fluid velocity in the duct.

What has been said for point A also applies to point B but, while z_{tot} remains unchanged in accordance with Bernoulli's theorem, z_s is less than in A as the velocity (and therefore the dynamic pressure) in B is greater due to the conservation of mass. It should be said that z_{tot} in B will be slightly less than in A due to the small viscous loss of load; this aspect will be explored in Sect. 5.9.

A suitable Pitot tube allows the measurement of the speed of aircraft, etc. Imagine a device composed of two separate chambers arranged parallel to the flow (Fig. 5.6).

The fluid velocity must cancel out on the front wall; therefore, the pressure that is established in the chamber with an opening facing the flow is the stagnation pressure p_{tot} (red points). On the other hand, in the chamber with an opening parallel to the flow (corresponding to the opening of the static port of Fig. 5.5) the static pressure p_s (blue points) is established. Applying Eq. (5.26) we get:

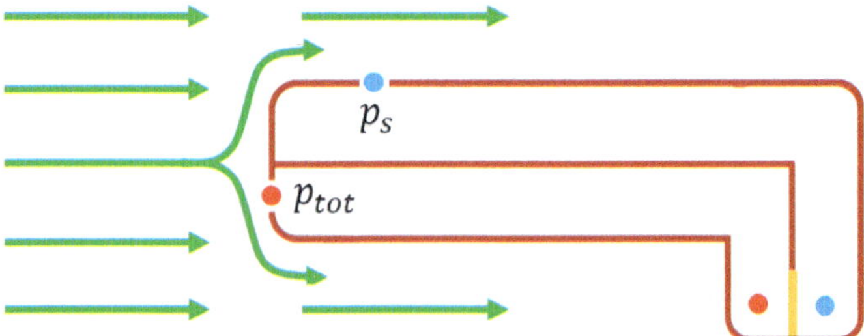

Fig. 5.6 Schematic illustration of the Pitot tube for the measurement of an aircraft's speed

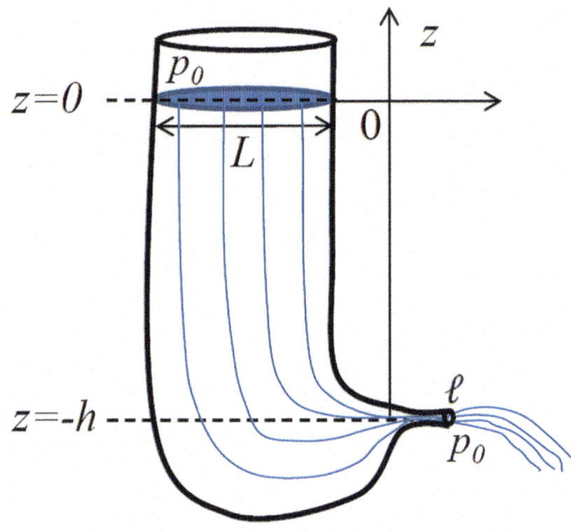

Fig. 5.7 Torricelli's flow

$$V = \sqrt{\frac{2(p_{tot} - p_s)}{\rho}}$$

Therefore, to obtain V it is enough to measure the difference $p_{tot} - p_s$ using a diaphragm (orange section) that separates the two chambers.

- **Application 3: Torricelli's theorem**

Consider a container of diameter L filled with liquid with free surface at $z = 0$ and equipped with a nozzle of diameter ℓ located at $z = -h$ from which the fluid can flow (Fig. 5.7). The quasi-stationarity of the flow is guaranteed by the condition $\ell \ll L$. Bernoulli's theorem can be applied in the form given by Eq. (5.25), as the flow is irrotational outside the thin boundary layer that adheres to the inner wall of the container (Sect. 8.1).

Applying the theorem at a point on the free surface (where $V \cong 0$ since $\ell \ll L$) and immediately outside the nozzle we get:

$$\frac{p_{atm}}{\rho} = \frac{1}{2}V^2 - gh + \frac{p_{atm}}{\rho}.$$

The exit speed of the liquid from the nozzle will therefore be:

$$V = \sqrt{2gh}.$$

This result goes under the name of Torricelli's theorem. It is interesting to note that V is the same speed that would be reached by an object at $z = -h$ if it fell from rest from $z = 0$.

5.8 Dynamics of Vorticity

In this paragraph, the evolution equation of vorticity for an incompressible fluid is derived. Taking into account the identity Eq. (5.14), the Navier–Stokes Eq. (5.8) is rewritten as follows:

$$\frac{\partial \mathbf{u}}{\partial t} + \frac{1}{2}\nabla|\mathbf{u}|^2 - \mathbf{u} \times \boldsymbol{\omega} = \nabla(gz) - \frac{1}{\rho}\nabla p + \nu\nabla^2\mathbf{u}. \tag{5.27}$$

Taking the curl of both members ($\nabla \times$) and remembering that $\nabla \times \nabla \cdot \equiv 0$ we get:

$$\frac{\partial \boldsymbol{\omega}}{\partial t} - \nabla \times (\mathbf{u} \times \boldsymbol{\omega}) = -\nabla \times \left(\frac{1}{\rho}\nabla p\right) + \nu\nabla^2\boldsymbol{\omega} \tag{5.28}$$

Now considering the identity

$$\nabla \times (\mathbf{A} \times \mathbf{B}) = (\mathbf{B} \cdot \nabla)\mathbf{A} - (\mathbf{A} \cdot \nabla)\mathbf{B} + \mathbf{A}\nabla \cdot \mathbf{B} - \mathbf{B}\nabla \cdot \mathbf{A} \tag{5.29}$$

with $\mathbf{A} = \mathbf{u}$ and $\mathbf{B} = \boldsymbol{\omega}$ and bearing in mind that the fluid is incompressible and that vorticity is a solenoidal field, we get:

$$\frac{d\boldsymbol{\omega}}{dt} = (\boldsymbol{\omega} \cdot \nabla)\mathbf{u} - \nabla \times \left(\frac{1}{\rho}\nabla p\right) + \nu\nabla^2\boldsymbol{\omega}. \tag{5.30}$$

As for the term containing the pressure we have:

$$\nabla \times \left(\frac{1}{\rho}\nabla p\right) = \frac{1}{\rho}\nabla \times \nabla p - \frac{1}{\rho^2}\nabla\rho \times \nabla p$$

(the first term on the right-hand side is trivially zero). In conclusion, the evolution equation of vorticity for an incompressible fluid is the following:

$$\frac{d\boldsymbol{\omega}}{dt} = (\boldsymbol{\omega} \cdot \nabla)\mathbf{u} + \nu\nabla^2\boldsymbol{\omega} + \frac{1}{\rho^2}\nabla\rho \times \nabla p. \tag{5.31}$$

The Lagrangian variation of vorticity is due to three effects. The first effect (term $(\boldsymbol{\omega} \cdot \nabla)\mathbf{u}$) goes under the name of *vortex stretching*, represents the main mechanism in turbulent dynamics and is responsible for the energy cascade between scales (Sect. 6.1). Note that this term is zero for two-dimensional flows, and this explains why turbulence is a strictly three-dimensional state of motion. The second effect (term $\nu\nabla^2\boldsymbol{\omega}$) describes the diffusion of vorticity: in this regard, see the discussion on the gradual diffusion of vorticity in a boundary layer in Sect. 8.1.

The third term describes the *baroclinic torque* mechanism, as a result of which vorticity is induced by the local misalignment of the isobaric and isopycnic surfaces

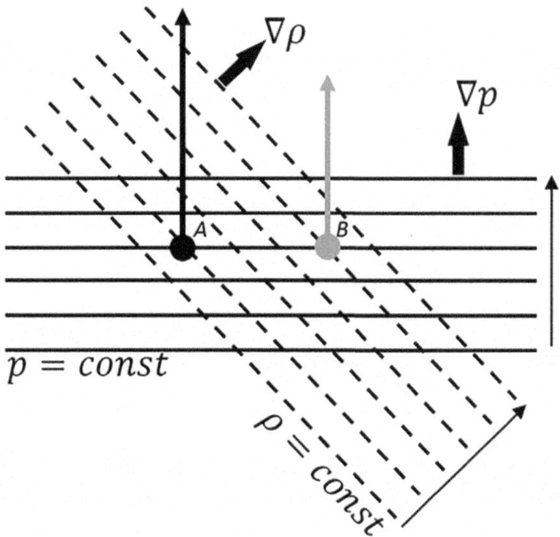

Fig. 5.8 Representation of the effect of the baroclinic torque on a plane. Solid (dashed) lines: isobars (isopycnals). The two fluid elements in A and B are subject to different accelerations, with resulting induction of vorticity

(in which case $\nabla\rho \times \nabla p \neq 0$). In Fig. 5.8 two fluid elements, black and grey, lie on the same isobar but on different isopycnals, with the density of the black element less than that of the grey element. The first fluid element will therefore be subject to a greater pressure gradient force per unit of mass; the different acceleration impressed on the two fluid elements will produce a rotation, and therefore vorticity. In this case, it is said that the fluid is in a *state of baroclinic motion*. Conversely, if $\nabla\rho \times \nabla p = 0$, the fluid is said to be in a *state of barotropic motion*. This second case occurs if the fluid is homogeneous ($\rho = const$) but also if ρ is variable but only depends on the pressure: $\rho = \rho(p)$. For example, as will be seen in Sect. 14.2, at low latitudes in the free atmosphere, geostrophic winds are substantially barotropic, while at medium and high latitudes the strong variation of temperature with latitude on an isobar ($\rho = \rho(p, T)$) makes the flow baroclinic.

5.9 Balance of Forces and Energy in Poiseuille Flow

This chapter concludes with the analysis of a viscous flow that will allow us to derive an instructive integral balance of forces and energy, thus complementing with a simple concrete example the dynamic concepts presented so far.

Consider a liquid flowing in a horizontal cylindrical tube, like the one schematised in Fig. 5.3 and 5.9.

The dynamics are described by Eq. (5.12) with $\rho = const$, which we rewrite for convenience:

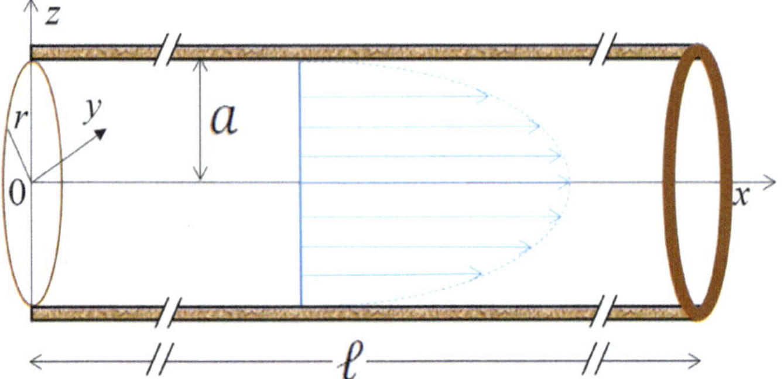

Fig. 5.9 Poiseuille flow diagram

$$\begin{cases} \rho \frac{d\mathbf{u}}{dt} = -\nabla p + \mu \nabla^2 \mathbf{u} \\ \nabla \cdot \mathbf{u} = 0 \end{cases} \tag{5.32}$$

where p is the modified pressure introduced in Sect. 5.6. Consider a stationary flow that, moreover, as is evident from Fig. 5.9, can be assumed to be *unidirectional*:

$$v = w = 0, \ \frac{\partial u}{\partial x} = 0 \tag{5.33}$$

(of course, the longitudinal extension ℓ of the conduit must be sufficient to exclude edge effects). With this position the continuity equation is identically satisfied and also $(\mathbf{u} \cdot \nabla)\mathbf{u} = 0$; therefore, in unidirectional flows the nonlinear terms are identically null. This is obviously a very important property for obtaining an exact solution of the equations. Note that Eq. (5.33) necessarily require:

$$\frac{\partial p}{\partial y} = \frac{\partial p}{\partial z} = 0. \tag{5.34}$$

Finally, imagine that a constant pressure difference Δp is maintained at the ends of the conduit, to which is associated a pressure gradient inside the conduit also constant:

$$-\frac{\partial p}{\partial x} = \frac{\Delta p}{\ell} = P = \cos t. \tag{5.35}$$

Considering Eqs. (5.33–5.35), the first of Eq. (5.32) is reduced to the classic *Poisson equation* in two dimensions:

$$\frac{\partial^2 u}{\partial y^2} + \frac{\partial^2 u}{\partial z^2} = -\frac{P}{\mu}. \tag{5.36}$$

Equation (5.36) can be easily solved by invoking the cylindrical symmetry of the problem. It is immediate to verify that

$$u(r) = -\frac{Pr^2}{4\mu} + A\ln r + B \tag{5.37}$$

(where r is the radial distance from the axis of the cylinder) is a solution of Eq. (5.36). As for the integration constants A and B, firstly, it is observed that A must be zero in order to eliminate the singularity at $r = 0$; B, on the other hand, must be determined by imposing the no-slip condition on the inner wall of the conduit of radius a:

$$u|_{r=a} = 0.$$

The speed profile thus obtained is a paraboloid with the maximum at the centre of the duct:

$$u(r) = \frac{P}{4\mu}\left(a^2 - r^2\right). \tag{5.38}$$

This is known as *Poiseuille flow*. Note that the flow does not depend on the density of the fluid: this is due to the fact that acceleration is identically zero in a unidirectional flow.

It should be noted that the applications of Bernoulli's theorem (valid for perfect fluids) relating to cylindrical ducts discussed in the previous paragraph would seem not to be applicable to real cases in light of the flow just obtained, for which viscous effects play a fundamental role. In fact, the flow given by Eq. (5.38) is the exact solution of the Navier–Stokes equation under the only assumption of unidirectional flow (and therefore implicitly of laminar flow, see the next chapter); so, how can a non-viscous flow exist in a cylindrical duct of practical interest? This aspect will be clarified in Sect. 8.2.

In the following, three applications of great conceptual interest will be considered: (i) the analogy between the mass flow rate of Poiseuille flow and Ohm's law, (ii) the integral balance of forces and energy and (iii) another category of unidirectional flows, called Couette flows.

- **Analogy with Ohm's law**

An interesting parameter to calculate is the volume flux Q (or mass flux ρQ, called also *mass flow rate*) through the generic section E orthogonal to the duct. Exploiting the cylindrical symmetry of the flow (Fig. 5.10), from Eq. (5.38) we get:

$$Q = \iint_E \mathbf{u}\cdot\mathbf{n}ds = \int_0^a u2\pi r dr = \frac{P\pi}{2\mu}\int_0^a \left(a^2 - r^2\right) r dr = \frac{\pi P a^4}{8\mu}.$$

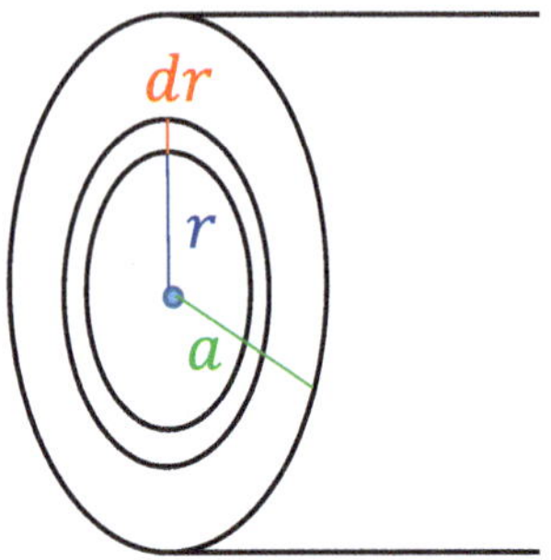

Fig. 5.10 Cylindrical symmetry of the flow

Therefore, we can write:

$$Q = \frac{\Delta p}{R}; R = \frac{8\mu\ell}{\pi a^4}. \tag{5.39}$$

This relationship expresses the *head loss* (pressure decrease) that occurs between two points at distance ℓ in the duct.

It is worth noting that the first relationship of Eq. (5.39) is perfectly analogous to Ohm's law which expresses the direct electric current I flowing in a conductor of electrical resistance R_{el} subjected to the electric potential difference ΔV :

$$I = \frac{\Delta V}{R_{el}}.$$

Even the expression of R presents some analogies with the resistance of a cylindrical conductor: ℓ corresponds to the length of the conductor and 8μ to the electrical resistivity; the dependence on the radius instead is different (in the electrical analogue at the denominator we have a^2).

- **Integral balance of forces and energy**

It is very instructive to evaluate the integral balance of the forces in play and the energy of the system. As for the forces, taking into account the cylindrical symmetry of the problem, the tangential stress acting on the inner surface of the duct is (Eqs. 2.12 and 5.38):

$$\mu \left.\frac{du}{dr}\right|_{r=a} = -\frac{1}{2}Pa.$$

Therefore, the total friction force turns out to be:

$$\mathcal{F}_{viscous} = 2\pi a\ell\left(-\frac{1}{2}Pa\right) = -\pi a^2 \Delta p$$

(the resultant of the internal stresses in the fluid is zero for the third law of dynamics). On the other hand, the total pressure gradient force is $\mathcal{F}_{pressure} = \pi a^2 \Delta p$, so a total zero balance of forces is obtained, as it must be for a non-accelerated flow like Poiseuille's:

$$\mathcal{F}_{viscous} + \mathcal{F}_{pressure} = 0.$$

Regarding energy, on one hand there is the work done by pressure forces dL_p, on the other hand the mechanical energy dissipated due to molecular viscosity $d\varepsilon_v$, which are calculated per unit of time. Indicating with $S = \pi a^2$ the area of the cylinder section, we have:

$$\frac{dL_p}{dt} = \Delta p S \frac{\langle dx \rangle}{dt} = \Delta p S \langle u \rangle = \Delta p Q,$$

where $\langle u \rangle$ is the velocity averaged over the section. On the other hand, applying the formula for the viscous dissipation rate Φ (Eq. 5.21) to the present problem with cylindrical symmetry, we obtain:

$$\frac{d\varepsilon_v}{dt} = \ell \int_0^a \Phi 2\pi r dr = 2\pi \ell \mu \int_0^a \left(\frac{du}{dr} \right)^2 r dr = \frac{\pi \ell P^2 a^4}{8\mu} = \ell P Q = \Delta p Q.$$

Therefore,

$$\frac{dL_p}{dt} = \frac{d\varepsilon_v}{dt}$$

or, the work done by pressure forces per unit of time is equal to the energy dissipated by viscosity, which is found in the form of heat irreversibly transferred to the fluid; in this way the flow is stationary.

- **Couette Flows**

As a simple extension of Poiseuille flow, we consider two other examples of unidirectional stationary flows known as *Couette flows*, which differ from the first in that the motion occurs between two parallel flat plates, as shown in Figs. 5.11 and 5.12.

Under these conditions the equation for velocity is as follows:

$$\frac{d^2 u}{dy^2} = -\frac{P}{\mu}.$$

Considering also the possibility that the upper plate is in motion with velocity U, the following solution is obtained:

$$u = \frac{P}{2\mu} y(D - y) + \frac{Uy}{D}.$$

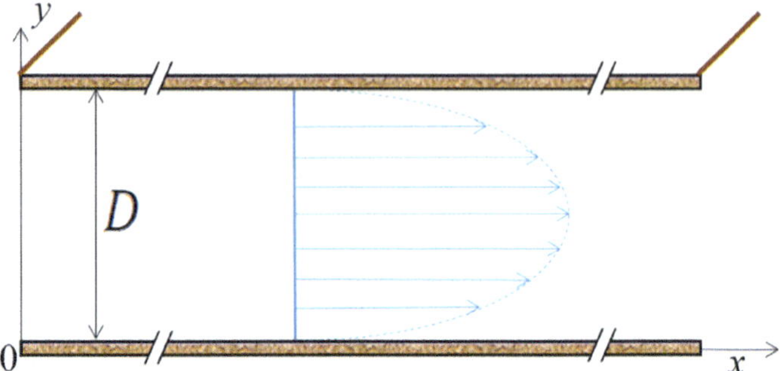

Fig. 5.11 Couette flow diagram with $U = 0$ and $P \neq 0$.

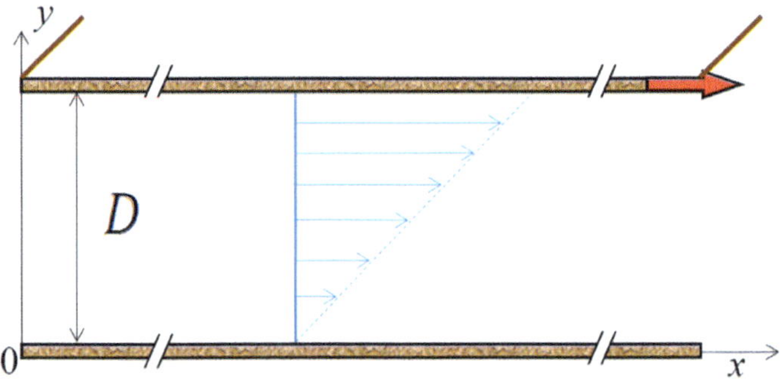

Fig. 5.12 Couette flow diagram with $U \neq 0$ e $P = 0$.

If $U = 0$ a paraboloid flow is obtained (Fig. 5.11). If instead $U \neq 0$ but $P = 0$, the linear shear flow shown in Fig. 5.12 is obtained.

This is the shear flow invoked in the argument used in Sect. 2.4 to experimentally determine the relationship between **d** and **e** for Newtonian fluids.

Recommended Reading

Batchelor, G.K.: An Introduction to Fluid Dynamics. Cambridge University Press, Cambridge (1967)

Çengel, Y.A., Cimbala, J.M.: Fluid Mechanics, Fundamentals and Applications. McGraw-Hill, Boston (2006)

Kundu, P.K., Cohen, I.M., Dowling, D.R.: Fluid Mechanics. Elsevier, Amsterdam (2012)

Landau, L.D., Lifshitz, E.M.: Fluid Mechanics. Pergamon Press, Oxford (1987)

Shapiro, A.H.: Illustrated Experiments in Fluid Mechanics. The MIT Press, Cambridge, Massachusetts (1972)

Tritton, D.J.: Physical Fluid Dynamics. Van Nostrand Reinhold, New York (1977)

White, F.M.: Fluid Mechanics. McGraw-Hill, New York (2011)

Chapter 6
Turbulence and Turbulent Viscosity

This chapter introduces the fundamental concepts related to the state of turbulent motion in a fluid. The peculiar characteristics of turbulence are described and the dimensionless Reynolds number, which governs the transition from a state of laminar motion to a turbulent one, is analysed. The concept of turbulent (eddy) viscosity, of fundamental importance in meteorological and oceanographic applications, is also introduced.

6.1 Phenomenology of Turbulence

Motion in a fluid can occur in two profoundly different modes: *laminar* or *turbulent motion*. In laminar motion the flow lines present a character of spatial and temporal regularity; however complex it may be, such a motion field varies gradually and predictably, that is, small variations in initial and boundary conditions produce small variations in the flow, with the consequent possibility, in principle, of predicting—for example by solving the equations of motion with numerical methods—the evolution of the flow itself. Figure 6.1a gives a schematic idea of this situation.

On the other hand, the state of turbulent motion is characterised by the presence of a wide variety of strongly variable motions (which can schematically be called *eddies*), which encompass a wide range of spatial and temporal scales (Figs. 6.1 and 6.2). Their most distinctive feature is to be unpredictable, with consequent *chaoticity* of the flow: small variations in initial and boundary conditions produce huge variations in the evolution of the flow, from which derives the impossibility of predicting the latter beyond a certain predictability time (but, as we will see in Sects. 6.3 and 6.4, it is possible to estimate the evolution of the average flow).

Other important characteristics of turbulence concern mixing and vorticity. Turbulent motion produces a mixing of fluid masses whose effect is in some ways similar to that due to molecular viscosity, but enormously more intense (Sect. 6.4). This

S. Pierini, *Oceanic and Atmospheric Fluid Dynamics*, UNITEXT for Physics,
https://doi.org/10.1007/978-3-031-77991-6_6

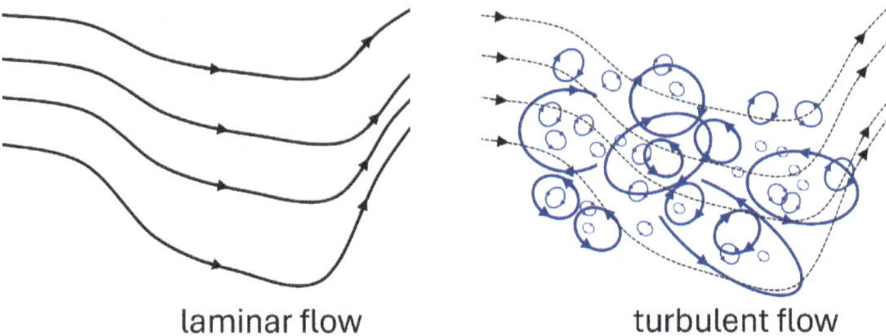

Fig. 6.1 Schematic representation of a laminar flow and a turbulent one. In the turbulent flow the dashed lines represent the average flow

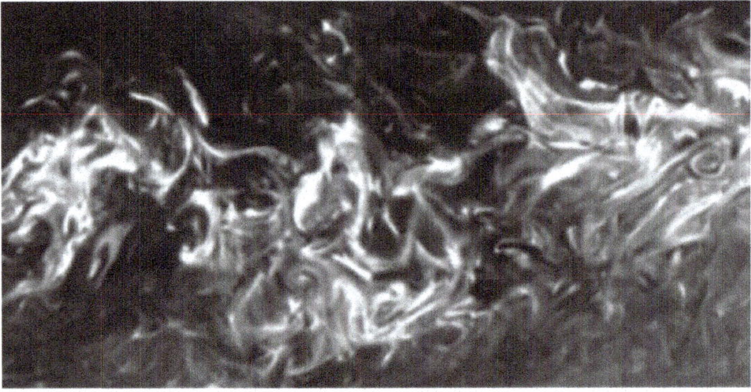

Fig. 6.2 Turbulent flow in a channel observed using "Planar laser-induced fluorescence". Figure adapted from Ishigami et al. (2022, Processes 10, 860, MDPI. Work distributed under a Creative Commons Attribution 4.0 International License)

phenomenon is associated with the stretching and folding of fluid elements; Fig. 6.3 shows an exhaustive example of this phenomenon. There is also a complex field of highly variable vorticity associated with the various scales of motion.

Turbulence is always present in the atmosphere and in the oceans. It manifests with very different modes and intensities, but never fails to accompany the dynamics of air masses and marine waters. Figure 6.4 shows an example of a time series of a horizontal component of the wind speed measured with a fast-response ultrasonic anemometer: the complexity of the evolution and the richness of the temporal scales at play is clearly represented.

The most interesting state of turbulent motion is that of the so-called *fully developed turbulence*. The most distinctive properties of this dynamic condition are the presence of an *energy cascade* from larger to smaller scales (Fig. 6.5) and the existence of a range of spatial scales within which this cascade is regulated by a universal law (there is also another form of pseudo-turbulence of a two-dimensional and geostrophic nature with much larger scales, for which the energy cascade is in the opposite direction; for hints see Sects. 17.5, 20.1 and 20.2).

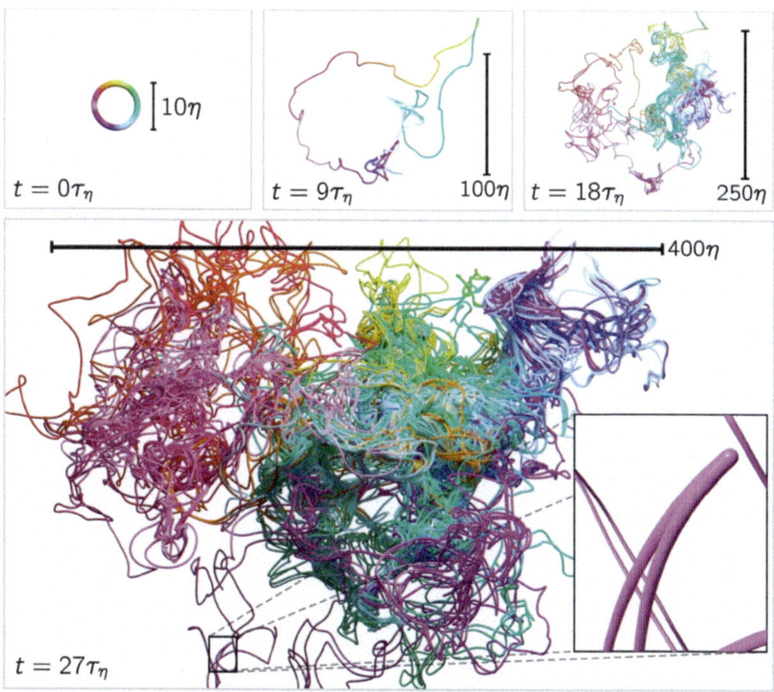

Fig. 6.3 Material evolution of an initial loop (first figure in the top left) produced by the advection of a turbulent motion (Bentkamp et al. 2022). The result is obtained through a *direct numerical simulation* (DNS) based on the Navier–Stokes equations in a situation of developed turbulence (τ_η is the Kolmogorov time scale). A Supplementary Movie of the same process is available: https://doi.org/10.1038/s41467-022-29,422-1. Figure taken from Bentkamp et al. (2022, Nat. Commun. 13, 2088; Springer Nature. Work distributed under a Creative Commons Attribution 4.0 International License)

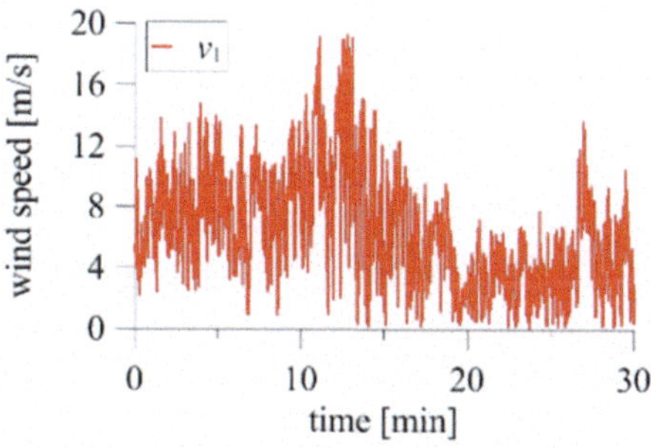

Fig. 6.4 Example of wind speed measured by a sonic anemometer. Figure adapted from Lipecki et al. (2020, Sensors 20, 5640, MDPI. Work distributed under a Creative Commons Attribution 4.0 International License)

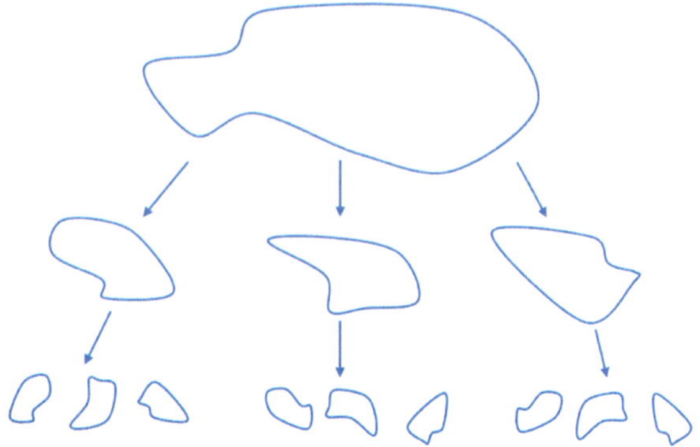

Fig. 6.5 Diagram of energy cascade from larger to smaller scales in three-dimensional turbulence

The schematic spectrum of Fig. 6.6 valid for fully developed turbulence shows how on the larger scales there is an input of energy coming from outside the system; on the smaller scales, however, the effects of molecular viscosity are manifested, with consequent dissipation of mechanical energy in favour of heat. As for the universal behaviour mentioned, in the 1940s the Russian mathematician Andrey Kolmogorov proposed a theory based purely on scale arguments, which indeed accurately describes the transition of energy through the various spatial scales. In the range, called inertial, the spectral energy density depends only on the wave number k according to the law $k^{-5/3}$, hence called *Kolmogorov's law*. This feature is also called *self-similar* and is associated with a so-called *fractal* behaviour.

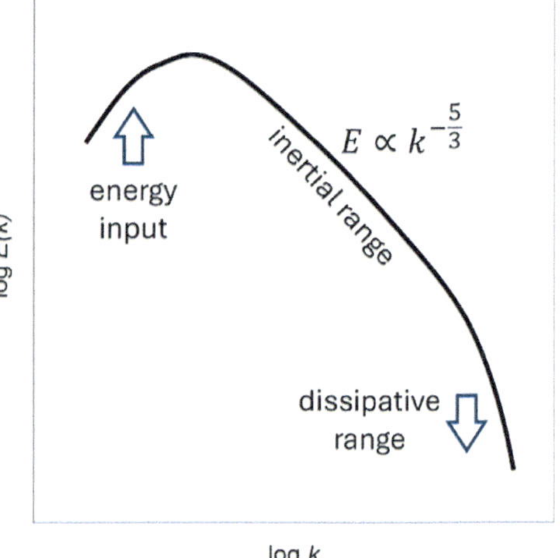

Fig. 6.6 Diagram of the energy spectrum of three-dimensional turbulence according to Kolmogorov's theory

6.2 Transition to Turbulence, Reynolds Number

One can now ask the following question: what determines the state of motion of a fluid? What dynamic characteristics discriminate between laminar and turbulent motion?

In Fig. 6.7a–c the schematic images of a flow (from left to right) impacting on a cylindrical obstacle as its speed increases are shown: the stages from (a) to (b) are characterised by a laminar motion of increasing complexity (in b there are recirculations downwind, which can also exhibit periodic or quasi-periodic oscillations). Conversely, stage (c) shows the presence of a turbulent motion downwind, certainly different from the developed one described in the previous paragraph, but still with chaotic characteristics (a further significant increase in speed would lead to the onset of fully developed turbulence). In this case, the increase in the overall flow speed upwind beyond a certain critical value leads to the transition from a laminar motion to a turbulent one.

In Fig. 6.7d, e the transition in the fluid exiting a nozzle occurs, instead, by varying the fluid's viscosity without changing the flow rate. The liquid introduced into the duct can be a mixture of water and a much more viscous fluid, such as glycerine. This has a kinematic molecular viscosity three orders of magnitude greater than that of water and, thanks to its three hydroxyl groups, is miscible with the latter. Therefore, by varying the proportions between the two liquids it is possible to control the viscosity of the mixture, changing its value very significantly. As can be seen, the transition from laminar to turbulent flow in this case is regulated by the decrease in ν.

The English engineer and scientist Osborne Reynolds (1842–1912) was the first to systematically study the transition to turbulence. He concluded that the transition from a laminar to a turbulent motion is determined by a dimensionless number,

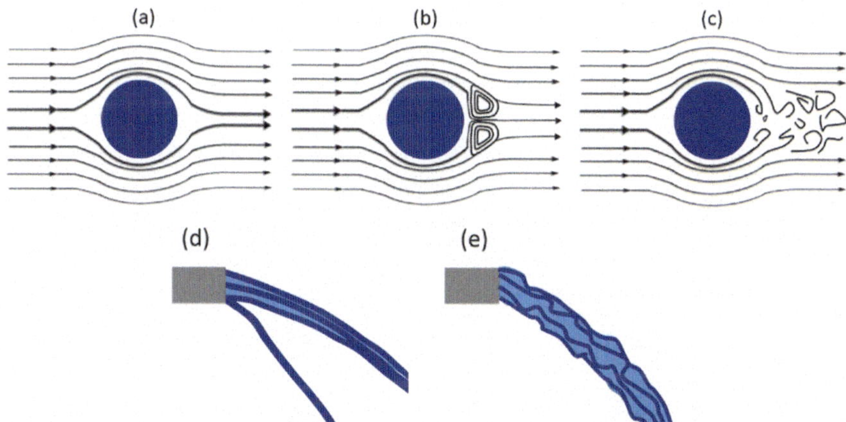

Fig. 6.7 Transition from a laminar flow to a turbulent one as the flow speed increases (**a–c**) or as the fluid viscosity decreases (**d, e**)

known as the Reynolds number, defined as:

$$Re = \frac{UL}{\nu}, \tag{6.1}$$

where ν is the kinematic viscosity of the fluid and U and L are respectively the typical scales of speed and length characterising the variations of the flow. The transition occurs, for given geometric conditions, at a critical value Re_c dependent on the case in question, but always much greater than unity (typically, $Re_c > O(10^3)$). As already noted in the previous paragraph, the transition does not involve a sudden change from a laminar situation to one of fully developed turbulence, but to a turbulent state that only gradually, with the increase of Re, reduces to Kolmogorov turbulence.

To understand the physical meaning of Re, it is necessary to switch to dimensionless variables and equations. Consider the Navier–Stokes equation for a homogeneous and incompressible fluid in which, for simplicity, p represents the modified pressure. The equation is made dimensionless, transforming all variables, both dependent and independent, into variables without physical dimension (indicated with an apostrophe), scaling them with appropriate dimensional parameters representing the typical scales of the flow variations:

$$\mathbf{u} = U\mathbf{u}'; \mathbf{x} = L\mathbf{x}'; t = \frac{L}{U}t'; p - p_0 = \rho U^2 p'$$

where p_0 is a representative value of the modified pressure. Note that the parameters that scale time and pressure depend on U and L; this is due to the constraint constituted by the same dynamic equations. After simple steps, we obtain:

$$-\nabla' p'_m = \frac{d\mathbf{u}'}{dt'} - \frac{1}{Re}\nabla'^2\mathbf{u}'. \tag{6.2}$$

This dimensionless version of the governing equations is also at the basis of the concept of *dynamic similarity* (see Sect. 20.5), which however cannot be explored in this discussion. The pressure gradient force (deliberately put in the left-hand side) plays a balancing role of the sum of the inertial (acceleration) and viscous effects. Thanks to the scaling carried out it is possible to state that:

$$O(|d\mathbf{u}'/dt'|) = (|\nabla'^2\mathbf{u}'|) = O(1).$$

Consequently, if $Re \ll 1$ the inertial terms are negligible compared to the viscous ones and vice versa if $Re \gg 1$. The same result can be reformulated in terms of dimensional variables as follows:

$$Re \approx O\left(\left|\frac{\rho d\mathbf{u}}{dt}\right| / |\mu\nabla^2\mathbf{u}|\right). \tag{6.3}$$

Thus, the Reynolds number expresses the ratio between the weight of the inertial and viscous terms. In Chaps. 7 and 8 some essential aspects of low and high Reynolds number flows will be discussed.

At this point, we are able to understand why turbulence necessarily requires $Re \gg 1$ to manifest. Without going into the delicate problem of the mechanisms that generate turbulent motion, it is understood how the most salient characteristic of turbulence, namely the transfer of energy between the various scales of motion, is associated with the nonlinear terms $(\mathbf{u} \cdot \nabla)\mathbf{u}$ present in the acceleration. Therefore, for turbulence to occur those terms must be predominant over the viscous ones, therefore $Re \gg 1$ must result. Of course, this condition is necessary but not sufficient for the flow to be turbulent; in fact, if $1 \ll Re < Re_c$ the flow is still laminar.

We conclude by noting that the meteorological and oceanographic dynamics is always characterised by a turbulent state of motion, albeit with very variable intensity in space and time. For example, considering the values of ν for air and water (Sect. 5. 3) extremely high values of Re are obtained for any reasonable value of U and L :

$$Re_{air} \sim UL \times 6 \times 10^4 \text{ m}^2\text{s}^{-1}; \; Re_{water} \sim UL \times 10^6 \text{ m}^2\text{s}^{-1}. \tag{6.4}$$

This refers to the necessity, particularly felt for meteorological and oceanographic problems, to resort to adequate parameterisations of turbulence, as discussed in the following paragraph.

6.3 Equations for the Average Fields

How to approach the study of fluid dynamics in the presence of turbulence? It has already been said about the intrinsic unpredictability of turbulent motions, so that even in principle it would not be possible to describe or model them in detail. Moreover, in the vast majority of cases there is not even an interest in solving turbulence deterministically. What is interesting, and also possible, is a statistical approach based on a methodology that once again (and it will not be the last time, as we will see in the next paragraph) recalls the name of Osborne Reynolds: the *Reynolds averaging*.

Let's define the average value (generally dependent on time) of a generic variable A as that signal obtained by applying the following *moving average* to the original variable:

$$\langle A \rangle(t) = \frac{1}{T} \int_{t-\frac{T}{2}}^{t+\frac{T}{2}} A(t')dt' \tag{6.5}$$

The parameter T represents the time window on which the average is calculated and must be chosen in such a way as to eliminate unwanted residual fluctuations (in our case the turbulent fluctuations) while preserving the lower frequency variability of which we are interested in knowing the evolution. All the variables of the turbulent fluid dynamics problem (which we consider here in the case of a homogeneous and incompressible fluid, cf. Eq. 5.32) can therefore be decomposed into an average part defined by Eq. (6.5) and a turbulent fluctuation (indicated with an apostrophe):

$$\begin{cases} \mathbf{u} = \langle \mathbf{u} \rangle + \mathbf{u}' \\ p = \langle p \rangle + p' \end{cases}$$

Naturally $\langle \mathbf{u}' \rangle = \langle p' \rangle = 0$ since $\langle\langle\cdot\rangle\rangle = \langle\cdot\rangle$. Substituting $\mathbf{u}$ thus decomposed into the continuity equation $\nabla \cdot \mathbf{u} = 0$, taking the average of both members and considering that $\partial \langle A \rangle / \partial x = \langle \partial A / \partial x \rangle$ we get:

$$\nabla \cdot \langle \mathbf{u} \rangle = 0; \ \nabla \cdot \mathbf{u}' = 0. \tag{6.6}$$

Proceeding similarly for the Navier–Stokes equations, after a series of simple mathematical steps we get:

$$\frac{D\langle u_i \rangle}{Dt} = -\frac{1}{\rho}\frac{\partial \langle p \rangle}{\partial x_i} + \nu \nabla^2 \langle u_i \rangle + \frac{1}{\rho}\sum_j \frac{\partial \tau_{ij}}{\partial x_j} \tag{6.7}$$

where

$$\frac{D}{Dt} = \frac{\partial}{\partial t} + (\langle \mathbf{u} \rangle \cdot \nabla)$$

is the Lagrangian derivative in which the advective field is given by the average field $\langle \mathbf{u} \rangle$ and where the matrix elements of the tensor $\boldsymbol{\tau}$, called *Reynolds stresses*, are defined as:

$$\tau_{ij} = -\rho \left\langle u_i' u_j' \right\rangle. \tag{6.8}$$

If the velocity fluctuations were, in this statistical approach, independent random variables, we would have $\tau_{ij} = 0$ for $i \neq j$; it is precisely the nonlinear character of turbulence that instead makes the Reynolds stresses generally non-zero.

It is important to note that Eq. (6.7) for the average fields are, net of the summation on the second member (the divergence of the tensor $\boldsymbol{\tau}$), formally identical to those for the total fields, or for the fields of a laminar flow. Therefore, the terms $\sum_j \partial \tau_{ij} / \partial x_j$ can be seen as the effect of turbulence on the mean fields. The next paragraph is dedicated to the analysis of these terms.

6.4 Reynolds Stress, Turbulent Viscosity

The matrix elements of the Reynolds stress tensor τ_{ij} represent further unknowns of the system. It is possible to derive the prognostic equations for such variables, but these in turn will contain other unknown terms and so on. For a fluid in a state of turbulent motion, a hierarchy of infinite prognostic equations is obtained whose solution necessarily requires the imposition of empirical assumptions to limit the number of equations: this is the so-called *turbulence closure problem*. Here the first order closure is introduced under particularly simplified, albeit significant, assumptions.

Referring to the so-called *flux-gradient theory* based on a series of analogies with Eq. (2.12) between shear stress and velocity shear for Newtonian fluids, the Reynolds stresses can be parameterised by the position:

$$\tau_{ij} = A_{ij} \frac{\partial \langle u_i \rangle}{\partial x_j} \tag{6.9}$$

The elements of the tensor A_{ij} (and of $K_{ij} = A_{ij}/\rho$), generally dependent on $\mathbf{x}$ and t, are called *eddy viscosity coefficients*.

In boundary layers (Sect. 8.1 and Chaps. 15 and 16), Ludwig Prandtl (German physicist, one of the founders of aerodynamics, 1875–1953) introduced the concept of *mixing length* ℓ, establishing an analogy with the mean free path Δr of the kinetic theory of gases (Sect. 1.1). Just as Δr gives the distance that on average a molecule travels between two successive collisions with other molecules, ℓ gives the average distance that a turbulent vortex travels before mixing with the surrounding vortices and thus losing its individuality. Referring to the application of this concept to the atmospheric and oceanic boundary layers (Chaps. 15 and 16), the eddy viscosity coefficient K_{13} that parameterises the vertical flux of horizontal momentum satisfies the empirical relationship

$$K_{13} = \ell^2 \frac{\partial \langle u \rangle}{\partial z}, \tag{6.10}$$

therefore, thanks to Eq. (6.9) the following relationship holds:

$$\tau_{13} = \rho \ell^2 \left(\frac{\partial \langle u \rangle}{\partial z} \right)^2. \tag{6.11}$$

Returning to Eq. (6.9), under the further assumptions that K_{ij} is constant and does not depend on the indices (isotropy),

$$K_{ij} = K = const,$$

Equation (6.7) is reduced to:

$$\frac{D\langle \mathbf{u} \rangle}{Dt} = -\frac{1}{\rho}\nabla \langle p \rangle + (\nu + K)\nabla^2 \langle \mathbf{u} \rangle. \tag{6.12}$$

These equations for the average fields of a turbulent flow differ from those for a laminar flow only by the presence of the terms $K\nabla^2\langle \mathbf{u} \rangle$, which are similar to those describing the molecular viscosity: for this reason their effect is called *turbulent* (or *eddy*) *viscosity*. The mathematical similarity reflects a similar physical effect that turbulent viscosity has compared to molecular viscosity, but with at least three major differences:

- While molecular viscosity is associated with the diffusion of momentum produced by molecular motion, in turbulent viscosity such diffusion is associated with the turbulent motion of macroscopic masses of fluid.
- While ν is practically constant (depending only weakly on T and p), K can depend substantially on space and time, as it represents a parameterisation of turbulent motions.
- The numerical values of ν and K differ enormously. For example, in the ocean $\nu \cong 10^{-6}\,\mathrm{m^2 s^{-1}}$ while the coefficient K_H associated with the effects of horizontal turbulence typical of the mesoscale dynamics can assume values $K_H = O\left(10 - 10^3\,\mathrm{m^2 s^{-1}}\right)$. For the effects of turbulence on the vertical, typical orders of magnitude are $K_\nu = O\left(10^{-3} - 10^{-2}\,\mathrm{m^2 s^{-1}}\right)$ for the oceanic surface mixed layer and $K_\nu = O\left(10\,\mathrm{m^2\,s^{-1}}\right)$ for the atmospheric boundary layer.

Therefore, in meteorological and oceanographic applications, molecular viscosity is negligible compared to turbulent viscosity; the term $\nu\nabla^2\langle \mathbf{u} \rangle$ is therefore eliminated from the force balance. In the same context, in order to consider the strong anisotropy present between horizontal and vertical motions in the dynamics from the mesoscale to the large scale, it is often used to introduce two distinct coefficients of eddy viscosity, K_H e K_ν (third point in the previous discussion). With this position, restoring for simplicity the symbol d/dt for the Lagrangian derivative, implying the average operator $\langle \cdot \rangle$ and reintroducing the force of gravity (for which now p is the total pressure), the following (vector) Navier–Stokes equation will be obtained for the average fields of a fluid in a state of turbulent motion, with closure at the first order and constant eddy viscosity for the parametrisation of turbulence:

$$\frac{d\mathbf{u}}{dt} = \mathbf{g} - \frac{1}{\rho}\nabla p + K_H \nabla_H^2 \mathbf{u} + K_\nu \frac{\partial^2 \mathbf{u}}{\partial z^2} \tag{6.13}$$

where $\nabla_H^2 = \partial^2/\partial x^2 + \partial^2/\partial y^2$. This is the equation (with the inclusion of the Coriolis force) to which reference will be made in Part II of the text.

Bibliography

Bentkamp, L., Drivas, T.D., Lalescu, C.C., Wilczek, M.: The statistical geometry of material loops in turbulence. Nat. Commun. **13**, 2088 (2022)

Ishigami, T., Irikura, M., Tsukahara, T.: Machine learning to estimate the mass-diffusion distance from a point source under turbulent conditions. Processes **10**, 860 (2022)

Lipecki, T., Jaminska-Gadomska, P., Sumorek, A.: Influence of ultrasonic wind sensor position on measurement accuracy under full-scale conditions. Sensors **20**, 5640 (2020)

Further Recommended Readings

Batchelor, G.K.: An Introduction to Fluid Dynamics. Cambridge University Press, Cambridge (1967)

Çengel, Y.A., Cimbala, J.M.: Fluid Mechanics, Fundamentals and Applications. McGraw-Hill, Boston (2006)

Kundu, P.K., Cohen, I.M., Dowling, D.R.: Fluid Mechanics. Elsevier, Amsterdam (2012)

Landahl, M.T., Mollo-Christensen, E.: Turbulence and Random Processes in Fluid Mechanics. Cambridge University Press, Cambridge (1986)

Landau, L.D., Lifshitz, E.M.: Fluid Mechanics. Pergamon Press, Oxford (1987)

Shapiro, A.H.: Illustrated Experiments in Fluid Mechanics. The MIT Press, Cambridge, Massachusetts (1972)

Townsend, A.A.: The Structure of Turbulent Shear Flow. Cambridge University Press, Cambridge (1976)

Tritton, D.J.: Physical Fluid Dynamics. Van Nostrand Reinhold, New York (1977)

White, F.M.: Fluid Mechanics. McGraw-Hill, New York (2011)

Chapter 7
Low-Reynolds Number Flows

This chapter discusses the general problem of flows at low Reynolds number. An interesting example of meteorological and oceanographic relevance is also considered.

7.1 Approximate Governing Equations for Low-Reynolds Number Flows

In Sect. 6.2 it was shown how the dimensionless Reynolds number *Re* provides an estimate of the ratio between the weight of the inertial terms and that of the viscous terms that appear in the Navier–Stokes equation. On the other hand, it was also noted (Eq. 6.4) that the viscous effects are so weak as to be comparable with acceleration, or to prevail over it, only if $UL \lesssim 10^{-5}$ $\mathrm{m^2s^{-1}}$ for air and $UL \lesssim 10^{-6}$ $\mathrm{m^2s^{-1}}$ for water (naturally in other applications referring to fluids with a much greater ν, UL can be much larger). This means that normally the Reynolds number is much greater, or even enormously greater than unity (*high-Reynolds number flows*, $Re \gg 1$); only if the motion is characterised by very small velocities and linear dimensions can viscosity prevail over inertia (*low-Reynolds number flows*, $Re \ll 1$). However, it should be noted that, as we will see in the next chapter, molecular viscosity plays a fundamental role even if, away from solid bodies, $Re \gg 1$; in fact, within the thin boundary layers along the solid boundaries, both U and L are drastically reduced to make the viscous effects prevalent.

Therefore, the case $Re \ll 1$ is beyond the scope of this text. However, for completeness and by way of example, this chapter will deal with a case that, despite extremely small velocity and length scales, is of undoubted meteorological and oceanographic interest: the flow induced by a small sphere moving in a fluid. But, first of all, in this paragraph we introduce the approximate governing equations based on the condition $Re \ll 1$.

S. Pierini, *Oceanic and Atmospheric Fluid Dynamics*, UNITEXT for Physics,
https://doi.org/10.1007/978-3-031-77991-6_7

For a homogeneous and incompressible fluid ($\rho = const$, $\nabla \cdot \mathbf{u} = 0$, it is, once more, convenient to switch to the modified pressure), under the assumption $Re \ll 1$ the acceleration in the Navier–Stokes equation can be neglected:

$$\nabla p = \mu \nabla^2 \mathbf{u}. \tag{7.1}$$

Multiplying both members scalarly by ∇ and taking into account the incompressibility, we obtain:

$$\nabla^2 p = 0. \tag{7.2}$$

Multiplying both members of Eq. (7.1) vectorially by ∇, we obtain:

$$\nabla^2 (\nabla \times \mathbf{u}) = 0 \tag{7.3}$$

(here it is convenient to leave the vorticity expressed in terms of $\mathbf{u}$, this being the unknown dependent variable). If the problem requires imposing boundary conditions that involve only velocity, Eq. (7.3) will need to be solved together with the continuity equation $\nabla \cdot \mathbf{u} = 0$; the pressure can then be determined using Eq. (7.1). If, on the other hand, the boundary conditions involve only pressure, Eq. (7.2) will need to be solved; the velocity can then be determined using Eq. (7.1) with $\nabla \cdot \mathbf{u} = 0$.

A case of great interest concerns the flow induced by a solid body of volume Ω (bounded by the surface $\partial\Omega$) moving translationally with constant velocity $\mathbf{U}$ in an otherwise static fluid. In this case, the conditions to be imposed will be:

$$\begin{cases} \mathbf{u}|_{\partial\Omega} = \mathbf{U} \\ \lim\limits_{|\mathbf{x}| \to \infty} \mathbf{u} = 0 \\ \lim\limits_{|\mathbf{x}| \to \infty} p = p_0 \end{cases} \tag{7.4}$$

The first is the no-slip condition on the surface of the body, the second requires that the fluid velocity cancels out far from the body and the third being that the pressure converges to a reference value p_0 at infinity. In this case, the velocity field $\mathbf{u}$ will depend *parametrically* on time, in the sense that, for example, the streamlines photographed at time t will be found translated by $d\mathbf{x}=\mathbf{U}dt$ at time $t + dt$. On the other hand, in the reference system fixed to the body (the option adopted in the following paragraph), there is a stationary flow ($\partial \mathbf{u}/\partial t = 0$) with $\mathbf{u} \to \mathbf{U}$ at infinity and $\mathbf{u} \to 0$ on $\partial\Omega$; in this case, the streamlines coincide with the trajectories of the fluid elements.

7.2 Flow Induced by a Sphere at Low Reynolds Number

Consider now the flow induced by a sphere in translational motion for $Re \ll 1$. The dynamics are governed by Eqs. (7.3) and (7.4) with $\nabla \cdot \mathbf{u} = 0$. The derivation of the flow requires rather complex calculations; considering the specific nature of the topic which is beyond the scope of this text, for a detailed derivation of the solution, refer to Batchelor (1967, section 4.9 therein).

The stress on the surface of the sphere is found to be:

$$\mathbf{S}|_{\partial\Omega} = -p_0\mathbf{n} - \frac{3\mu}{2a}\mathbf{U}, \tag{7.5}$$

where $\mathbf{n}$ is the normal to the generic point on $\partial\Omega$ and a is the radius of the sphere. It is interesting to note that the stress due to viscosity is the same at every point on $\partial\Omega$ (this is not a general property but is only valid for spherical bodies). Moreover, the pressure is simply represented by the normal stresses that would be at infinity. The resistance $\mathbf{D}$ (called *drag*) that the fluid opposes to the motion of the body is obtained by integrating $\mathbf{S}|_{\partial\Omega}$ on $\partial\Omega$:

$$\mathbf{D} = \oiint_{\partial\Omega} \mathbf{S} ds = -p_0 \oiint_{\partial\Omega} \mathbf{n} ds - \frac{3\mu}{2a} 4\pi a^2 \mathbf{U} = -6\mu\pi a \mathbf{U}. \tag{7.6}$$

This expression is known as *Stokes' law*.

In general, the force exerted by the fluid on a moving body is expressed using a dimensionless coefficient called the *drag coefficient*, obtained by scaling D with a parameter that has the dimension of a stress (for homogeneous fluids $\rho U^2/2$ is used) multiplied by the area of the projection of the body on the plane perpendicular to $\mathbf{U}$, which in this case is πa^2:

$$C_D = \frac{2D}{\rho U^2 \pi a^2}. \tag{7.7}$$

If the typical length scale of the motion is defined as $L = 2a$ and, as is natural, the body's velocity U is adopted for the typical velocity scale, the corresponding Reynolds number will be

$$Re = \frac{2aU}{\nu}.$$

Therefore, the drag coefficient for Stokes' law can be expressed in terms of Re as follows:

$$C_D = \frac{24}{Re}. \tag{7.8}$$

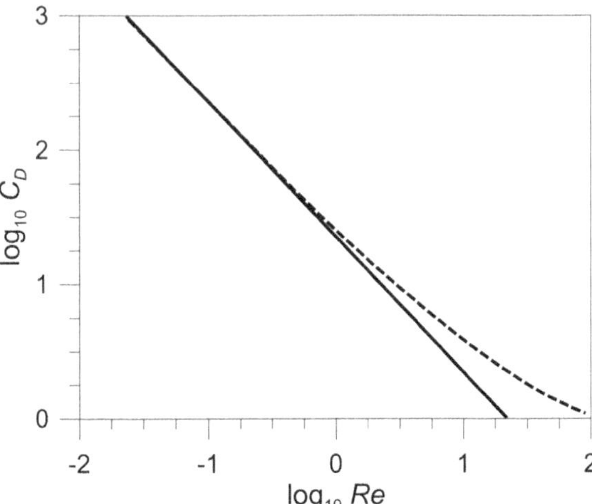

Fig. 7.1 Comparison between the drag coefficient C_D of Stokes' law (solid line) and that obtained experimentally (dashed line)

In relation to this, one must ask to what extent Re must be less than unity for Eq. (7.8) to be considered valid. Figure 7.1 shows the comparison between the theoretical C_D coefficient and that obtained from experimental data: as can be seen, the agreement is good even for $Re \lesssim 1$.

7.3 Meteorological and Oceanographic Applications

An important application of Stokes' law of meteorological and oceanographic interest concerns the determination of the limit speed U_{lim} reached (i) by cloud droplets formed by the condensation of water vapour on condensation nuclei in free fall in the atmosphere and (ii) by particulate matter in free fall in water. Given the very small linear scale of these particles (of the order of hundredths of a mm), it can be assumed that the flow from them induced can verify the condition $Re \lesssim 1$; naturally this will need to be verified a posteriori. One of the most delicate and relevant aspects of atmospheric physics is the description and modelling of microphysical phenomena of clouds and rain formation which, by their very nature, must be treated using appropriate parameterisations; in this context, the estimate of U_{lim} is of fundamental importance. An analogous role is played by the same parameter in the study of particulate deposition in the sea, of geophysical and environmental relevance. In this context, we will now determine the speed U_{lim} in the two cases in question.

Consider a sphere assumed rigid with density $\overline{\rho}$ in free fall in a fluid of density $\rho < \overline{\rho}$. The particle will be subject to reduced gravity directed downwards (Sect. 3. 3) and to the force of drag D given by Eq. (7.6) directed upwards (Fig. 7.2). The constant deposition speed U_{lim} is reached when the two forces balance:

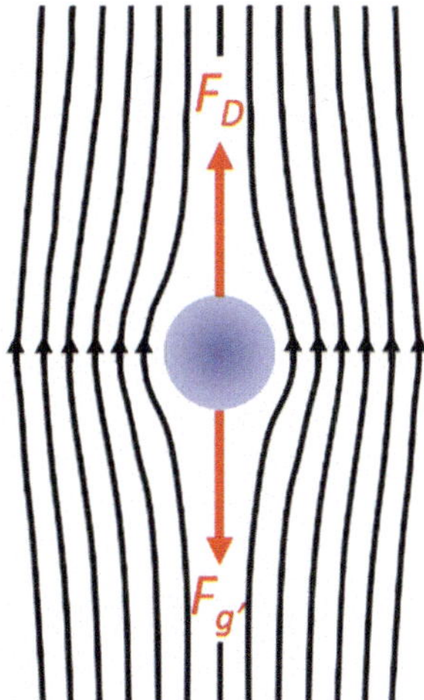

Fig. 7.2 Balance of forces for a sphere in free fall in a fluid

$$6\mu\pi a U_{lim} = \frac{4}{3}\pi a^3(\overline{\rho} - \rho)g.$$

Therefore, the expression for the limit speed is as follows:

$$U_{lim} = \frac{2}{9}\frac{a^2 g}{\nu}\left(\frac{\overline{\rho}}{\rho} - 1\right). \tag{7.9}$$

The corresponding limit Reynolds number is consequently:

$$Re_{lim} = \frac{4}{9}\frac{a^3 g}{\nu^2}\left(\frac{\overline{\rho}}{\rho} - 1\right). \tag{7.10}$$

It is first necessary to verify the consistency a posteriori of the assumption $Re_{lim} \lesssim 1$. We start by considering a cloud droplet (assumed rigid but considered composed of liquid water) in fall in the air, in which case $\overline{\rho}/\rho \cong 780$ and $\nu_{air} \cong 1.5 \times 10^{-5}\,\mathrm{m^2 s^{-1}}$. Substituting these values into Eq. (7.10) we get $Re_{lim} \cong 1$ if $\overline{a} = 40\,\mu\mathrm{m}$; therefore, the assumption of low Reynolds number motion is satisfied if $a \lesssim \overline{a}$. This condition is actually satisfied in cumulus clouds, for which a varies in a range $\sim 0.5 - 20\mu m$. For a reference radius $a = 10\mu m$ we get $Re_{lim} \cong 0.015$ and $U_{lim} \cong 11 mm s^{-1}$.

Collision and coalescence processes involving the droplets during their fall lead to the formation of raindrops. The relative dynamics is completely outside the range considered here, resulting in $Re \gg 1$; in fact, linear dimensions and precipitation speeds come into play that are respectively $O(100)$ and $O(1000)$ times greater than those of the droplets.

To complete the analysis just presented, it is useful to briefly discuss this case here, despite it being related to the different range $Re \gg 1$. The drag coefficient for a sphere for $Re \gg 1$ depends on the range of Re. From values $Re \approx 10$ the boundary layer separation begins (Sect. 8.1) downwind of the sphere, with the release of vortices (*vortex shedding*). The drag now depends essentially on the pressure rather than the viscosity (*pressure drag*), with C_D decreasing significantly up to $Re \approx 10^3$. In the range $10^3 < Re < 10^5$, C_D settles on a nearly constant value $C_D \cong 0.4 - 0.5$. At the transition to turbulence (for $Re_c \approx 2 \times 10^5$) there is finally a sudden decrease in C_D, reaching values around $C_D \cong 0.1$. In light of this, it is possible to calculate the limit speed of a sphere for $Re \gg 1$ by equating the drag D, expressed in terms of C_D from Eq. (7.7), to the reduced gravity,

$$\frac{1}{2}\overline{\rho}U_{lim}^2 \pi a^2 C_D = \frac{4}{3}\pi a^3 (\overline{\rho} - \rho) g,$$

obtaining

$$U_{lim} = \sqrt{\frac{8ag}{3C_D}\left(\frac{\overline{\rho}}{\rho} - 1\right)}. \tag{7.11}$$

For a raindrop of radius $a = 1$ mm with $C_D = 0.4$, $U_{lim} \cong 7\,\text{m s}^{-1}$, corresponding to $Re_{lim} \cong 950$.

Finally, returning to the range $Re_{lim} \lesssim 1$, as far as the precipitation of particulate in water is concerned, $\overline{\rho}/\rho \cong 2$ and $\nu_{water} \cong 1.1 \times 10^{-6}\,\text{m}^2\text{s}^{-1}$. Substituting these values into Eq. (7.10) we get $Re_{lim} = 1$ if $\overline{a} = 65\mu$m; therefore, the assumption of motion at low Reynolds number is satisfied if $a \lesssim \overline{a}$. Also in this case the condition finds interesting application, as a wide class of solid particles have dimensions ranging from a few μm up to values comparable with $\overline{a}$. In the considered case, for a reference radius of $a = 50\,\mu$m, $Re_{lim} \cong 0.45$ and $U_{lim} \cong 5\,\text{mm s}^{-1}$.

Bibliography

Batchelor, G.K.: An Introduction to Fluid Dynamics. Cambridge University Press, Cambridge (1967)

Further Recommended Readings

Çengel, Y.A., Cimbala, J.M.: Fluid Mechanics, Fundamentals and Applications. McGraw-Hill, Boston (2006)
Landau, L.D., Lifshitz, E.M.: Fluid Mechanics. Pergamon Press, Oxford (1987)
Shapiro, A.H.: Illustrated Experiments in Fluid Mechanics. The MIT Press, Cambridge, Massachusetts (1972)

Tritton, D.J.: Physical Fluid Dynamics. Van Nostrand Reinhold, New York (1977)

Wallace, J.M., Hobbs, P.V.: Atmospheric Science, an Introductory Survey. Elsevier, Amsterdam (2006)

White, F.M.: Fluid Mechanics. McGraw-Hill, New York (2011)

Chapter 8
High-Reynolds Number Flows

This chapter discusses the general problem of flows at high Reynolds number. We start by considering the boundary layer, its structure in both laminar and turbulent cases, and its separation from the solid body. The transition of an internal flow from potential to viscous is then addressed. The irrotationality in perfect fluids is also analysed based on the classic Kelvin theorem. Finally, a series of potential external flows of aerodynamic and naval interest are derived.

8.1 The Boundary Layer: Structure and Separation

It has already been observed (Sects. 6.2 and 7.1) that, except in very particular cases (Chap. 7), flows occur with high Reynolds numbers ($Re \gg 1$); in other words, in the Navier–Stokes equation, the inertial effects ($d\mathbf{u}/dt$) have a predominant weight compared to the viscous ones ($\nu \nabla^2 \mathbf{u}$). However, in many cases this does not imply that the perfect fluid approximation can be adopted without any consideration of the effects of molecular viscosity. In fact, there will still be a region, however confined to a thin layer that wraps the solid boundaries that delimit the fluid, called *boundary layer*, in which, due to the no-slip condition, viscosity, together with inertial effects, plays a fundamental role with sometimes significant consequences for the overall flow.

Therefore, while it is true that the assumptions of perfect fluid and irrotational—and therefore potential—flow can be usefully adopted in many cases of great interest (as will be seen in this chapter), it is nevertheless necessary to analyse the main characteristics of the boundary layers to obtain a correct overall view of flows with $Re \gg 1$.

This paragraph will briefly address the problem of the laminar boundary layer, the turbulent one, and its possible separation from the solid boundary (a more detailed and quantitative treatment is beyond the scope of this text). In this chapter, it is

S. Pierini, *Oceanic and Atmospheric Fluid Dynamics*, UNITEXT for Physics,
https://doi.org/10.1007/978-3-031-77991-6_8

assumed everywhere that the fluid is homogeneous and incompressible. In Part II of the text, atmospheric and oceanic boundary layers (Chaps. 15 and 16) will also be analysed, in which the Coriolis force plays an important role.

- **Laminar boundary layer**

Figure 8.1 shows the boundary layer (white background) that forms in a uniform laminar flow with velocity U (Fig. 8.1a) that impinges on a flat plate E in the absence of a horizontal pressure gradient. Figure 8.1b shows how the vanishing of the fluid velocity immediately in contact with the plate is associated with the formation of a thin boundary layer in which the strong transverse velocity gradient implies a significant effect of molecular viscosity and the presence of intense shear vorticity. Conversely, outside the boundary layer (grey background), the viscous stresses are negligible and the flow is irrotational.

The thickness of the boundary layer δ is generally defined as the distance along y between the solid boundary and the level at which the velocity corresponds to a certain, high, percentage of that of the free flow U (usually 99%). Figures 8.1b-c show that δ grows along the plate, as the deceleration of the fluid gradually transfers between the layers due to viscosity. This corresponds to a decrease in the velocity shear on the plate $s = du/dy|_E$, and therefore of the corresponding shear stress $\tau = \mu s$, with the increase of x.

The cause of the growth of δ along x can be understood in terms of vorticity. First of all, note that the vorticity integrated within the boundary layer is constant along x. In fact, consider the flux C of $\boldsymbol{\omega}$ through the surface Ω located at x_A whose lower section (1) of length ℓ (within which the flow does not vary) corresponds to the solid boundary while the specular one (3) is well beyond the boundary layer (Fig. 8.1b):

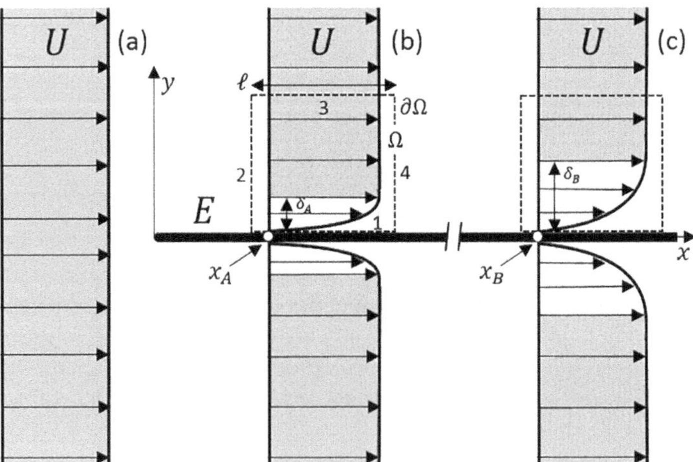

Fig. 8.1 Boundary layer that forms in a uniform flow with velocity U on a flat plate E. The boundary layer, of thickness δ, is with a white background; the free and uniform flow is with a grey background

$$C = \iint_{\Omega} \boldsymbol{\omega} \cdot \mathbf{n} ds = \iint_{\Omega} \omega ds.$$

According to Stokes' theorem, C is equal to the line integral of $\mathbf{u}$ along the boundary $\partial\Omega$:

$$C = \oint_{\partial\Omega} \mathbf{u} \cdot d\boldsymbol{\ell} = \int_{\partial\Omega(3)} \mathbf{u} \cdot d\boldsymbol{\ell} = -\ell U \tag{8.1}$$

Along section 1 the velocity vanishes and along the lateral sections 2 and 4 the velocity is perpendicular to $d\boldsymbol{\ell}$; therefore, C reduces to $-\ell U$ and is therefore independent of x (the calculation at x_B would give the same result). But the vorticity is all contained within the boundary layer, so the total vorticity of the boundary layer contained in a length interval ℓ is constant along x despite δ increasing with x. Moreover, since the flow before hitting the plate is devoid of vorticity, it can be inferred that C is generated by the interaction of the free flow with the leading edge of the plate (at $x = 0$ in the figure).

These conclusions can be summarised by stating that the vorticity contained in the boundary layer—and generated at the leading edge—is diffused transversely by the mechanism of molecular diffusion during the movement of the fluid along the plate. To understand this important statement, reference must be made to Eq. (5.31) of vorticity evolution. Since here we are considering a fluid not only incompressible but also homogeneous ($\rho = const$), therefore barotropic, and that the flow is two-dimensional in the plane (x, y) with $\boldsymbol{\omega} = (0, 0, \omega)$, Eq. (5.31) reduces to the classic *advection–diffusion equation* for the vorticity:

$$\frac{d\omega}{dt} = \nu \nabla^2 \omega. \tag{8.2}$$

The *advection* process is provided by the advective terms in the Lagrangian derivative and that of *diffusion* by the Laplacian term (this latter corresponds to the term $\nu \nabla^2 \mathbf{u}$ present in the Navier–Stokes equation). In light of Eq. (8.2), δ can be seen as a measure of the distance from the plate within which vorticity diffusion has taken place.

To obtain an estimate of the dependence of δ on x, one can invoke the *Fick's second law*, which states that the length scale of diffusion is proportional to $\sqrt{Kt}$, where K is the diffusion coefficient. In this case, setting $K = \nu$ and considering the time $t \cong x/U$ that the fluid element takes to reach the position x, we get:

$$\delta \propto \sqrt{\frac{\nu x}{U}}. \tag{8.3}$$

Therefore, the thickness of the boundary layer grows as $\delta \propto x^{1/2}$. It is also interesting to note that it decreases as $\delta \propto U^{-1/2}$ at the same distance from the leading edge, as with a higher U the boundary layer has less time to grow in thickness.

What has been said has a general character, being indicative of the structure of the laminar boundary layer that is established around a generic solid body in the absence of pressure gradients. In a case like this, in which the boundary layer remains everywhere adherent to the solid surface, the main effect of viscosity is that of generating a friction resistance D_f (*skin friction drag*) due to the integrated effect of the shear stresses acting on the surface.

- **D'Alembert's paradox**

In this regard, it is worth mentioning the so-called *D'Alembert's paradox* (from the mathematician, physicist and French philosopher Jean le Rond D'Alembert, 1717–1783). D'Alembert demonstrated that, under the hypothesis of stationary and potential flow in an incompressible and inviscid fluid, a body in uniform motion does not register any resistance, in clear contradiction with experience; this can be intuitively understood as follows. Let $\mathbf{U}$ be the constant speed of the body in the fluid and let $\mathbf{D}$ be the resistance that the fluid opposes to the motion of the body: based on experimental evidence this must have the same direction as $\mathbf{U}$ and opposite direction ($\mathbf{D} \propto -\mathbf{U}/U$). However, the force$-\mathbf{D}$ that, conversely, the body would exert on the fluid would perform work with a consequent continuous input of energy: this should in turn be dissipated by viscosity to allow the flow to be stationary. But if we reduce to consider an inviscid fluid, it necessarily follows that $\mathbf{D} = 0$. Note that this does not exclude the presence of a force perpendicular to $\mathbf{U}$, the *lift* (see Sect. 8.4), as the latter does not perform work.

The paradox is resolved by considering the resistance produced by molecular viscosity due to the shear stresses acting on the surface of the body. Consider that when the paradox was stated (1752) the theory of fluid dynamics could not yet count on the fundamental concept of boundary layer. It was only at the beginning of the twentieth century that Ludwig Prandtl experimentally studied and mathematically described the boundary layers, definitively resolving D'Alembert's paradox.

It is important to emphasise that the resistance that the fluid exerts on a solid body in motion also depends, and sometimes substantially, on the possible separation of the boundary layer. This aspect will be subsequently analysed in this same paragraph and, for further study, refer to Sects. 8.4 and 8.5.

- **Turbulent boundary layer**

The boundary layer analysed in the previous paragraph is laminar, however in most cases of practical relevance the boundary layer exhibits a turbulent dynamics (Chap. 6). This implies that locally the motion will fluctuate in an irregular and three-dimensional manner.

Figure 8.2a shows some velocity profiles corresponding to different instants in a turbulent boundary layer with stationary free flow U (grey arrow); the dashed line represents the average of the profiles. Figure 8.2b shows the comparison between the average turbulent and laminar profile. It can be observed how the turbulent mixing,

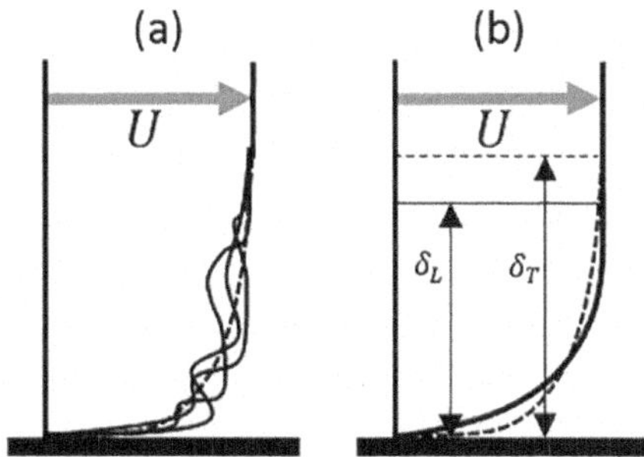

Fig. 8.2 **a** Instantaneous velocity profiles in a turbulent boundary layer (solid lines) and average profile (dashed line). **b** Comparison between turbulent and laminar boundary layer

producing a homogenisation of the motion, provides a velocity profile that differs little from U up to a small distance from the surface, where the transition $\mathbf{u} = 0$ occurs suddenly; the laminar profile is instead more gradual. This implies that the shear stress on the surface in a turbulent boundary layer is greater than the corresponding laminar one: $\tau_T > \tau_L$. Note, furthermore, that the thickness of the turbulent boundary layer δ_T is systematically greater than the laminar one δ_L.

The expression of the integrated vorticity C given by Eq. (8.1) is also valid in the turbulent case; therefore, a laminar boundary layer and a turbulent one corresponding to the same free flow U contain the same total vorticity per unit length along the solid surface. However, in the turbulent case the vorticity is more concentrated near the surface than in the laminar case.

Figure 8.3 shows that, in the case of a flat plate discussed earlier, the transition from a laminar boundary layer to a turbulent one from the leading edge can occur in different ways; in any case the thickness of the turbulent boundary layer grows more rapidly than the laminar one. If we define the local Reynolds number of the boundary layer as $Re_x = Ux/\nu$, the transition from laminar to turbulent boundary layer occurs for values of the order of $Re_{xc} \sim 10^6$.

- **Separation of the boundary layer**

The boundary layer separates from the solid boundary as a result of the slowing down—up to reversal—of the flow due to an adverse pressure gradient, that is, a pressure that increases in the direction of the flow itself. The analysis of the causes

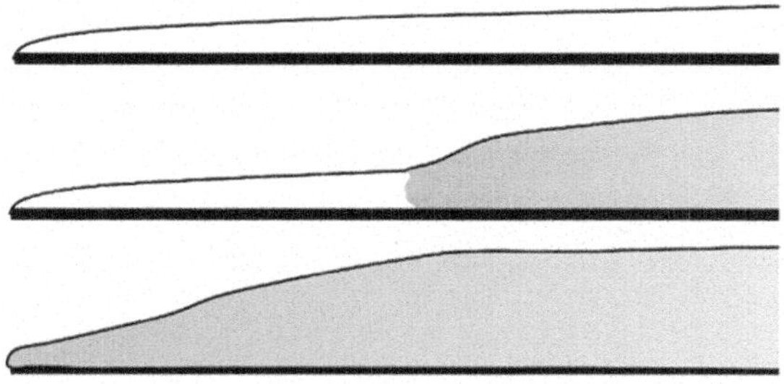

Fig. 8.3 Various modes of transition from a laminar boundary layer to a turbulent one, with a grey background (dimensions not to scale)

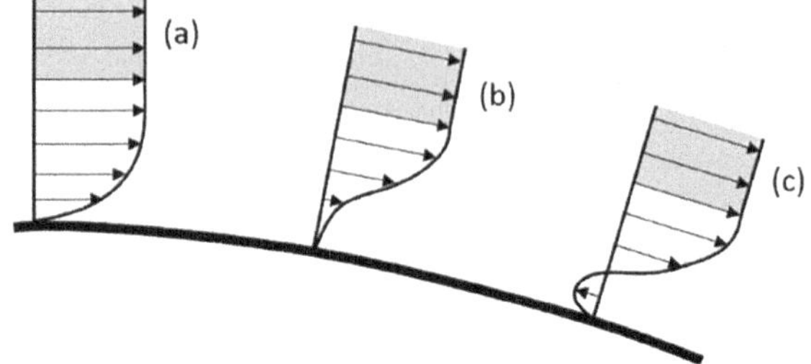

Fig. 8.4 Separation of the boundary layer caused by an adverse pressure gradient. The grey colour indicates the free flow

that produce such a pressure gradient and of the specific separation mechanism are beyond the scope of this discussion; here it is enough to note that the separation occurs when, as schematically shown in Fig. 8.4, the flow along the solid boundary changes sign. Figure 8.4b shows the beginning of this transition; the reversal of the flow (Fig. 8.4c) corresponds to the separation. In all this, the free flow is not affected by the separation except for the distance between its lower limit and the solid boundary.

8.2 Transition of an Internal Flow from Potential to Viscous

In Sect. 5.9 an apparent contradiction was noted between the existence of an exact solution of the viscous flow in a cylindrical tube (the Poiseuille flow) and the manifestation of inviscid flows governed by Bernoulli's theorem in the same geometric context. Now, after having studied the boundary layers, we are able to clarify the problem.

Figure 8.5 shows an initially potential flow (a) with a uniform velocity profile except within the boundary layer. It was seen in Sect. 8.1 how the thickness δ of the boundary layer along a flat plate increases due to the diffusion of vorticity by molecular viscosity (Eq. 8.3); naturally this behaviour is quite general and also in this case the boundary layer that delimits the area B within which the flow is inviscid will grow with the displacement of the fluid along the duct. At the distance ℓ from the mouth of the duct (the *entrance length*) the boundary layer will have invaded the entire section, connecting with the Poiseuille solution (b) (solid line); from this point the flow will be independent of x and the pressure drop will be linear, as predicted by the theory. For completeness, a schematic average profile of a turbulent flow (dashed line in (b)) is also shown, which highlights a greater shear stress on the surface of the duct and a more uniform velocity profile.

The entrance length can be estimated from the following empirical formula:

$$\ell \approx 0.06\, d\, Re \tag{8.4}$$

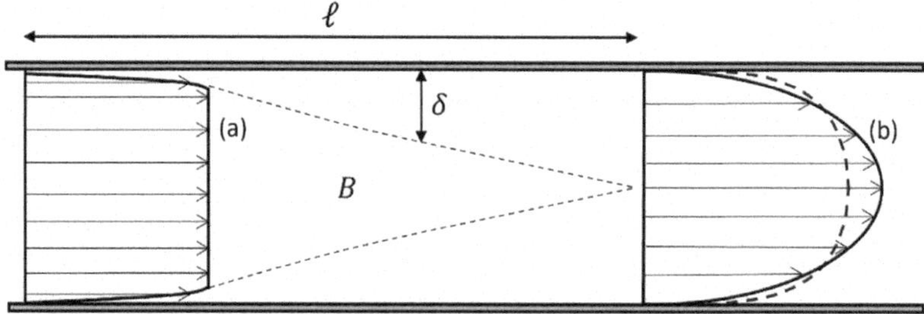

Fig. 8.5 Transition of a flow in a cylindrical tube from potential (**a**) to viscous (**b**) as a result of the gradual increase in the thickness δ of the boundary layer. The dashed velocity profile in (b) refers to a turbulent flow. The dimensions are not to scale

where d is the diameter of the duct. For internal flows of this type the transition to turbulence occurs for values $Re_c \approx 10^3 - 10^4$; if we choose, for example $Re = 2000$ for a laminar flow, we get $\ell \approx 120\, d$. This shows that in many practical applications the Poiseuille flow may never occur and, consequently, Bernoulli's theorem can be applied to the real flow (it should be noted that in a turbulent flow the boundary layer grows more rapidly; in this case ℓ/d turns out to be smaller).

In conclusion, a reduced duct length can be advantageous in various contexts. For example, in a wind tunnel in which it is intended to simulate (through an appropriate dynamical scaling) the inviscid and irrotational conditions of an aircraft's flight, there is an interest in acting at the beginning of area B: for example, for a tunnel of 1 m in diameter, this condition is normally largely satisfied.

8.3 Irrotationality in Perfect Fluids, Kelvin's Theorem

There is clear experimental evidence that, for barotropic motions in inertial reference frames, flows at high Reynolds number outside the boundary layers—and outside any wakes formed by the separation of the same—are irrotational, and therefore potential (Sect. 4.5). In Sect. 8.5 it will be shown how this property allows for obtaining exact two-dimensional solutions of Euler's equations in idealised but nonetheless significant cases; more generally, the irrotationality of flows outside the boundary layers under the conditions mentioned above is an essential dynamic characteristic in countless applications. But what is the reason why such flows are irrotational?

The vorticity equation provides a first answer. Imagine a fluid element that, to fix ideas, is initially to the left of an obstacle (like, for example, the wing represented in Fig. 8.9) in the absence of motion: its initial vorticity is therefore trivially zero ($\boldsymbol{\omega} = 0$). With the establishment of a flow from left to right, the fluid element will move towards the obstacle until it bypasses it (above or below it depending on its initial position), but its vorticity will remain zero as, from Eq. (5.31), $d\boldsymbol{\omega}/dt = 0$ if initially $\boldsymbol{\omega} = 0$ and if the baroclinic term is zero. This result is known as the *persistence of irrotationality* in perfect fluids: in such case the forces acting on the fluid element are due to pressure alone and these, being normal, do not apply any couple that can induce rotation, and therefore vorticity. In reality, vorticity will be generated in the boundary layer that wraps the wing, and will be there diffused by molecular viscosity as discussed in the previous paragraph (see the following Sect. 8.4), but this does not affect the irrotationality of the free flow outside the boundary layer.

A rigorous integral explanation equivalent to the semi-qualitative one just provided is based on the concepts of vorticity flow through a surface A and circulation along its boundary ∂A introduced in Sect. 4.6. It is recalled that these quantities are linked by Stokes' theorem:

$$C = \iint_{\Gamma} \boldsymbol{\omega} \cdot \mathbf{n} ds = \oint_{\partial \Gamma} \mathbf{u} \cdot d\boldsymbol{\ell}.$$

Kelvin's circulation theorem demonstrated in Appendix D on the basis of the equations of motion of fluid dynamics asserts the following:

- For barotropic motions $[\rho = \rho(p)]$ in a perfect fluid (absence of viscous effects), *the circulation C along a material line is invariant*:

$$\frac{dC}{dt} = 0. \tag{8.5}$$

This property is illustrated in Fig. 8.6: if $\partial\Gamma(t')$ is the evolution of the material curve $\partial\Gamma(t)$, then $C(t') = C(t)$, and this holds for every choice of Γ. Therefore, if at time t the vorticity flux is zero through every surface Γ, as is trivially true in the case of a fluid initially at rest, the same can be said at every subsequent time: this is equivalent to stating that the flow remains irrotational. Kelvin's theorem will be used in Sect. 8.6 to explain the origin of the lift acting on a wing.

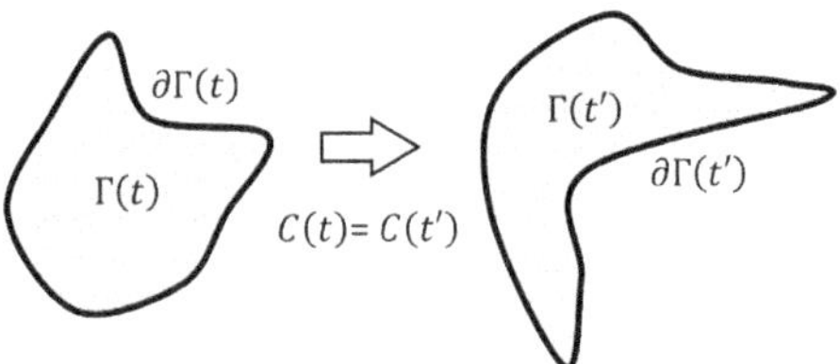

Fig. 8.6 Illustration of Kelvin's theorem. The surface $\Gamma(t')$ is the material evolution of the surface $\Gamma(t)$ and the same applies to the two curves $\partial\Gamma(t')$ and $\partial\Gamma(t)$. In this case the circulation is invariant under the assumptions of Kelvin's theorem: $C(t) = C(t')$

8.4 External Flows: Lift and Drag

In this paragraph and the next, we will consider some particularly significant potential external flows. By *external*, we mean a flow in which the fluid completely surrounds a solid body. Here we begin by briefly and qualitatively discussing the most salient aspects of the flow around an airfoil, which is the element that best represents the flight conditions of an aircraft.

Figure 8.7 shows a symmetrical wing at rest, infinitely extended in the longitudinal direction: this assumption allows us to reduce the problem to two dimensions (the more realistic case of a three-dimensional wing will not be considered in this synthetic treatment). The wing is immersed in a stationary two-dimensional flow on the vertical plane which, at a sufficient distance from it, is uniform with speed U parallel to the x-axis (the so-called *free stream*): this, however, is represented as horizontal in the figure only for convenience (the x-axis can in turn be inclined). The dashed line that joins the *leading edge* with the *trailing edge*, called the *chord*, forms an angle α with the direction of the flow, called the *angle of attack* or *incidence*. The vector $\mathbf{F}$ indicates the resultant of the surface forces (per unit length) applied to the wing (the *aerodynamic force*) while its projections in the x direction and in the perpendicular direction are respectively called *drag* $\mathbf{D}$ and *lift* $\mathbf{L}$. Of course, the wing is also subject to a moment of force, the effect of which will not be considered here.

Lift and drag are traditionally represented by the following empirical formulas:

$$L = \frac{1}{2}\rho U^2 S C_L \tag{8.6}$$

$$D = \frac{1}{2}\rho U^2 S C_D \tag{8.7}$$

where ρ is the density of the air far from the wing, S is the area of the wing surface considered and C_L and C_D are suitable dimensionless coefficients respectively for lift and drag. For a wing of a certain shape, these coefficients depend on the Reynolds number Re, the local Mach number $M = V/c_s$ (where V and c_s are respectively the local fluid and sound speeds) and the angle of attack α. Limiting ourselves to subsonic

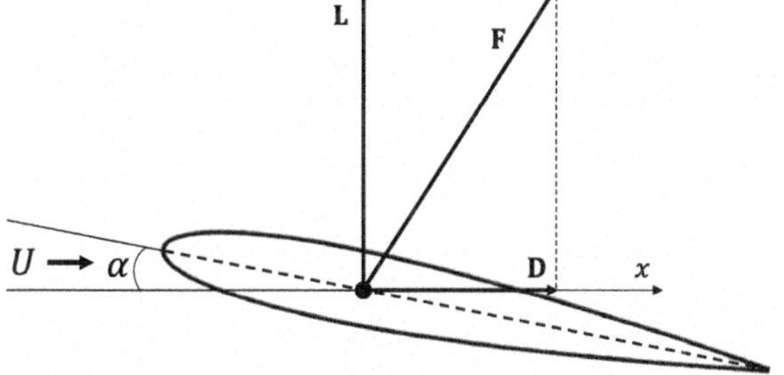

Fig. 8.7 Airfoil profile immersed in a stationary and uniform flow with speed U at a sufficient distance from the wing

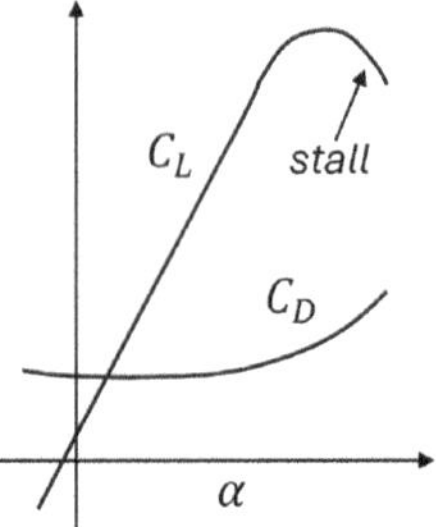

Fig. 8.8 Schematic representation of the dependence of C_L and C_D on the angle of attack α

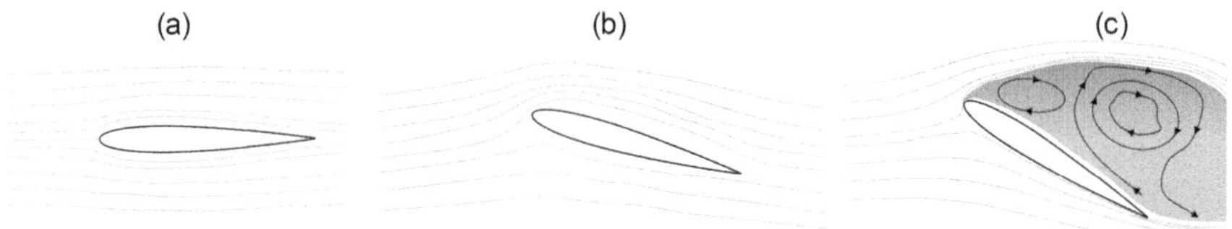

Fig. 8.9 Flow around the wing as the angle of attack varies. In cases (**a**, **b**) the boundary layer adheres to the wing along its entire profile. In case (**c**) the boundary layer separates immediately after the leading edge and forms an extensive turbulent wake (in grey); for intermediate angles, the detachment occurs between the leading and trailing edges

flows ($M < 1$ at every point), for a given airfoil profile, C_L and C_D essentially depend on α, as schematically indicated in Fig. 8.8.

For a given speed U, the lift increases linearly with α while, for sufficiently high values of α, L reaches a maximum value and then decreases abruptly. This reversal is associated with the separation of the boundary layer, with the formation of a turbulent wake, as shown in Fig. 8.9c. Note that corresponding to the decrease in lift there is an increase in drag due to the birth of a new component beyond the previously defined skin friction D_f, called *pressure drag* (or *form drag*) D_p: this drag is caused by the pressure drop that is established in the wake. The total resistance in a separated flow can therefore be expressed as follows:

$$D = D_f + D_p. \tag{8.8}$$

The phenomenon of the sudden decrease in lift and increase in drag, called *stall*, has drastic consequences on flight dynamics and can be prevented by resorting to mechanical devices, such as flaps etc., which, for example, are used during take-off and landing to increase the lift of the wing at low speeds.

To conclude, let's now consider the effect of turbulence on separation and drag. As shown in Fig. 8.2b, the average speed in a turbulent boundary layer in the immediate vicinity of the solid boundary is greater than that of a laminar boundary layer corresponding to the same free flow. This means that a turbulent boundary layer is less likely to separate due to an adverse pressure gradient compared to the laminar case. Therefore, to delay separation—and thus stall—in order to have sufficient lift for higher angles of attack, it is preferable for the boundary layer to be turbulent. For this purpose, some wings are equipped with appropriate vortex generators which, when necessary, can ensure a high level of turbulence in the boundary layer.

On the other hand, while it is true that D_p is greater for a laminar boundary layer compared to a turbulent one (for which separation is delayed), it is also true that D_f is greater for a turbulent boundary layer compared to a laminar one (it has already been noted that $\tau_T > \tau_L$ on the solid boundary). Identifying the optimal compromise in specific cases is a delicate problem in aeronautical engineering.

8.5 Elementary and Composite Potential External Flows

In this paragraph, some simple exact solutions of two-dimensional external flows at high Reynolds numbers outside the boundary layer will be derived; such flows must therefore be potential. To clarify, one can think of naval applications, for which the flow describes the motion of water around a vessel; however, the mathematical formulation of the problem and some relevant properties obtained in this paragraph, and expanded upon in the next, also apply to flows around a wing in motion in the air like those qualitatively discussed in the previous paragraph. Therefore, these examples have a rather general conceptual and applicative value.

Consider an incompressible ($\nabla \cdot \mathbf{u} = 0$) and homogeneous ($\rho = const$) fluid. As seen, this assumption, obviously valid for a liquid, also applies to air for subsonic motions (Sect. 4.2). Furthermore, consider an irrotational flow ($\nabla \times \mathbf{u} = 0$), therefore potential ($\mathbf{u} = \nabla\varphi$) and harmonic ($\nabla^2\varphi = 0$). In conclusion, we consider solenoidal and irrotational velocity fields (it is interesting to note that a similar mathematical problem is encountered in electrostatics; in fact, in that case Eq. (A3) and (A5) give $\nabla \cdot \mathbf{E} = 0$ in vacuum and $\nabla \times \mathbf{E} = 0$). Finally, we limit ourselves to stationary and two-dimensional flows on the horizontal plane ($\mathbf{u} = \mathbf{u}(x, y)$; in the aerodynamic extension of the problem the plane would be vertical: (x, z)).

Under these conditions, the relationships that link the velocity $\mathbf{u}$ to the velocity potential φ and the stream function ψ are:

$$\begin{cases} u = \dfrac{\partial\varphi}{\partial x} = -\dfrac{\partial\psi}{\partial y} \\ v = \dfrac{\partial\varphi}{\partial y} = \dfrac{\partial\psi}{\partial x} \end{cases} \tag{8.9}$$

From these it follows that also ψ satisfies Laplace's equation: $\nabla^2\psi = 0$; moreover, since $\nabla\varphi \cdot \nabla\psi = 0$, the isolines of φ and those of ψ (the streamlines) are perpendicular at every point. Finally, the free-slip boundary conditions along the solid boundary $\partial\Omega$ of a domain Ω containing the fluid are:

$$\begin{cases} \left.\frac{\partial \varphi}{\partial \mathbf{n}}\right|_{\partial\Omega} = 0 \\ \psi|_{\partial\Omega} = \text{const} \end{cases} \tag{8.10}$$

where $\mathbf{n}$ is the unit vector at each point normal to $\partial\Omega$.

It is worth noting that the relationships between φ and ψ expressed by Eq. (8.9) coincide with the *Cauchy-Riemann conditions* of the theory of analytic functions of a complex variable. It can be demonstrated that the complex function

$$w = \varphi + i\psi, \tag{8.11}$$

where i is the imaginary unit and φ and ψ satisfy Eq. (8.9), depends on x and y only through the complex variable

$$z = x + iy$$

In this case, $w = w(z)$ (which can be considered the *complex potential* of the problem) is called an *analytic* (or *holomorphic*) *function*. For functions of this type, a number of important properties hold: for example, for an analytic function it is possible to define the derivative with respect to z, it is possible to reconstruct its imaginary part from the knowledge of its real part (and vice versa), a theorem, called *residue theorem*, allows for the simple and elegant calculation of many complicated integrals, etc.. Moreover, starting from any real function $w(\xi)$ of real variable ξ, a flow satisfying Eq. (8.9) can be obtained simply by performing the *analytic continuation* of the function from the real axis to the complex plane: $w(\xi) \rightarrow w(z)$. The technique of *conformal mapping* should also be mentioned, of particular interest for fluid dynamics (as well as for electromagnetism). For a deeper understanding of this fascinating, but rather complex, field of investigation in fluid dynamics, it is recommended to consult the texts of Lamb (1895), Batchelor (1967) and Lavrentiev and Chabat (1972).

In the present discussion, some simple external potential flows will be obtained by exploiting the linearity of Laplace's equation, which both φ and ψ must satisfy. Four elementary harmonic external flows will be appropriately combined, obtaining more complex—yet still harmonic—flows with elements of realism. The nonlinearity of the fluid dynamics equations manifests itself (but only in algebraic terms, not differential) in the calculation of pressure using Bernoulli's Eq. (5.26), which is here again reported for convenience:

$$\frac{1}{2}\rho|\mathbf{u}|^2 + p_s = \text{const.} \tag{8.12}$$

The first elementary flow is the *uniform flow*, for which

$$u = U; \quad v = 0; \quad \varphi = Ux; \quad \psi = -Uy. \tag{8.13}$$

For the other three flows, it is convenient to switch to polar coordinates for the independent variables

$$\begin{cases} x = r\cos\theta \\ y = r\sin\theta \end{cases}$$

and to the radial and tangential components for the velocity field:

$$\begin{cases} u_r = \frac{\partial\varphi}{\partial r} = -\frac{1}{r}\frac{\partial\psi}{\partial\theta} \\ u_\theta = \frac{1}{r}\frac{\partial\varphi}{\partial\theta} = \frac{\partial\psi}{\partial r} \end{cases} \tag{8.14}$$

The second elementary flow is a *point source*

$$u_r = \frac{Q}{2\pi r};\ u_\theta = 0;\ \varphi = \frac{Q}{2\pi}\ln r;\ \psi = -\frac{Q}{2\pi}\theta \tag{8.15}$$

with Q constant, the third is a *point sink*

$$u_r = -\frac{Q}{2\pi r};\ u_\theta = 0;\ \varphi = -\frac{Q}{2\pi}\ln r;\ \psi = \frac{Q}{2\pi}\theta \tag{8.16}$$

and the fourth is the *irrotational vortex* (free vortex) already encountered in Sect. 4.4

$$u_r = 0;\quad u_\theta = \frac{\Gamma_0}{2\pi r};\quad \varphi = \frac{\Gamma_0}{2\pi}\theta;\quad \psi = \frac{C_0}{2\pi}\ln r. \tag{8.17}$$

The constant C_0 has a particular meaning. In fact, if the circulation (Eq. 4.26) along a circle L_R of radius R centred at the origin is calculated, one obtains:

$$C = \oint_{L_R} \mathbf{u}\cdot d\boldsymbol{\ell} = \oint_{L_R} u_\theta d\ell = \frac{C_0}{2\pi R}2\pi R = C_0.$$

So, C does not depend on R. More generally, given the irrotationality of the vortex, the same result would be obtained along any curve that contains the origin: in fact, calculate the circulation along the closed curve $L = L_R \cup L_1 \cup L' \cup L_2$ bounding the surface B as shown in Fig. 8.10. Given the irrotationality of the flow and the cancellation of the sum of the two line integrals along L_1 and L_2 (evaluated in opposite directions), we obtain $C_{L'} = C_{L_R}$:

$$C = \oint_L \mathbf{u}\cdot d\boldsymbol{\ell} \cong \oint_{L_R} \mathbf{u}\cdot d\boldsymbol{\ell} - \int_{L'} \mathbf{u}\cdot d\boldsymbol{\ell} = \iint_B \boldsymbol{\omega}\cdot\mathbf{n}ds = 0. \tag{8.18}$$

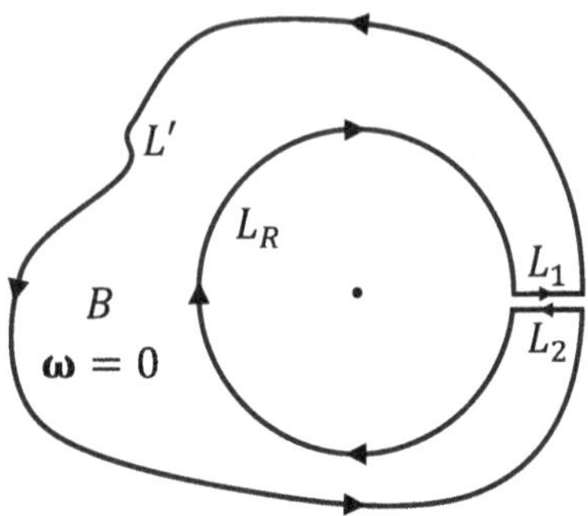

Fig. 8.10 Independence of circulation from the closed line along which it is calculated in an irrotational flow

In relation to the demonstration in Eq. (8.18), it is useful to note an apparent paradox easily resolved by considering the applicability of Stokes' theorem. In an irrotational vortex, the circulation calculated along any curve that encloses the origin is $C = C_0$; on the other hand, if the curve does not enclose the origin, $C = 0$, and yet, one might object, in both cases $\boldsymbol{\omega} = 0$ inside them. This statement is not correct, as at the origin of the vortex $\boldsymbol{\omega}$ is not only non-zero, it is not even defined as $r = 0$ represents a singularity point of the flow: in this case Stokes' theorem is not applicable. The independence of the circulation from the closed curve including an airfoil profile in an irrotational field, will be used in the next paragraph in relation to an important theorem.

We now move on to consider composite potential flows. Consider the flow obtained by adding a source and a sink symmetrically arranged on the $x-$ axis at a distance δ superimposed on a uniform flow (from now on it is convenient to express the fields in terms of ψ as its isolines represent the streamlines, thus providing an immediate image of the velocity field):

$$\psi = -\frac{Q}{2\pi}\theta_1 + \frac{Q}{2\pi}\theta_2 - Uy = \frac{Q}{2\pi}\arctan\left(\frac{2y\delta}{x^2+y^2-\delta^2}\right) - Uy \tag{8.19}$$

where θ_1 and θ_2 are the angles formed by the position vector with the x-axis for source and sink. An example of the flow thus obtained is shown in Fig. 8.11a. The main characteristic of this flow lies in the existence of a closed streamline L_E in the shape of an ellipse that encloses the two singular points of source and sink. This means that the flow external to L_E corresponds to an irrotational velocity field that satisfies the free-slip condition along the same curve: this therefore represents a realistic flow, also by virtue of the exclusion from the domain of the singular points. This solution is known as Rankine flow, in honour of the Scottish physicist William Rankine (1820–1872) who, by combining various elementary flows differently, was able to obtain flows around closed profiles that more realistically approximated naval sections.

For very small values of δ, from Eq. (8.19) we get:

$$\psi \cong \frac{Q\delta}{\pi}\left(\frac{y}{x^2+y^2}\right) - Uy = \left(\frac{\chi}{r} - Ur\right)\sin\theta \tag{8.20}$$

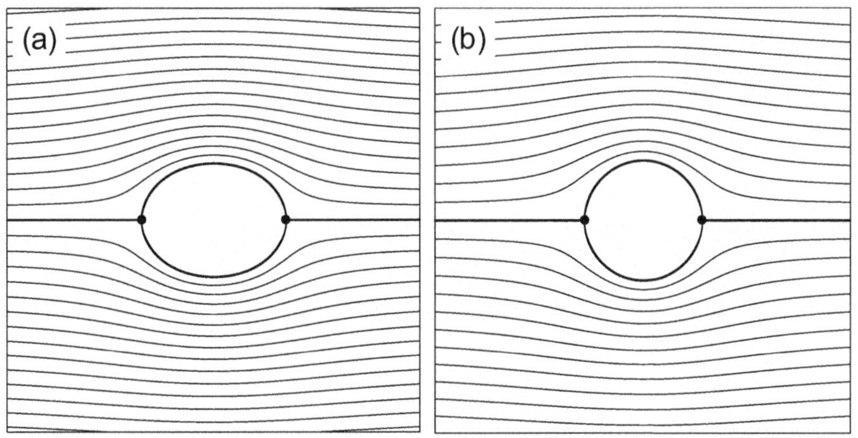

Fig. 8.11 Potential flow (from left to right, $U > 0$) obtained by adding a source, a sink and a uniform flow. **(a)** Example obtained from Eq. (8.19). **(b)** Example obtained from Eq. (8.20)

where $\chi = Q\delta/\pi$, from which it follows that the streamline $\psi = 0$ is a circle (therefore the section of a cylinder) of radius $R = \sqrt{\chi/U}$ (Fig. 8.11b). The tangential speed along the circle is therefore:

$$u_\theta|_{r=R} = \left.\frac{\partial\psi}{\partial r}\right|_{r=R} = -2U\sin\theta. \tag{8.21}$$

From Eq. (8.21) it follows that, as is obvious, the two points indicated by dots in Fig. 8.11b are stagnation points:

$$u_\theta|_{r=R;\theta=0,\pi} = 0.$$

On the other hand, at the two points on the y-axis for $x = 0$, the speed is double that far from the body:

$$u_\theta|_{r=R;\theta=\pm\pi/2} = \mp 2U.$$

Moving away from those points along y, the speed tends to U as y^{-2}:

$$u_\theta|_{\theta=\pm\frac{\pi}{2}} = \left.\frac{\partial\psi}{\partial r}\right|_{\theta=\pm\frac{\pi}{2}} = \mp U\left(1+\frac{R^2}{y^2}\right).$$

We now move on to calculate the pressure on the surface of the cylinder. The static pressure p_s is broken down into the sum of the pressure that would be present in the absence of the obstacle, therefore very far from it (p_∞), and that associated with the motion in the vicinity of the obstacle (p):

$$p_s = p_\infty + p. \tag{8.22}$$

From Eq. (8.12) we have

$$\frac{1}{2}\rho|\mathbf{u}|^2 + p_\infty + p = \frac{1}{2}\rho U^2 + p_\infty$$

from which we obtain

$$p = \frac{1}{2}\rho U^2\left(1 - \frac{|\mathbf{u}|^2}{U^2}\right) \tag{8.23}$$

where $|\mathbf{u}|^2 = \psi_x^2 + \psi_y^2$. Moving to the dimensionless pressure $p' = p/(\rho U^2/2)$ we have

$$p' = 1 - \frac{|\mathbf{u}|^2}{U^2} \tag{8.24}$$

Figure 8.12 shows the field of dimensionless pressure thus obtained.

Substituting in Eq. (8.23) the expression of the tangential speed for $r = R$ given by Eq. (8.21), we obtain the pressure on the surface of the cylinder:

$$p|_{r=R} = \frac{1}{2}\rho U^2(1 - 4\sin^2\theta). \tag{8.25}$$

The corresponding dimensionless pressure,

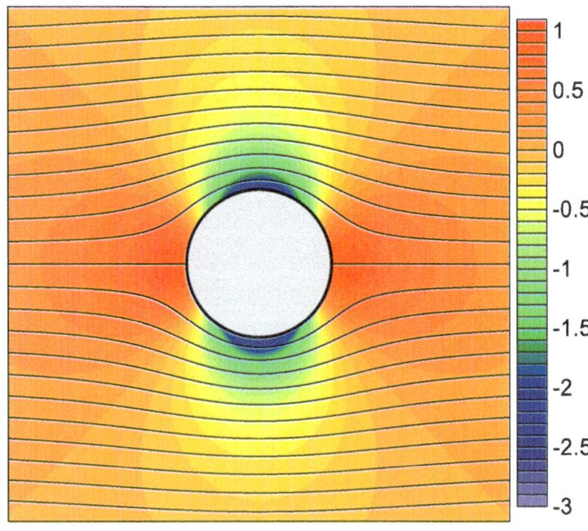

Fig. 8.12 Field of pressure p' obtained from Eq. (8.24) with ψ given by Eq. (8.20). The streamlines are the same as in Fig. 8.11b

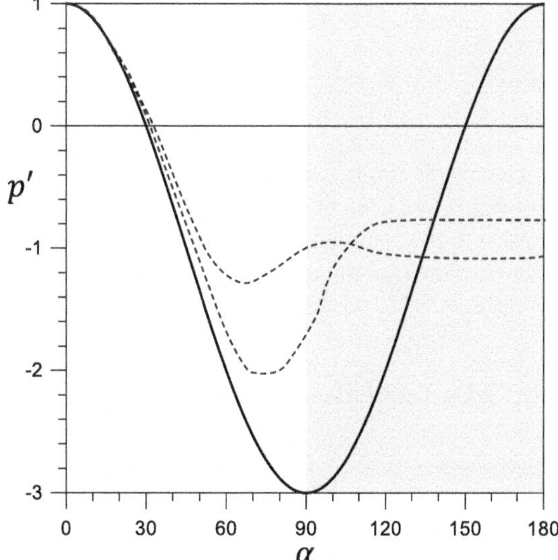

Fig. 8.13 Solid line: pressure p' of Fig. 8.12 on the surface of the cylinder (Eq. 8.26; $\alpha = \pi - \theta$). Dashed lines: examples of pressure in real cases. The white (grey) area indicates the upstream (downstream) side of the cylinder

$$p'\big|_{r=R} = 1 - 4\sin^2\theta \tag{8.26}$$

is shown by the solid line in Fig. 8.13. The dashed lines show examples of pressure on the cylinder in real cases with $Re \sim 10^5 - 10^6$, for which there is separation of the boundary layer (Sect. 8.1) with a turbulent wake downstream. Near the upstream stagnation point ($\alpha = 0$), the pressure in a real fluid is well approximated by the pressure of the potential flow but for larger angles, already on the upstream side, there is a significant reduction in pressure; downstream (grey area), the average pressure is significantly lower than the pressure of the potential flow. This pressure imbalance between the two areas of the cylinder produces a form drag D_p that far exceeds the friction drag D_f (Eq. 8.8, see also Sect. 8.1).

The resultant $\mathbf{P}$ of the pressure forces on the surface of the cylinder ($r = R$) per unit of transverse length is calculated as follows:

$$\mathbf{P} = \oint p\mathbf{n}ds = R\int_0^{2\pi} p|_{r=R}\mathbf{n}d\theta.$$

The components along x and y are

$$\begin{cases} P_1 = R\int_0^{2\pi} p|_{r=R}\cos\theta d\theta \\ P_2 = R\int_0^{2\pi} p|_{r=R}\sin\theta d\theta \end{cases} \tag{8.27}$$

Substituting the expression of $p|_{r=R}$ given by Eq. (8.25) and solving the integrals, we get

$$P_1 = 0; \quad P_2 = 0. \tag{8.28}$$

This result is quite obvious as the pressure is symmetrical with respect to both the x- and y-axis, but this calculation procedure will be useful in the next paragraph. In general, the component parallel to the motion P_1 would be null for any stationary potential flow (see the discussion on D'Alembert's paradox in Sect. 8.1).

8.6 Magnus Effect, Kutta-Žukovskij Theorem

Now modify the flow given by Eq. (8.20) by superimposing an irrotational vortex (Eq. 8.17):

$$\psi = \left(\frac{\chi}{r} - Ur\right)\sin\theta + \frac{C_0}{2\pi}\ln r. \tag{8.29}$$

The streamlines are shown in Fig. 8.14.

The flow thus obtained retains its symmetry with respect to the y-axis but loses it with respect to the x-axis due to the asymmetry introduced by the direction of rotation of the vortex. It is therefore natural that the resultant of the pressure forces on the surface of the cylinder along y is not null. It is immediate to verify that in this case Eq. (8.25) transforms into

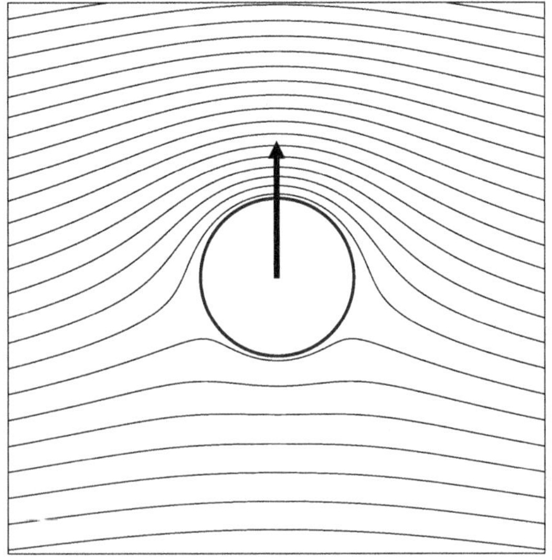

Fig. 8.14 Potential flow (from left to right, $U > 0$) obtained by adding a source, a sink, a uniform flow and an irrotational vortex. Example obtained from Eq. (8.29)

$$p|_{r=R} = \frac{1}{2}\rho U^2\left[1 - 4\left(\sin\theta + C'\right)^2\right], \tag{8.30}$$

where

$$C' = \frac{C_0}{4\pi RU}.$$

Substituting this expression into Eq. (8.27), we get

$$P_1 = 0; \quad P_2 = \rho U C_0. \tag{8.31}$$

Therefore, the presence of a circulation around the cylinder produces a resultant of the pressure forces in the direction perpendicular to the motion (the arrow in Fig. 8.14). In reality, this circulation is produced by the rotation of the cylinder which, due to the viscosity acting in the boundary layer, drags the fluid producing a lift similar to that discussed in Sect. 8.4. This phenomenon, known as the *Magnus effect*, manifests itself in various circumstances, also applying to spherical bodies. Think, for example, of tennis or soccer balls to which, in addition to an initial impulse, a rotation has also been imparted: the resulting lift determines the curvature of the trajectory. It is important to underline that, despite the result contained in Eq. (8.31) deriving from a purely potential (therefore inviscid) theory, the phenomenon requires the action of viscosity to manifest itself.

That lift is associated with the circulation around a body is a more general property than the case just studied. Consider, for example, an airfoil profile like that in Fig. 8.7. The typical flow around it in the absence of boundary layer separation (Fig. 8.9b) can be seen schematically as the result of the sum of a uniform flow (Fig. 8.15a) and one with clockwise circulation around the wing (Fig. 8.15b) which causes the downward deviation of the streamlines (the *downwash*). This latter effect already suggests the necessity of the existence of lift: in fact, this deviation is determined by a downward force that the wing imparts to the fluid but, according to the third law of dynamics, the same force—with the opposite direction—is exerted by the fluid on the wing itself, the lift indeed.

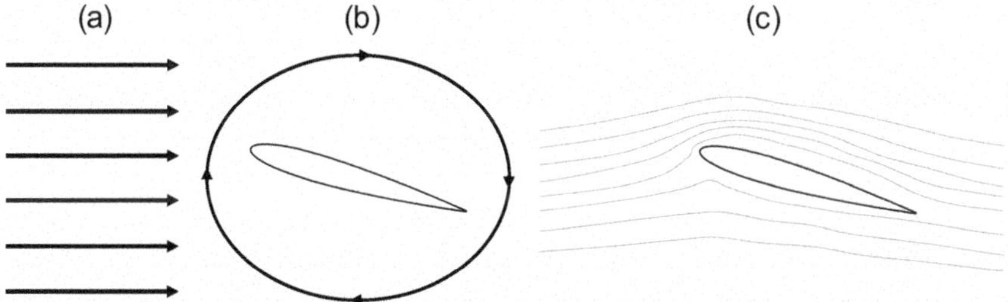

Fig. 8.15 Schematic representation of the effect of the superposition of a uniform flow (**a**) and a circulation around a wing profile (**b**). The result is the typical flow (**c**)

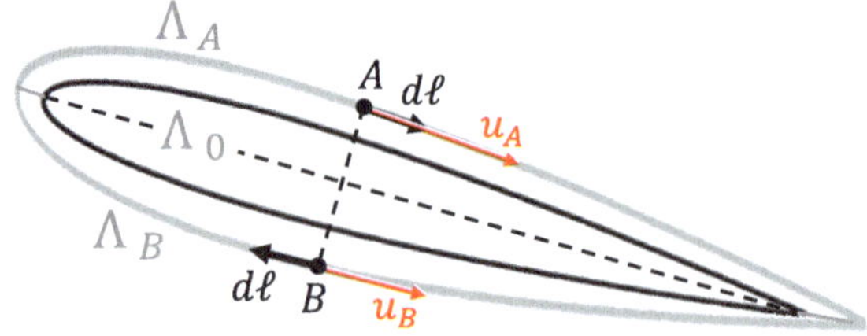

Fig. 8.16 Demonstration of the Kutta-Žukovskij theorem for a thin wing and a small angle of attack. The wing profile is indicated by the black line; the grey line wraps the wing just outside the boundary layer. Note that $u_B < u_A$, so $p_B > p_A$ according to the Bernoulli theorem

The general mathematical relationship that links circulation and lift is expressed by the *Kutta-Žukovskij theorem* (from the German mathematician Martin Kutta, 1867–1944, and the Russian mathematician and engineer Nikolay Žukovskij, 1847–1921). The theorem states that the lift L (per unit of transverse length) exerted on a two-dimensional body (like a wing profile, a cylinder like the one just considered, etc.) in a stationary potential flow in the absence of boundary layer separation, is given by the formula:

$$L = \rho U C \tag{8.32}$$

where C is the circulation around the wing (the argument used in the discussion of Fig. 8.10 ensures that C does not depend on the closed curve chosen). This expression is in fact identical to that of P_2 in Eq. (8.31). It should be added that, although the theorem is derived for purely potential flows, its validity extends with good approximation also to real flows in typical aerodynamic applications.

The Kutta-Žukovskij theorem can be easily demonstrated in a heuristic way in the particular case of a thin airfoil slightly inclined with respect to the direction of motion (Fig. 8.16).

Consider points A and B, symmetric with respect to the wing chord Λ_0 and located respectively on the upper and lower part of the wing just outside the boundary layer (it is emphasised that the pressure jump between these points and those in direct contact with the wing—where the pressure that determines the lift is applied—is negligible). Using Bernoulli's law Eq. (8.12), the pressure difference between A and B turns out to be

$$p_B - p_A = -\frac{1}{2}\rho\left(u_B^2 - u_A^2\right) = \rho\frac{1}{2}(u_A + u_B)(u_A - u_B) \cong \rho U(u_A - u_B), \tag{8.33}$$

where, thanks to the thinness of the wing and the small angle of attack, it can be assumed that the average value of the two speeds is approximately equal to the flow speed far from the wing: $\overline{u} \cong U$. Therefore, the lift acting on the wing is

$$L = \int_{\Lambda_B} p_B d\ell - \int_{\Lambda_A} p_B d\ell \cong \int_{\Lambda_0} (p_B - p_A) d\ell = \rho U \int_{\Lambda_0} (u_A - u_B) d\ell \tag{8.34}$$

where, thanks to the thinness of the wing, the first two integrals can be calculated along the wing chord Λ_0 with the values of the integrand functions corresponding to the points along Λ_A and Λ_B. On the other hand, the circulation C around the closed curve $\Lambda = \Lambda_A \cup \Lambda_B$ is

$$C = \oint_{\Lambda} u d\ell \cong \int_{\Lambda_0} (u_A - u_B) d\ell. \tag{8.35}$$

Combining Eqs. (8.34) and (8.35), the theorem is demonstrated:

$$L \cong \rho U C.$$

What remains is to identify the cause that produces the circulation around the airfoil. This certainly cannot be the rotation of the body as in the case of the Magnus effect, but rather it must be sought in the transient phase that connects the initial state of rest of the wing to its asymptotic uniform translational motion.

In this phase, one observes the formation of a vortex (counterclockwise in Fig. 8.17b), called the *starting vortex*, which detaches from the wing starting from the trailing edge, then completely separating from it (see specialist texts, such as those by Anderson (2011); Houghton et al. (2013), for further study). The circulation around the wing, for example along the line L_A of Fig. 8.17a, is initially null ($\mathbf{u} = 0$).

Let L_B now be the material evolution of L_A (in Fig. 8.17b, $L_B = L_{B1} \cup L_{B2}$ minus the two adjacent lines L_{C1} and L_{C2}). According to Kelvin's theorem (Eq. 8.5, Appendix D) the circulation along L_B must therefore be null:

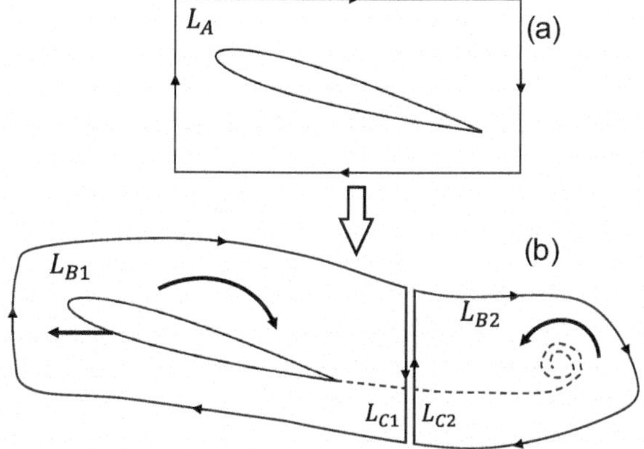

Fig. 8.17 Explanation of the formation of circulation around a wing by applying Kelvin's theorem. In **(a)** the wing is at rest, in **(b)** it is moving to the left

$$C_{L_A} = C_{L_B} = \oint_{L_{B1}} \mathbf{u} \cdot d\boldsymbol{\ell} + \oint_{L_{B2}} \mathbf{u} \cdot d\boldsymbol{\ell} = 0$$

Consequently, we will have

$$C_{L_{B1}} = -C_{L_{B2}}. \tag{8.36}$$

Therefore, the circulation associated with the starting vortex produces a circulation of the opposite sign around the wing and, therefore, the lift that allows it to sustain flight. Note that this equation expresses the conservation of the fluid's angular momentum implied by the third law of dynamics.

Bibliography

Anderson, J.D.J.: Fundamentals of Aerodynamics. McGraw-Hill, New York (2011)
Batchelor, G.K.: An Introduction to Fluid Dynamics. Cambridge University Press, Cambridge (1967)
Houghton, E.L., Carpenter, P.W., Collicott, S.H., Valentine, D.T.: Aerodynamics for Engineering Students. Elsevier, Amsterdam (2013)
Lamb, H.: Hydrodynamics. Cambridge University Press, Cambridge (1895)
Lavrentiev, M., Chabat, B.: Méthodes de la théorie des fonctions d'une variable complexe. Editions MIR, Moscou (1972)

Further Recommended Reading

Çengel, Y.A., Cimbala, J.M.: Fluid Mechanics, Fundamentals and Applications. McGraw-Hill, Boston (2006)
Kundu, P.K., Cohen, I.M., Dowling, D.R.: Fluid Mechanics. Elsevier, Amsterdam (2012)
Landau, L.D., Lifshitz, E.M.: Fluid Mechanics. Pergamon Press, Oxford (1987)
Tritton, D.J.: Physical Fluid Dynamics. Van Nostrand Reinhold, New York (1977)
White, F.M.: Fluid Mechanics. McGraw-Hill, New York (2011)

Chapter 9
Surface Gravity Waves: General Aspects

This chapter introduces the main aspects of surface gravity waves, initially considering linear waves and subsequently weakly nonlinear waves. For linear waves, the velocity field associated with a regular sinusoidal wave and the dispersion relation that links the angular frequency to the wave number are derived, delving into the two limits of deep water and shallow water waves and providing hints on the related phenomenology. The nonlinear extension for periodic waves is then considered. Finally, the solitonic solution for shallow water waves is discussed.

9.1 Linear Surface Gravity Waves

The surface gravity waves that manifest at the air-sea interface—or sea surface—(Fig. 9.1) originate from the restoring force due to gravity, which tends to restore the original position of fluid elements displaced from their equilibrium position by a peculiar transformation of energy from kinetic to potential form, and vice versa. Such waves are mainly generated by wind, but also by the movement of boats, by earthquakes and volcanic phenomena and by astronomical forces (tides).

These waves constitute a fascinating fluid dynamic problem, but they also represent a phenomenon of great importance for a series of application aspects ranging from maritime navigation to the dispersion of pollutants, from bathing to coastal protection, with strong impacts on biological, ecological, social aspects, etc. Moreover, the energy of wave motion is now considered one of the alternative energy sources. On the other hand, in addition to the typically oceanic surface gravity waves there are also internal gravity waves, which affect both the ocean and the atmosphere. Therefore, the more general problem of gravity waves is of utmost relevance for oceanic and atmospheric fluid dynamics. These aspects will be addressed in this Chap. 9 and in the subsequent Chaps. 10 and 11.

S. Pierini, *Oceanic and Atmospheric Fluid Dynamics*, UNITEXT for Physics,
https://doi.org/10.1007/978-3-031-77991-6_9

Fig. 9.1 Surface gravity waves

We start by discussing the linear approximation and its implications. The Fourier analysis allows to represent any field as an appropriate linear combination of terms containing circular functions (every mathematical detail, that is assumed to be known to the reader, is omitted). For example, suppose that, at a certain instant t_0, the deviation of the sea surface from its equilibrium position $\eta_0(\mathbf{x})$ (Fig. 9.2a; in this treatment, with $\mathbf{x}$ we will refer only to the horizontal coordinates: $\mathbf{x} = (x, y)$) can be expressed by the following combination of a finite number of Fourier components:

$$\eta_0(\mathbf{x}) = \sum_{i=1}^{N} a_i \cos(\mathbf{k}_i \cdot \mathbf{x} + \phi_i) \tag{9.1}$$

where a_i, $\mathbf{k}_i$, and ϕ_i are, respectively, amplitudes, wave numbers and phases. It is recalled that the vector $\mathbf{k} = (k_1, k_2)$ has as its modulus the *wave number* $k = 2\pi/\lambda$, where λ is the *wavelength*, while its direction is perpendicular to the wave fronts (Fig. 9.2b), and therefore indicates the direction and the sense of propagation of the sinusoidal wave.

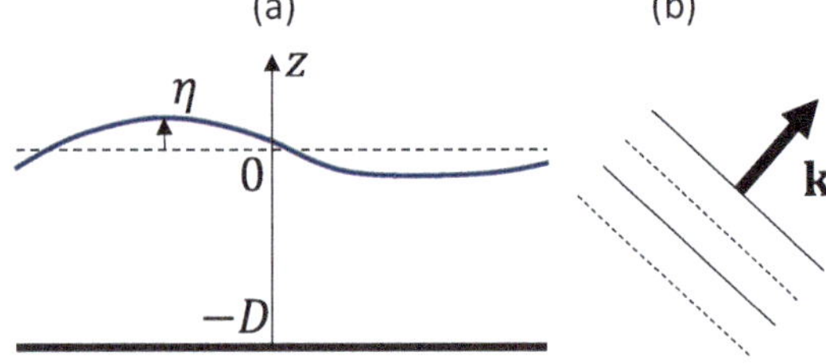

Fig. 9.2 (**a**) Definition of the deviation η of the sea surface from its equilibrium position. (**b**) The wave number vector $\mathbf{k}$

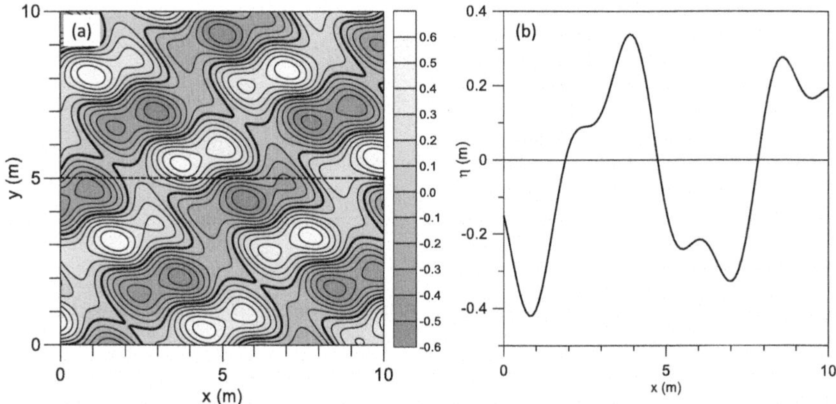

Fig. 9.3 Example of superposition of linear waves. Amplitude η (**a**) and relative profile along the section $y = 5m$ (**b**). In Eq. (9.1), $N = 3$; $a_1 = 0.1$ m, $\lambda_1 = 1.5$ m, $\gamma_1 = 45°$; $a_2 = 0.2$ m, $\lambda_2 = 2.5$ m, $\gamma_2 = 90°$; $a_3 = 0.3$ m, $\lambda_3 = 4$ m, $\gamma_3 = 135°$; $\phi_1 = \phi_2 = \phi_3 = \pi/2$

In Fig. 9.3 an example of superposition of three Fourier components is reported (this example is extremely idealised both for its simplicity and for the infinite extension of the field, and yet it is useful to clarify the role of the linearisation of the governing equations). It is interesting to note how, even with such a limited number of components, the structure of the field is rather irregular.

Now, due to the nonlinear terms $(\mathbf{u} \cdot \nabla)\mathbf{u}$ in the Navier–Stokes equation, the temporal evolution $\eta(x, t)$ of the initial condition Eq. (9.1) will generally produce a broad spectrum in terms of wave numbers, with an N much larger than the initial one. However, if the nonlinear terms are negligible, two properties of great dynamic relevance occur that greatly simplify the problem:

1. There is a real function that links the *angular frequency* $\sigma = 2\pi\nu$ (where $\nu = 1/T$ is the *frequency* and T is the *wave period*) to the wave number $\mathbf{k}$ ($\sigma = \sigma(\mathbf{k})$, called *dispersion relation*), such that

$$\eta(\mathbf{x}, t) = a\cos[\mathbf{k} \cdot \mathbf{x} - \sigma(\mathbf{k})t + \phi] \tag{9.2}$$

   is a solution of the dynamic equations for every choice of a, $\mathbf{k}$ and ϕ.
2. Any linear combination of functions of the type of Eq. (9.2),

$$\eta(\mathbf{x}, t) = \sum_{i=1}^{N} a_i \cos[\mathbf{k}_i \cdot \mathbf{x} - \sigma(\mathbf{k}_i)t + \phi_i] \tag{9.3}$$

   is in turn a solution of the dynamic equations (*superposition principle*).

Therefore, for example, under the linear hypothesis, the evolution of the initial condition Eq. (9.1) will be given by Eq. (9.3) with the same a_i, $\mathbf{k}_i$, and ϕ_i Naturally,

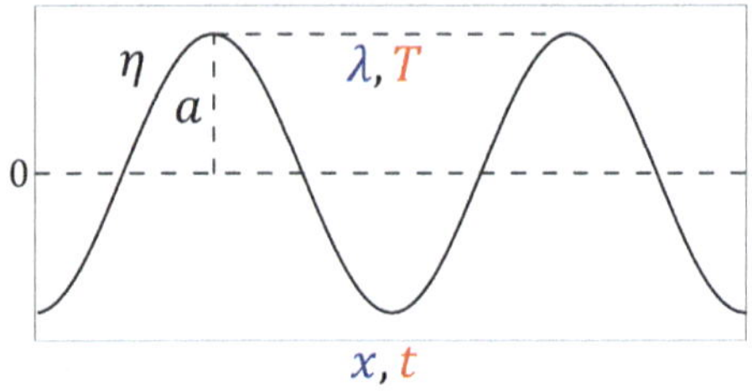

Fig. 9.4 Profile of a sinusoidal wave of amplitude a, wavelength λ and period T

these properties are not limited to the specific case but apply to a wide class of partial differential evolution equations with hyperbolic characteristics. For example, limited to the topics covered in this text, in the study of linear Rossby waves (Chap. 19) the same properties will be found despite the great diversity of the two wave fields.

Figure 9.4 summarises the characteristics of the *sinusoidal wave* η given by Eq. (9.2). The wave-like character of the solution lies in the fact (easily verifiable) that the profile at time t is found shifted by $\Delta x = c_p \Delta t$ at time $t + \Delta t$, where

$$c_p = \sigma/k = \lambda/T$$

is called the *phase velocity* of the wave.

When can it be assumed that the nonlinear terms are negligible in surface gravity waves? In the limit of infinitesimal wave amplitude this is certainly true, as the quadratic character of the nonlinear terms makes them tend to zero faster than the others. To justify the linear approximation for small but finite amplitude waves in a quantitative way, it is necessary to analyse the dimensionless dynamic equations; a general analysis of this aspect is beyond the scope of this discussion. However, it is worth anticipating some considerations on nonlinear effects: these manifest in two different contexts.

The first nonlinear effect concerns the shape of the regular wave, which constitutes the basic building block of the theory. The generalisation of the linear solution Eq. (9.2) to the case of nonlinear waves will be briefly outlined in Sect. 9.7, where it will be shown that these have steeper crests and wider troughs and also propagate with a phase velocity c_p that, unlike linear waves, depends on the wave amplitude. In Sects. 9.7 and 9.8, a dimensionless number (*Ursell's number*) will also be obtained that will clarify how, in addition to the amplitude, the wavelength also plays a fundamental role in determining the weight of the nonlinear terms.

The second nonlinear effect concerns the validity of a decomposition such as that expressed by Eq. (9.3). It is important to note that, for short waves (Sect. 9.3) of small but finite amplitude, the nonlinear interaction between the various regular components is rather weak (Hasselmann 1962, 1963a,b), so that an expression like Eq. (9.3) can be adopted as a first approximation for the spectral description of the wave field.

In the following, the linearisation of the equations will be assumed, so as to develop a simple and powerful theory that can then be generalised to the case of non-negligible nonlinear effects. In addition to linearisation, the other assumptions that can be adopted with excellent approximation are those of (i) homogeneous and incompressible fluid, (ii) perfect fluid and (iii) irrotational flow.

(i) Stratification, although present in the ocean, has no influence on the dynamics of surface waves due to the small variations in density involved (but these are of utmost importance for internal waves, see Chap. 11), therefore, it can be assumed:

$$\nabla \cdot \mathbf{u} = 0;\ \rho = \text{ const.} \tag{9.4}$$

(ii) Molecular viscosity is negligible for the motions in question (rather, for very short waves—of the order of *cm*—surface tension comes into play), therefore reference will be made to the linearised Euler equation:

$$\frac{\partial \mathbf{u}}{\partial t} = \mathbf{g} - \frac{1}{\rho}\nabla p. \tag{9.5}$$

Since we are looking for a solution that on the sea surface is of the form Eq. (9.2), the velocity field $\mathbf{u}$ is assumed to be the product of a circular function with the same argument as Eq. (9.2) (but with a phase γ that can be different from ϕ) multiplied by a function of depth z alone:

$$\mathbf{u}(\mathbf{x}, z, t) = \mathbf{U}(z)\cos(\mathbf{k} \cdot \mathbf{x} - \sigma t + \gamma). \tag{9.6}$$

(iii) As for irrotationality, under the assumptions described above Eq. (5.31) provides

$$\frac{\partial \boldsymbol{\omega}}{\partial t} = 0. \tag{9.7}$$

But in oscillatory motions, the condition that $\boldsymbol{\omega}$ is constant over time necessarily requires that the flow be irrotational: $\boldsymbol{\omega} = 0$. This implies (Sect. 4.5) that there will be a velocity potential φ (such that $\mathbf{u} = \nabla\varphi$) which, for Eq. (9.4), will have to satisfy the Laplace equation

$$\nabla^2\varphi = 0 \tag{9.8}$$

This equation will have to be accompanied by appropriate boundary conditions, which are derived below.

The first boundary condition is that of free-slip on the sea floor. Here we consider a layer of fluid of constant thickness D and fix the origin of the z-axis at the undisturbed sea surface level (Fig. 9.2a). Therefore, imposing $w = 0$ on the bottom will give:

$$\frac{\partial \varphi}{\partial z} = 0; \quad z = -D. \tag{9.9}$$

The derivation of the (*dynamic*) condition on the free surface is more laborious. First of all, substituting $\mathbf{u} = \nabla \varphi$ in Eq. (9.5) gives a time-dependent generalisation (but in the linear approximation) of Bernoulli's Eq. (5.25),

$$\frac{\partial \varphi}{\partial t} + gz + \frac{p}{\rho} = 0 \tag{9.10}$$

where the constant on the right-hand side is arbitrary and can be chosen to be zero since φ is defined up to a constant with respect to the spatial coordinates (for example, we go from a constant C to 0 through the transformation $\varphi \to \varphi + Ct$). On the free surface ($z = \eta$) we have

$$\frac{\partial \varphi}{\partial t} + g\eta + \frac{p_0}{\rho} = 0; \quad z = \eta, \tag{9.11}$$

where p_0 is the atmospheric pressure assumed to be constant; following the same argument above, this term can therefore be removed. Moreover, for the linearity assumption, the equation essentially holds in $z \cong 0$ (from now on we will simply indicate $z = 0$). Ultimately, Eq. (9.11) can be rewritten as:

$$\frac{\partial \varphi}{\partial t} + g\eta = 0; \quad z = 0. \tag{9.12}$$

Differentiating with respect to time we get

$$\frac{\partial^2 \varphi}{\partial t^2} + g\frac{\partial \eta}{\partial t} = 0; \quad z = 0. \tag{9.13}$$

But η_t is nothing more than the vertical component of the velocity w at $z = 0$, so $\eta_t = \varphi_z|_{z=0}$. Substituting this expression into the previous equation eliminates η in favour of φ, thus obtaining the following dynamic boundary condition on the sea surface:

$$\frac{\partial^2 \varphi}{\partial t^2} + g\frac{\partial \varphi}{\partial z} = 0; \quad z = 0 \tag{9.14}$$

See Sect. 4.5 for a discussion on why, in general, irrotational flows require a moving boundary: here that boundary is the free surface, whose motion, however, is not prescribed but is determined by the dynamics itself. Also note that it is precisely in this condition that the forces (gravity) and the time derivative of the unknown variable, both absent in the Laplace equation, intervene.

In conclusion, the complete differential problem for linear surface gravity waves with constant depth D is given by Eqs. (9.8, 9.14 and 9.9). which are reported here for completeness:

$$\begin{cases} \text{(a) } \nabla^2\varphi = 0 \\ \text{(b) } \frac{\partial^2\varphi}{\partial t^2} + g\frac{\partial\varphi}{\partial z} = 0; \quad z = 0 \\ \text{(c) } \frac{\partial\varphi}{\partial z} = 0; \quad z = -D \end{cases} \tag{9.15}$$

9.2 Derivation of the Velocity Field and the Dispersion Relation

For simplicity, and without loss of generality, consider a wave travelling along the x-direction, for which $\mathbf{k} = (k, 0, 0)$. We are therefore looking for a solution of the type:

$$\begin{cases} (a)\ \eta = a\cos(kx - \sigma t) \\ (b)\ \varphi = \Phi(z)\sin(kx - \sigma t) \end{cases} \tag{9.16}$$

(the choice of $\cos(\cdot)$ in (a) and $\sin(\cdot)$ in (b) is imposed by the condition $\eta_t = \varphi_z|_{z=0}$). Substituting φ into Eq. (9.15a) we get the ordinary differential equation for Φ

$$\Phi'' = k^2\Phi,$$

whose most general solution is given by

$$\Phi = Be^{kz} + Ce^{-kz} \tag{9.17}$$

with B and C arbitrary constants. For Eq. (9.15c) we have:

$$\Phi'(-D) = kBe^{-kD} - kCe^{kD} = 0.$$

This condition reduces the number of arbitrary constants from two to one (which we call A):

$$B = \frac{A}{2}e^{kD}; \quad C = \frac{A}{2}e^{-kD}.$$

Substituting into Eq. (9.17) we get:

$$\Phi = A\cosh[k(z + D)]. \tag{9.18}$$

The constant A is related to the amplitude a in the expression of η in Eq. (9.16a) by matching w at $z = 0$:

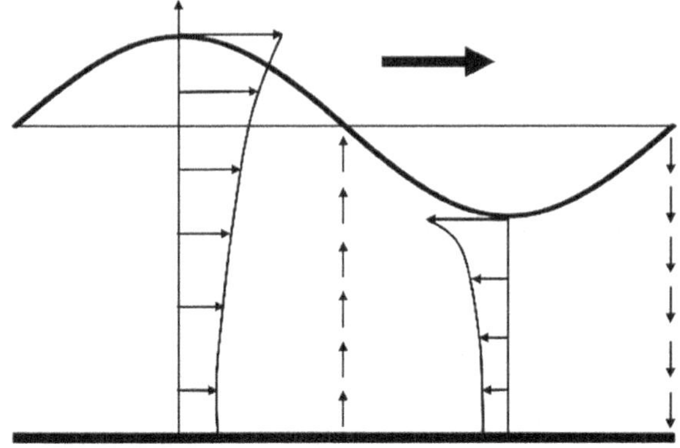

Fig. 9.5 Schematic representation of the surface elevation and velocity field in a sinusoidal surface wave. The arrow indicates the direction of propagation

$$\frac{\partial \eta}{\partial t} = \left. \frac{\partial \varphi}{\partial z} \right|_{z=0} \rightarrow a\sigma \sin(kx - \sigma t) = kA \sinh(kD) \sin(kx - \sigma t)$$

from which

$$A = \frac{a\sigma}{k \sinh(kD)}$$

Substituting this expression into Eq. (9.18) and then into Eq. (9.16b) we get:

$$\varphi = \frac{a\sigma \cosh[k(z + D)]}{k \sinh(kD)} \sin(kx - \sigma t). \tag{9.19}$$

This expression provides the three-dimensional velocity field $\mathbf{u} = \nabla\varphi$. A detailed analysis of $\mathbf{u}$ is omitted, but it is clear that $u = \partial\varphi/\partial x$ and η (Eq. 9.16a) oscillate in phase and, in turn, are out of phase by 90° with respect to $w = \partial\varphi/\partial z$, as schematically shown in Fig. 9.5.

Finally substituting Eq. (9.19) into the dynamic condition on the surface (Eq. 9.15b) we get:

$$\sigma^2(k) = gk \tanh(kD). \tag{9.20}$$

Without any loss of generality, we can consider a wave travelling along the positive x-direction, in which case Eq. (9.20) provides the *dispersion relation for surface gravity waves* in a fluid with constant thickness D:

$$\sigma(k) = \sqrt{gk \tanh(kD)}. \tag{9.21}$$

It is worth noting that, given the isotropy of the problem in question, σ depends on $\mathbf{k}$ only through its modulus: $\sigma(\mathbf{k}) \equiv \sigma(k)$. We will see that this property does not apply to Rossby waves (Chap. 19), as in that case the planetary beta effect (Sect. 12.2) breaks the isotropy of the problem.

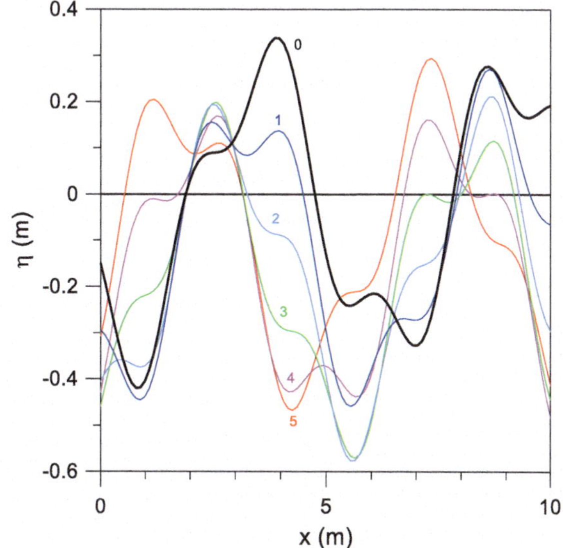

Fig. 9.6 Evolution of the η profile at $t = 0$ along the section $y = 5$ m shown in Fig. 9.3b (black line) for a depth $D = 100$ m. The lines marked by 1, ..., 5 correspond, respectively, to the instants $t = 0.1, ..., 0.5$ s and are obtained from Eq. (9.22)

Equation (9.21) will be discussed in detail in the next paragraph. Here, to emphasise the potential of a wave field's dispersion relation to provide the evolution of the corresponding linear problem, in Fig. 9.6 the evolution of the field composed of three sinusoidal waves described in Fig. 9.3, considered as the initial condition, is shown.

The elevation $\eta(\mathbf{x}, t)$ is obtained from Eq. (9.3) by adopting for σ the dispersion relation Eq. (9.21) with $D = 100$ m, chosen as a reference depth on the continental shelf:

$$\eta(\mathbf{x}, t) = \sum_{i=1}^{3} a_i \cos[\mathbf{k}_i \cdot \mathbf{x} - \sigma(k_i)t + \phi_i]. \tag{9.22}$$

As will be seen in the next paragraph, the three sinusoidal waves considered (see the caption of Fig. 9.3) fall into the category of deep water waves considering the chosen depth. Moreover, their respective amplitudes are sufficiently small to be compatible with the linear approximation.

9.3 Deep and Shallow Water Waves

Figure 9.7 shows the dispersion relation Eq. (9.21) (black line) for the reference depth $D = 100$ m. There are two important approximations related to the product $kD = 2\pi D/\lambda$. If the waves are shorter than the depth ($kD > 1$, in practice if $\lambda \lesssim 2D$) we have with excellent approximation:

$$\sigma \cong \sqrt{gk}. \tag{9.23}$$

In this case we speak of *deep water waves* (red line in Fig. 9.7): for these waves the fluid behaves as if it were at infinite depth. Figure 9.8a shows the corresponding orbital motions obtainable from the study of the velocity field: the orbits are circles whose amplitudes decrease exponentially with depth. Since the orbits are closed there is no net mass transport (but see Sect. 9.7 for a modification due to nonlinear effects).

If instead the waves are much longer than the depth ($kD \ll 1$, in practice if $\lambda \gtrsim 20D$), we have with excellent approximation:

$$\sigma \cong \sqrt{gDk}. \tag{9.24}$$

In this case we speak of in *shallow water waves* (blue line in Fig. 9.7). Figure 9.8b shows the corresponding orbital motions: at small depths the orbits are ellipses with increasing eccentricity with increasing depth, until they reduce to segments corresponding to purely horizontal motions (if $\lambda \gg D$ this happens practically along the

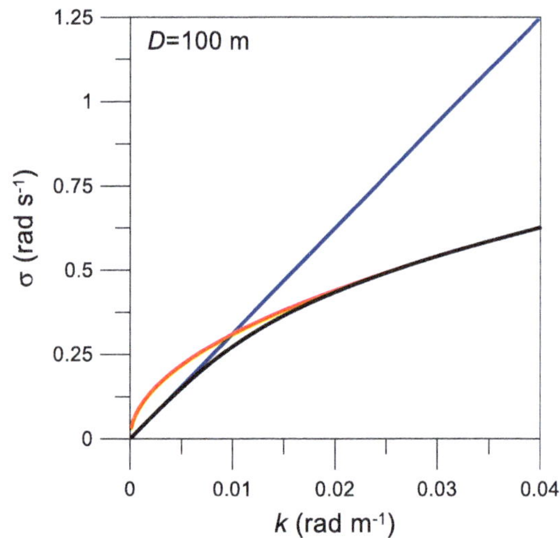

Fig. 9.7 Black line: dispersion relation for surface gravity waves with $D = 100$ m. Red (blue) lines: approximate relation for deep (shallow) water waves

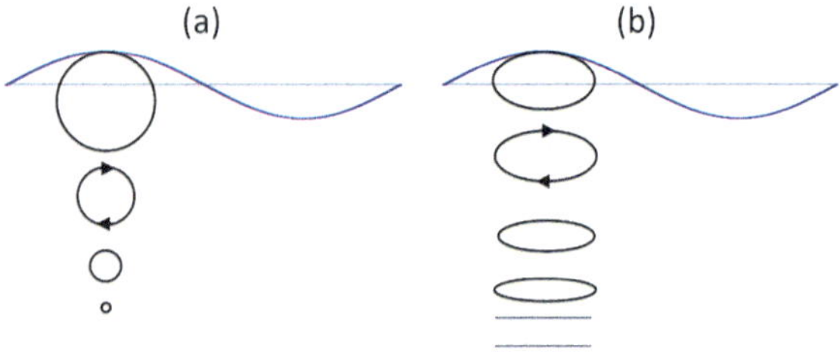

Fig. 9.8 Trajectories of fluid elements in a sinusoidal wave (wave amplitude and depth not to scale). In **(a)** $\lambda < 2D$; in **(b)** $\lambda \gtrsim D$

entire thickness of the fluid). On the seabed, which for these waves is involved in the wave motion, such horizontal motion is required by the free-slip conditions.

We have already seen that the speed with which the profile of the sinusoidal wave translates is given by the phase velocity $c_p = \sigma/k$. For surface gravity waves we have:

$$c_p = \sqrt{\frac{g}{k}\tanh(kD)}. \tag{9.25}$$

The solid black line in Fig. 9.9 shows this dependence.

In the limit of deep water waves we have (solid red line),

$$c_p \cong \sqrt{\frac{g}{k}}. \tag{9.26}$$

Note that for such waves the speed increases with the wavelength like $c_p \propto \lambda^{1/2}$, as results from the following useful formulas valid in the same limit:

$$c_p = \frac{g}{\sigma} = \frac{gT}{2\pi};\ \lambda = \frac{gT^2}{2\pi}.$$

In the limit of shallow water waves we have (solid blue line),

$$c_p \cong \sqrt{gD}. \tag{9.27}$$

Therefore, for such waves c_p does not depend on the wavelength. This case is analogous to electromagnetic waves in vacuum, which travel at the speed of

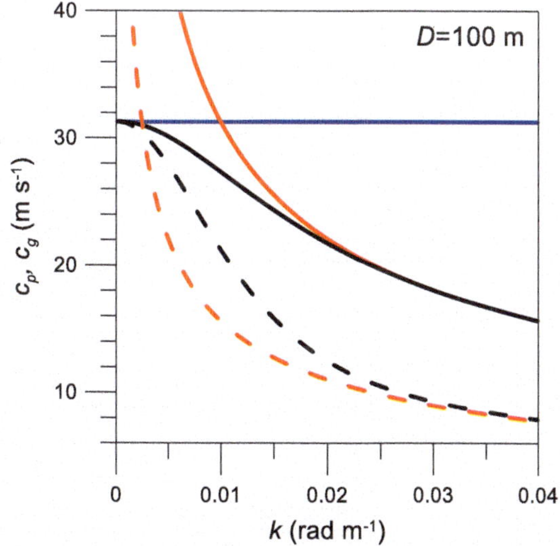

Fig. 9.9 Solid black line: phase velocity for surface gravity waves with $D = 100$ m. Solid red (blue) lines: approximate phase velocity for deep (shallow) water waves. Dashed lines: corresponding group velocities (the dashed blue line coincides with the solid line)

light $c = 299792\,\text{km s}^{-1}$ regardless of their length (and regardless of the inertial reference frame considered! See Appendix E). Note that the speed of shallow water waves represents the upper limit for gravity waves.

9.4 Group Velocity

Deep and shallow water waves are respectively called *dispersive* and *non-dispersive*. This has to do with the evolution of a *group*, or *packet* of linear waves, as we now proceed to analyse.

Suppose we know the wave profile (for simplicity dependent only on x) at time $t = 0$: $\eta(x, 0) = \eta_0(x)$. As already seen in the discussion of Figs. 9.3 and 9.6, knowledge of the dispersion relation allows us to determine the evolution of a field of linear waves by applying the Fourier method. Express the initial condition η_0 as

$$\eta_0(x) = \frac{1}{2\pi}\int_{-\infty}^{+\infty} \tilde{\eta}(k)e^{ikx}dk \tag{9.28}$$

where

$$\tilde{\eta}(k) = \int_{-\infty}^{+\infty} \eta_0(x)e^{-ikx}dx$$

is the *Fourier transform* of $\eta_0(x)$. Since, thanks to the linearity of the problem, the principle of superposition applies, the evolution of η will be obtained by multiplying each Fourier component in Eq. (9.28) by the phase factor $e^{-i\sigma t}$ (recall the Euler formula $e^{i\sigma t} = \cos(\sigma t) + i\sin(\sigma t)$, where i is the imaginary unit):

$$\eta(x, t) = \frac{1}{2\pi}\int_{-\infty}^{+\infty} \tilde{\eta}(k)e^{i[kx-\sigma(k)t]}dk \tag{9.29}$$

where $\sigma(k)$ is given by Eq. (9.21).

Now, imagine that $\tilde{\eta}$ is concentrated in a region of the spectrum containing only deep water waves. Each component will travel at a different speed from the others, resulting in a substantial modification of the initial wave packet profile. Over time, various wavelengths will emerge from an initial compact condition, resulting in a gradual widening of the wave packet. This phenomenon is expressed by saying that there is *phase dispersion* and that, therefore, such waves are *dispersive*.

Conversely, if the initial wave packet contains only shallow water waves, all Fourier components will travel at the same speed $c_p = \sqrt{gD}$. Consequently, the wave profile will propagate at this same speed without modification of its shape. This phenomenon is expressed by saying that there is no phase dispersion and that, therefore, such waves are *non-dispersive*. Naturally, if the wave packet is two-dimensional, its evolution will involve a modification of its shape, but this has nothing to do with phase dispersion.

This discussion introduces us to the concept of *group velocity*. In the evolution of a generic initial condition $\eta_0(x)$, limited areas (the *wave groups*) in which the wave amplitude is consistent will be distinguishable. The question we ask ourselves is: at what speed does the envelope of a wave group travel?

The simplest example of this phenomenon is obtained by superimposing two waves with the same amplitude, but with slightly different wavelengths. Figure 9.10 shows two sinusoidal waves of amplitude $a = 10$ cm with wavelengths respectively of $\lambda = 2$ m (green line) and $\lambda' = 2.2$ m (orange line). Using the first prosthaphaeresis formula, the sum of the two waves provides

$$\begin{aligned}\eta_{tot} &= a\sin(kx-\sigma t) + a\sin\left(k'x-\sigma' t\right)\\ &= 2a\cos\left(\frac{\Delta k}{2}x - \frac{\Delta\sigma}{2}t\right)\sin(\bar{k}x-\bar{\sigma}t) \equiv F(x,t)f(x,t)\end{aligned} \tag{9.30}$$

where

$$\Delta k = k-k';\ \Delta\sigma = \sigma-\sigma';\ \overline{k} = \frac{k+k'}{2};\ \overline{\sigma} = \frac{\sigma+\sigma'}{2}.$$

The blue lines in Fig. 9.11 show η_{tot} for gravity waves with $D = 100$ m at different time instants ($t = 0, 3, 6, 9, 12$ s). When the two waves are in phase the amplitude is maximum (*constructive interference*) while when they are in antiphase the amplitude is null. The black line shows the envelope $F(x, t)$ in the form of periodic wave groups that modulate the *carrier wave* $f(x, t)$. The latter has a wavelength $\overline{\lambda} = 2\pi/\overline{k}$ equal to the average of the two starting waves, while $F(x, t)$ has a much greater wavelength ($\Lambda = 2\pi/(\Delta k/2) = 2\lambda\lambda'/\left(\lambda'-\lambda\right) = 44\,m$).

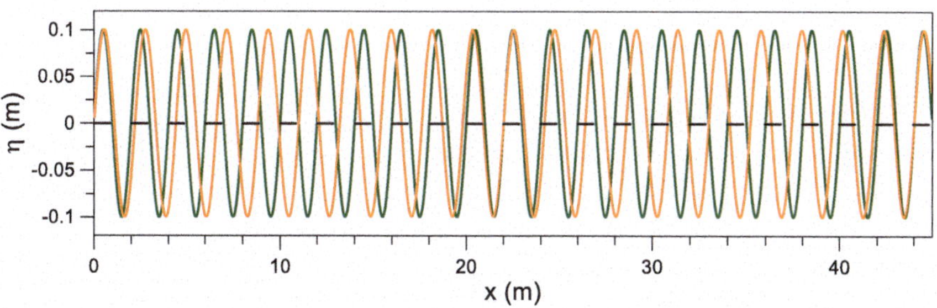

Fig. 9.10 Two sinusoidal waves of amplitude $a = 10$ cm with wavelengths $\lambda = 2$ m (green line) and $\lambda' = 2.2$ m (orange line)

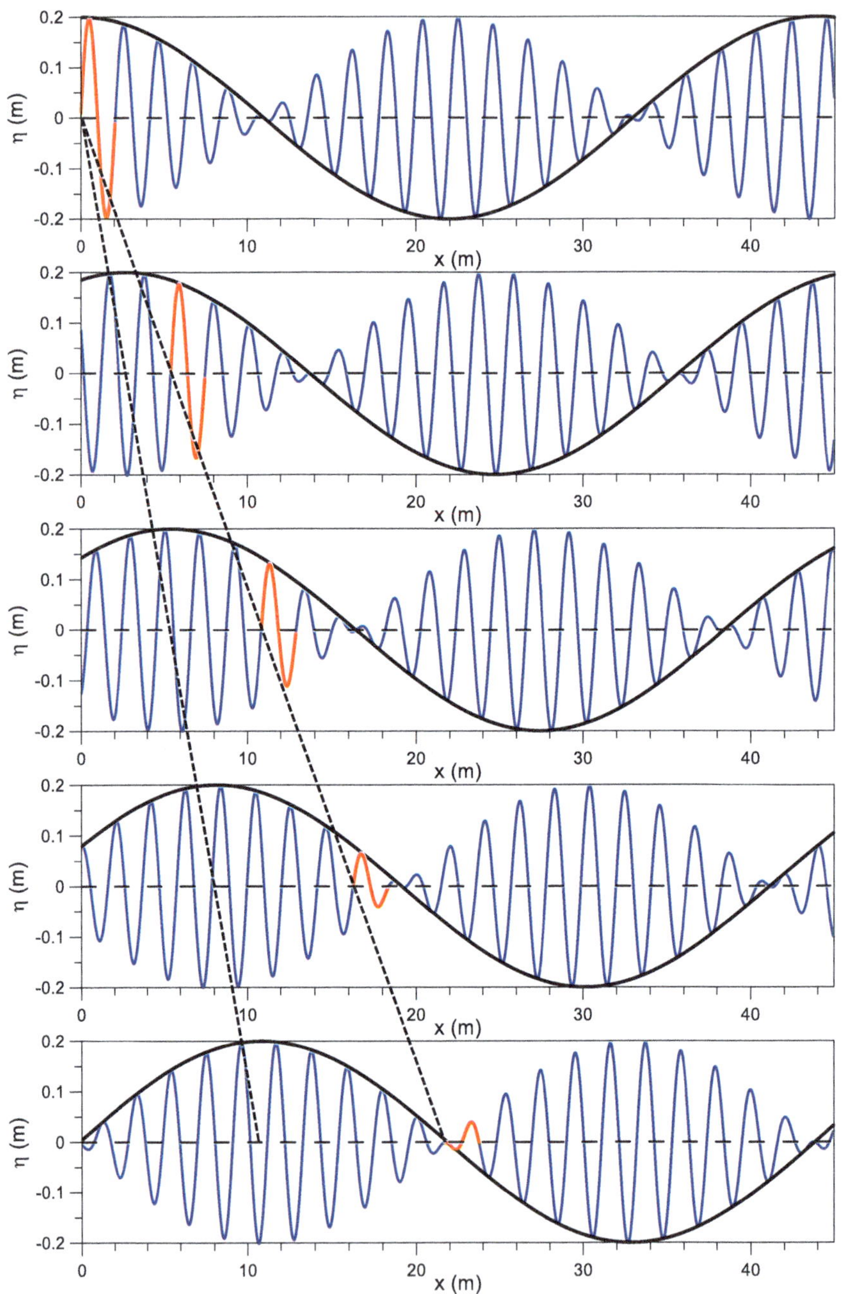

Fig. 9.11 Blue lines: η_{tot} defined by Eq. (9.30) with $\sigma = \sigma(k)$ and $D = 100$ m at different times ($t = 0, 3, 6, 9, 12$ s). Black lines: modulation $F(x, t)$ at different times. Red lines: one cycle of η_{tot} at different times

It is important to note that, while the carrier wave travels at the speed $c = \overline{\sigma}/\overline{k}$ (very close to the phase speed of the two original waves), the envelope travels at the speed:

$$c_g = \frac{\Delta\sigma}{2} / \frac{2}{\Delta k} \cong \frac{d\sigma}{dk}. \tag{9.31}$$

This is called the *group velocity* of the wave. This relationship is generally valid, no matter how complex the integral Fourier representation of η is, and therefore, for wave groups of any shape and structure (c_g will obviously have to be calculated for that k which competes with the carrier wave within the group). In Sect. 10.1 we will see how c_g describes the propagation of energy of an irregular wave field as well as of the single sinusoidal wave.

For gravity waves, the following expression for c_g is obtained from Eq. (9.25):

$$c_g = \frac{c_p}{2}\left[1 + \frac{2kD}{\sinh 2kD}\right]. \tag{9.32}$$

In the limit of deep water waves (red dashed line in Fig. 9.9)

$$c_g \cong \frac{c_p}{2}.$$

Therefore, the oscillations will overtake the wave group in which they are temporarily located. Figure 9.11 clearly shows this phenomenon: follow the evolution of a cycle of the carrier wave (in red): at $t = 12$ s in the example you can see (dashed lines) how the displacement undergone by that cycle is double that of the envelope wave.

Conversely, for shallow water waves (blue line in Fig. 9.9),

$$c_g \cong c_p \cong \sqrt{gD}.$$

In this case the two speeds coincide and the profile will translate, in one-dimensional propagation, without changing its shape.

9.5 Phenomenology of Surface Waves

Surface gravity waves are essentially generated by the wind (see Sect. 10.2, but see Sect. 9.6 for an interesting exception to this rule) and have periods ranging from a fraction of a second up to a maximum of a few tens of seconds. The highest frequency part of the wave field (with maximum periods of a few seconds and maximum wavelengths of a few metres) is called *sea*, while the lower frequency part (with periods even exceeding 10 s and with wavelengths even greater than 100 m) is called *swell*. These are obviously all to be considered waves in deep water if you are in the open sea off the continental shelf.

The sea and the swell have quite different propagation properties. The sea does not propagate for great distances from the generation area (the *wind fetch*) due to the high slope of the waves which facilitates their breaking; therefore, the sea is essentially localised in the fetch. On the contrary, thanks to the low slope of the waves, the swell can also propagate for tens or hundreds of km if generated in the open sea by strong and persistent winds on an extended fetch. This is the regular field of waves that can be observed, even in the absence of wind, along the coasts even hours after its generation.

It should be noted that swell waves, originally in deep water, approaching the coast can transition into a regime of waves in shallow water: this happens when the wavelength becomes comparable to the depth. From the moment when $\lambda \sim D$ the wave changes due to the phenomenon of *refraction*. Since the wave's frequency and the power transferred in the direction of propagation must remain constant, the wave records a decrease in λ and an increase of the amplitude as D decreases (and therefore as c_p decreases). The slope of the wave can also reach such high values that it breaks near the shore. Another effect due to refraction is the tendency of wave fronts to align along the bathymetric lines, resulting in a deviation of the propagation direction. Another important phenomenon that can substantially modify the wave fronts is that of *diffraction*, which occurs when a wave train hits a gap delimited by the mainland. For a detailed discussion of the phenomena of refraction and diffraction of surface gravity waves, refer to specific texts on the subject.

As an example, some indicative values of the parameters for sea and swell waves are reported:

- Indicative extreme values for deep water waves: $T =\lesssim 15\ s$; $c_p \lesssim 24\ ms^{-1}$; $\lambda \lesssim 360\ m$
- Example of sea waves: $T = 2\,\mathrm{s}$; $c_p \cong 3\,\mathrm{m\,s^{-1}}$; $\lambda \cong 6\,\mathrm{m}$
- Example of swell waves: $T = 10\,\mathrm{s}$; $c_p \cong 16\,\mathrm{m\,s^{-1}}$; $\lambda \cong 160\,\mathrm{m}$

Here is an example of parameters for shallow water waves near the coastline for $D = 20\,\mathrm{m}$:

- $c_p \cong 14\,\mathrm{m\,s^{-1}}$
- $\lambda \cong 200\,\mathrm{m}(\lambda/D = 10)$: $T = 14\,\mathrm{s}$
- $\lambda \cong 300\,\mathrm{m}(\lambda/D = 15)$: $T = 21\,\mathrm{s}$

9.6 Tsunami Waves

Surface gravity waves are essentially generated by the wind. However, there can also be waves in deep water generated by mechanisms that have nothing to do with the atmosphere: this is the case of *tsunami* (from the Japanese word meaning *harbour wave*). These waves result from the sudden displacement of a large sea surface area due to different causes, such as an underwater earthquake or volcanic eruption or, in less energetic phenomena, a landslide that sinks into the sea. The local displacement of the sea surface (which essentially reflects that of the seabed if this is—as often happens—very rapid) can also be modest, but the possible large extension of the generation area and the subsequent refraction near the coasts can make a tsunami extremely destructive.

In history, tsunamis have been documented that have caused tremendous catastrophes. Consider, for example, the tsunami produced by the Krakatoa volcano on the Indonesian island of Rakata in 1883, the Lisbon tsunami of 1755, that of Messina on 28 December 1908 (which in less than a minute severely damaged the cities of Messina and Reggio Calabria, representing, together with the earthquake, perhaps the most immense natural disaster in European history), the one that hit the Hawaiian islands in 1946 and the more recent ones of the Indian Ocean on 26 December 2004 (the most catastrophic natural disaster of the modern era originated off the island of Sumatra causing 250,000 deaths) and of Tohoku on 11 March 2011 (which produced, together with the earthquake, the serious accident at the Fukushima nuclear power plant, hit by a wave 14 m high).

Since tsunamis generally have a wavelength greater than the thickness of the deep sea, these represent waves in shallow water with very high speeds. For example, for $D = 4\,\mathrm{km}$, the speed is $c_p \cong 200\,\mathrm{m\,s^{-1}} \cong 720\,\mathrm{km\,h^{-1}}$, close to that of a cruising airliner. This makes the implementation of early warning systems problematic, which however in the great oceans are able to prevent damage to populations located thousands of km from the generation area (think, for example, of the Hawaiian islands, which can be alerted to the arrival of tsunamis generated in the Aleutian islands area). An efficient early warning system may never be implemented for tsunamis too close to the coast, like those of Messina and Tohoku.

9.7 Nonlinear Effects: Stokes and Cnoidal Waves

The analysis of nonlinear effects on surface gravity waves is very complex and varied and partly goes beyond the aims of this text. On the other hand, the topic is too relevant to be ignored in a discussion—albeit introductory—like the present one. Therefore, in this paragraph and the next, the most salient aspects of the problem will be described without, however, reporting the complex mathematical derivations that lead to those results. For a more in-depth treatment, refer to specialist texts, such as, for example, those by Whitham (1974), Le Blond and Mysak (1978) and Dean and Dalrymple (1984).

The consideration of the nonlinear effects produced by the terms $(\mathbf{u} \cdot \nabla)\mathbf{u}$ in the Navier–Stokes equations modifies the characterisation of the waves presented in Sect. 9.1 and summarised in Eqs. (9.2 and 9.3).

The expression in Eq. (9.2) will no longer be a solution to the equations of motion. There are periodic solutions also in the nonlinear case, however these are not sinusoidal as in the linear case but present, as will be seen, an asymmetry between crests and troughs. Moreover, the dispersion relation, which in the linear case links the frequency to the wave number alone, now also depends on the wave amplitude: $\sigma = \sigma(k, a)$. These aspects, analysed for the first time in a fundamental article by Stokes (1847), will be briefly discussed in the present paragraph.

As for the principle of superposition (Eq. 9.3), it will no longer be valid; however, for only *weakly nonlinear waves* it is possible to adopt a similar Fourier representation, but taking into account the nonlinear energy transfer between the various components of the wave field (Hasselmann 1962; 1963a,b). This aspect is too advanced and will not be discussed in this text.

We start by considering periodic waves in deep water (called *Stokes waves*). If $kD \gg 1$ Eq. (9.16a) generalises as follows ($\theta = kx - \sigma t$):

$$\eta = a\cos\theta + a^2\chi\cos(2\theta) + ... \tag{9.33}$$

where

$$\chi = \frac{1}{2}k\coth(kD)\left[1 + \frac{3}{2\sinh^2(kD)}\right]. \tag{9.34}$$

Figure 9.12 shows the profiles of nonlinear waves in deep water, as usual for a reference depth of $D = 100\,\text{m}$, for $\lambda = 5\,\text{m}$ and for three different amplitudes: $a = 0.1, 0.2, 0.3\,\text{m}$.

For $a = 0.1$ m (black line) the wave deviates very little from the sinusoidal shape, hence the nonlinear effects are negligible. Conversely, if $a = 0.2\,\text{m}$ (blue line) the wave presents a marked asymmetry between a narrower crest of maximum amplitude $\cong +0.23\,\text{m} > a$ and a wider trough of minimum amplitude $\cong -0.18\,\text{m} > -a$. This asymmetry is even more evident if $a = 0.3\,\text{m}$ (red line).

As for the dispersion relation, Eq. (9.23) becomes:

$$\sigma^2 = gk\left(1 + a^2k^2 + ...\right). \tag{9.35}$$

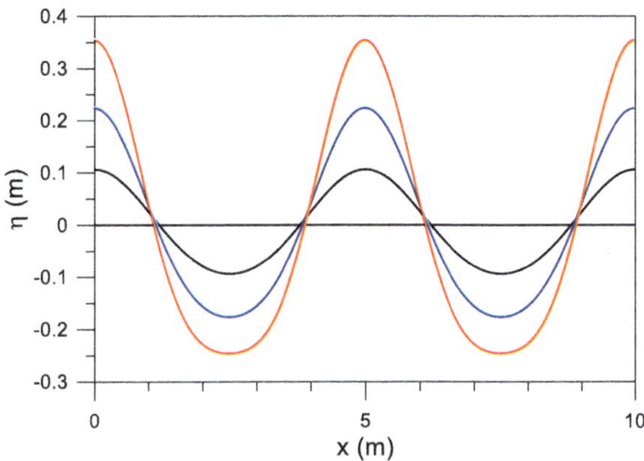

Fig. 9.12 Nonlinear periodic waves in deep water (Stokes waves) obtained from Eqs. (9.33 and 9.34) with $t = 0$, $\lambda = 5\,\text{m}$, $D = 100\,\text{m}$ and $a = 0.1\,\text{m}$ (black line), $a = 0.2\,\text{m}$ (blue line) and $a = 0.3\,\text{m}$ (red line)

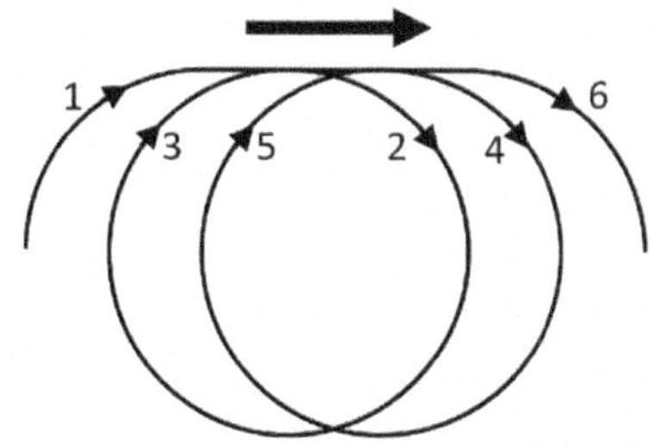

Fig. 9.13 The Stokes drift in deep water

The dependence of the frequency of a periodic wave on its amplitude is a distinctive characteristic of all nonlinear wave fields in fluid dynamics.

Another important nonlinear effect concerns the mass transport of the wave. In Sect. 9.3 it was observed that the orbits of the fluid elements are closed in a linear wave (Fig. 9.8) and this implies that there is no net mass transport. However, in waves of small but finite amplitude, for which nonlinear effects are not negligible, the orbits do not close exactly, as shown in Fig. 9.13: this results in a small net mass transport in the direction of wave propagation, known as *Stokes drift*.

The average speed (along x for a wave propagating in the same direction) associated with the Stokes drift as a function of depth is given by the formula:

$$u_{SD} = c_p a^2 k^2 e^{2kz}. \tag{9.36}$$

We now move on to consider periodic waves in shallow water. If $kD \ll 1$, Eq. (9.16a) generalises as follows:

$$\eta = a\cos\theta + \frac{3a^2}{4k^2D^3}\cos(2\theta) + \frac{27a^3}{64k^4D^6}\cos(3\theta) + \dots \tag{9.37}$$

Such waves are called *cnoidal*, as they can be expressed using the Jacobi elliptic function $cn(\cdot,\cdot)$. Figure 9.14 shows the profiles of cnoidal waves for a shallow coastal sea with $D = 2$ m for $\lambda = 20$ m and for three different amplitudes: $a = 0.1, 0.2, 0.3$ m. Similar considerations apply as those made for Fig. 9.12.

As for the dispersion relation, Eq. (9.24) generalises as follows ($c_0 = \sqrt{gD}$):

$$\sigma = c_0 k\left(1 - \frac{1}{6}k^2D^2 + \frac{9a^2}{16k^2D^4} + \dots\right). \tag{9.38}$$

Note that the first two terms

$$\sigma = c_0 k - \gamma k^3, \tag{9.39}$$

where

$$\gamma = c_0 D^2/6,$$

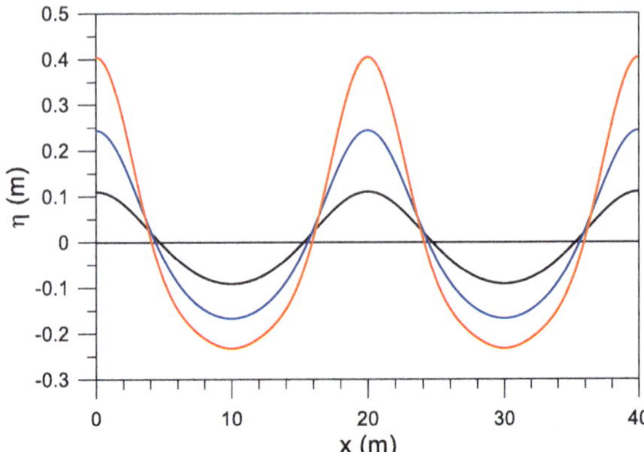

Fig. 9.14 Nonlinear periodic waves in shallow water (cnoidal waves) obtained from Eq. (9.37) with $t = 0$, $\lambda = 20$ m, $D = 2$ m and $a = 0.1$ m (black line), $a = 0.2$ m (blue line) and $a = 0.3$ m (red line)

can be obtained from the McLaurin series expansion of the exact dispersion relation (Eq. 9.21) and imply a weak phase dispersion for long but not infinitely long waves. The thin blue line in Fig. 9.15 shows this relation, which should be compared with the non-dispersive $\sigma = c_0 k$ indicated by the thick blue line. Finally, the third term in Eq. (9.38) represents a nonlinear correction that, as for waves in deep water, introduces a dependence on the amplitude a.

Note that Eqs. (9.37 and 9.38) can be rewritten as

$$\eta = a\left[\cos\theta + \frac{3}{16\pi^2}U_r\cos(2\theta) + \ldots\right], \tag{9.40}$$

$$\sigma = c_0 k\left(1 - \frac{1}{6}k^2D^2 + \frac{9}{64\pi^2}\varepsilon U_r + \ldots\right), \tag{9.41}$$

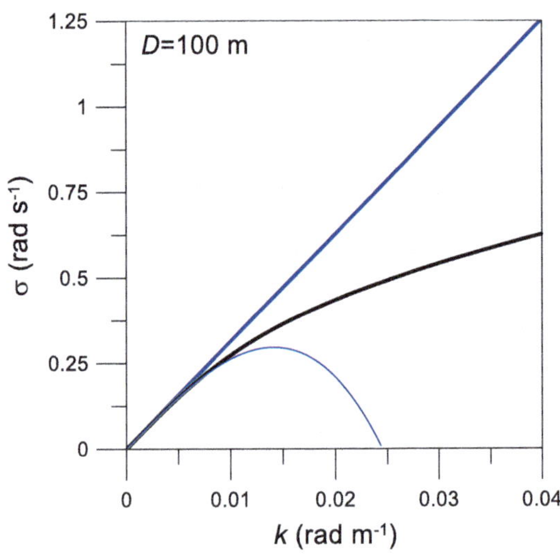

Fig. 9.15 Black line: dispersion relation for surface gravity waves with $D = 100$ m. Thick (thin) blue line: approximate relation for non-dispersive (weakly dispersive: Eq. 9.39) waves in shallow water

where the two dimensionless numbers Ur and ε are defined as follows:

$$U_r = \frac{a\lambda^2}{D^3}; \varepsilon = \frac{a}{D}. \tag{9.42}$$

The parameter ε plays an important role in the wave breaking process and it is required that $\varepsilon \ll 1$. Furthermore, it is clear that the so-called *Ursell number* U_r measures the weight of the nonlinearities. It is now clear why, as anticipated in Sect. 9.1, the degree of nonlinearity—and therefore the validity of any linear approximation—depends not only on the wave amplitude but also from its length λ. An analogous dimensionless number will appear in the next paragraph in relation to solitonic dynamics.

9.8 Nonlinear Effects: Korteweg-de Vries Equation, Solitons

We now move on to consider an important evolution equation that describes the propagation of *long, weakly dispersive and weakly nonlinear waves*: this will allow us to generalise the dynamics of cnoidal waves to non-periodic waves. The rigorous derivation of such an equation requires perturbative developments based on a differential problem that generalises Eq. (9.15) to the nonlinear case. Here we prefer to follow a heuristic procedure which, although not rigorous from a mathematical point of view, has the advantage of being intuitive and conceptually satisfying (for the rigorous derivation, refer to more specialised texts).

Note that the dispersion relation Eq. (9.27) for shallow water waves, but including both directions of propagation,

$$c_0 = \pm\sqrt{gD} \tag{9.43}$$

is perfectly equivalent to the following second-order, linear, hyperbolic partial differential equation (in this paragraph for simplicity we adopt the compact notation for partial derivatives):

$$\eta_{tt} - c_0^2 \eta_{xx} = 0. \tag{9.44}$$

Equation (9.43) immediately derives from Eq. (9.44) by looking for a solution of the type

$$\eta = a e^{i(kx - \sigma t)}. \tag{9.45}$$

Equation (9.44) is the classic equation of unidimensional and *bidirectional* linear non-dispersive waves; "bidirectional" refers to the fact that both directions of propagation along the x-axis are allowed.

On the other hand, the dispersion relation $c_0 = +\sqrt{gD}$ is equivalent to the equation

$$\eta_t + c_0\eta_x = 0. \tag{9.46}$$

This equation, describing the propagation of unidimensional and *unidirectional* linear non-dispersive waves, is physically rather banal as the evolution of any initial condition $\eta_0(x)$, which is intended to travel at speed c_0, will consist only in a translation at the same speed without modification of its own form; in fact, it is immediate to verify that $\eta = f(x - c_0 t)$ is solution of Eq. (9.46) (but also of Eq. 9.44) for every f.

More interesting is the equation equivalent to the dispersion relation Eq. (9.39):

$$\eta_t + c_0\eta_x + \gamma\eta_{xxx} = 0. \tag{9.47}$$

Now the evolution of an initial travelling condition shows the effect of (weak) phase dispersion. Figure 9.16 shows the evolution (thin blue line) of an initial Gaussian-shaped condition (thick blue line, whose amplitude a and width σ_G are compatible with the required conditions) obtained by solving Eq. (9.47) using an implicit finite difference numerical scheme (Pierini 1986); note that this solution (and all the others shown in this paragraph) refers to the reference frame moving with speed c_0. The wave front, where the energy of the waves with the longest wavelengths among those that make up the initial condition accumulates, is called the *Airy phase*, whose amplitude decreases over time as $a_{Airy} \propto t^{-1/3}$ while its width increases inversely, $\ell_{Airy} \propto t^{1/3}$, with consequent conservation of wave mass. The *dispersive trail* that develops behind the wave front clearly shows the effect of phase dispersion, with longer waves preceding the shorter ones in accordance with the dispersion relation for weakly dispersive waves.

Another non-trivial extension of Eq. (9.46) consists in considering non-dispersive but weakly nonlinear waves. In this case, considering the dependence of the phase velocity on the amplitude, it can be assumed that

$$\eta_t + c_0\left(1 + q_1\eta + q_2\eta^2 + ...\right)\eta_x = 0.$$

In fact, it can be shown that, at the lowest order of nonlinearity, η satisfies the nonlinear and non-dispersive wave equation

$$\eta_t + c_0\eta_x + \delta\eta\eta_x = 0 \tag{9.48}$$

with $q_1 = 3/(2D)$ and, therefore, where

$$\delta = \frac{3}{2}\sqrt{\frac{g}{D}}.$$

Figure 9.16 shows the evolution (thin red line) of an initial Gaussian-shaped condition (thick red line, whose amplitude a and width σ_G are compatible with

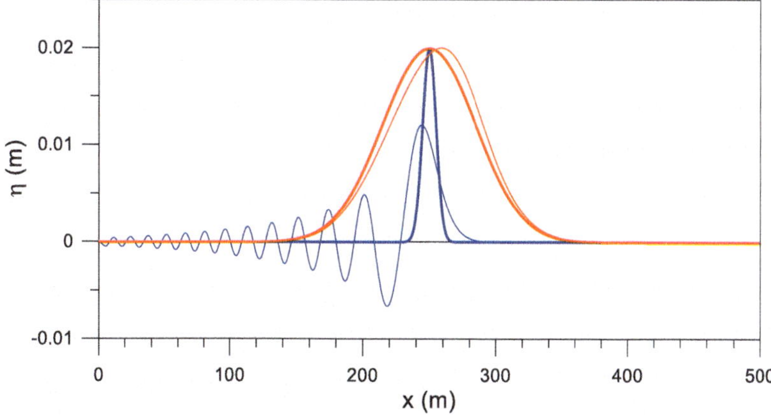

Fig. 9.16 Linear, weakly dispersive evolution with Airy phase and dispersive trail (blue lines) and nonlinear, non-dispersive evolution with progressive steepening of the wave front (red lines). Thick blue line: initial Gaussian condition with $a = 0.02$ m and $\sigma_G = 20$ m and, thin blue line, evolution determined by Eq. (9.47) in a shallow sea with $D = 2$ m. Thick red line: initial Gaussian condition with $a = 0.02$ m and $\sigma_G = 140$ m and, thin red line, evolution determined by Eq. (9.48) in a shallow sea with $D = 2$ m. The simulations were carried out in the reference frame moving with speed c_0 using an implicit finite difference numerical scheme (Pierini 1986)

the required conditions) obtained by solving Eq. (9.48) using the same numerical scheme mentioned above. The steepening of the wave front is due to the dependence of the speed of a point on the wave on its height relative to the undisturbed level $\eta = 0$: in fact, for example, the point of maximum amplitude of the initial condition will move with speed $c_0 + \delta a$ while a point on the wave front with $\eta \cong 0$ will move with speed $\sim c_0$.

At this point, a reflection is necessary. The asymptotic evolution of the blue line in Fig. 9.16 determined by the linear and weakly dispersive Eq. (9.47) will lead—as discussed above—to a complete dispersion of the initial wave packet, with the amplitudes of the Airy phase and the dispersive trail tending to zero. On the contrary, the evolution of the red line in Fig. 9.16 determined by the weakly nonlinear and non-dispersive Eq. (9.48) will lead to a vertical wave front, corresponding to the wave breaking. However, these asymptotic behaviours are in contradiction with the very assumptions on which Eqs. (9.47 and 9.48) are based, as we will now discuss.

On the one hand, we have seen that—for long periodic waves—the nonlinearity is weighted by the Ursell number defined in Eq. (9.42); an obvious extension to the case of the Airy wave front suggests that, for weakly dispersive and initially linear waves, the nonlinearity grows over time as $U_r \propto a_{Airy}\ell^2_{Airy} \propto t^{1/3}$. Therefore, for sufficiently large times, nonlinear effects will inevitably come into play. This is expressed by saying that Eq. (9.47) is not uniformly valid over time.

On the other hand, if the length scale ℓ_{nonlin} corresponding to the horizontal extension of the nonlinear wave front governed by Eq. (9.48) is initially large enough to exclude dispersive effects, it is also true that the progressive steepening will lead to a decrease in ℓ_{nonlin} until the phase dispersion is no longer negligible. Also in this case, it can therefore be stated that also Eq. (9.48) is not uniformly valid over time.

It seems plausible that an equation that takes into account both the effects of a weak phase dispersion included in Eq. (9.47), and of weak nonlinearity included in Eq. (9.48), can be uniformly valid over time:

$$\eta_t + c_0\eta_x + \delta\eta\eta_x + \gamma\eta_{xxx} = 0. \tag{9.49}$$

This famous equation, known as the *Korteweg-de Vries equation* (1895, abbreviated KdV) is indeed uniformly valid over time. In fact, regardless of the initial degree of nonlinearity and phase dispersion, these two effects will asymptotically tend to balance each other out, giving rise to solutions of invariant form called *solitons*. Moreover, the periodic cnoidal waves discussed in Sect. 9.7 turn out to be solutions of the same equation. It should be emphasised that, in addition to the case in question, the KdV equation also applies to other wave fields in fluid dynamics; for example, in Sect. 11.3, an application of it to long internal waves will be described.

The soliton (an example is shown by the thin line in Fig. 9.17) has the following mathematical expression:

$$\eta_s(x) = a\cosh^{-2}\left(\frac{x - c_s t}{\sigma_s}\right), \tag{9.50}$$

$$c_s = c_0\left(1 + \frac{a}{2D}\right); \; \sigma_s = 2D\sqrt{\frac{D}{3a}}.$$

The most salient aspects of this solution are as follows: (i) this wave is isolated, solitary, as it is not part of a wave train; (ii) its width is greater the lower the wave is:

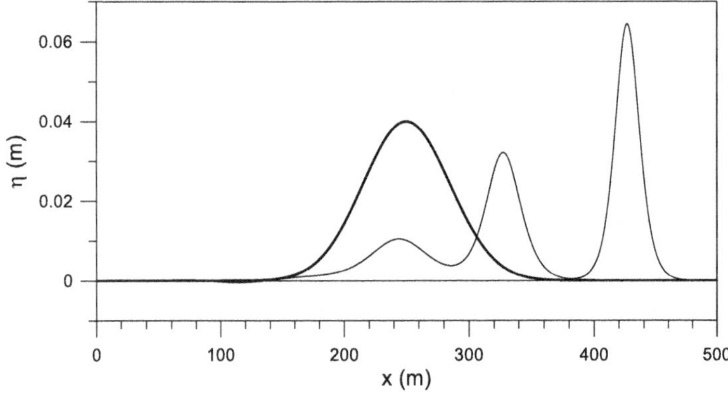

Fig. 9.17 Formation of solitons. Thick line: initial Gaussian condition with $a = 0.04$ m and $\sigma_G = 140$ m and, thin line, evolution determined by a regularised version of the KdV Eq. (9.49) in a shallow sea with $D = 2$ m. The simulation was carried out in the reference frame moving with speed c_0 using an implicit finite difference numerical scheme (Pierini 1986)

$\sigma_s \propto a^{-1/2}$; (iii) the wave translates without modifying its own shape as in a soliton there is a perfect and persistent balance between phase dispersion and nonlinearity; (iv) the translation speed c_s depends on the amplitude a through a correction to the non-dispersive phase speed c_0 equal to $\varepsilon/2$ (the parameter ε see Eq. (9.42)—must also in this case be $\varepsilon \ll 1$).

A fascinating aspect concerns the evolution governed by the KdV equation: every initial condition η_0 of positive volume gives asymptotically rise to a series of $n \geq 1$ solitons whose vertices will be aligned along a line, with an inclination that tends to infinity due to the nonlinear correction to the translation speed. An example of this behaviour is shown by the numerical simulation reported in Fig. 9.17. An initial Gaussian condition gives rise, in the specific case, to three solitons, the first of which is fully developed in the figure. The nonlinear correction to the soliton speed is well highlighted by the graph which, it is remembered, refers to the reference system moving with c_0.

As already observed, since the surface soliton is an elevation wave, the initial condition η_0 must also be so in order for one or more solitons to emerge (in fact the "mass" of the wave $\int \eta_0 dx$ must be conserved). What happens if this condition is not met? Fig. 9.18 shows the evolution of an initial depression condition equal and opposite to that of Fig. 9.17. The nonlinear effects now slow down the points on the wave of greatest negative amplitude ($\delta\eta < 0$); consequently, instead of the steepening of the wave front, an expansion of the latter and the formation of a substantial dispersive trail is observed.

Another aspect of great interest is the following: the interaction of two solitons (which can occur if, for example, a soliton of amplitude a_1 precedes the second of amplitude $a_2 < a_1$) results in the overtaking of the first soliton by the second, with the result of a small phase shift but without any modification of their respective shapes.

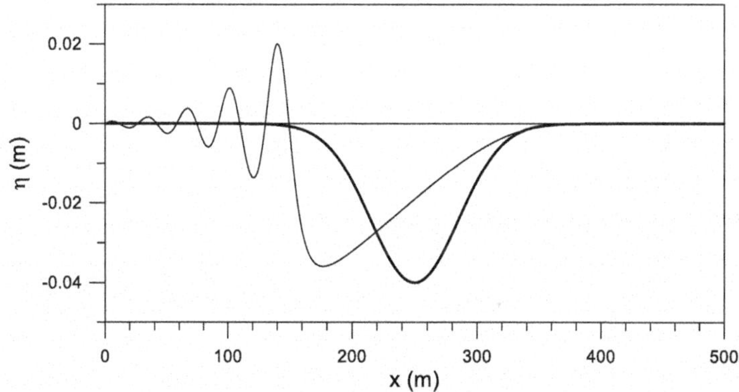

Fig. 9.18 KdV evolution of a negative initial condition equal and opposite to that of Fig. 9.17. Instead of the steepening of the wave front, with the consequent formation of solitons, a progressive widening of it and the formation of a dispersive trail is observed. The simulation was carried out in the reference system moving with speed c_0 using an implicit finite difference numerical scheme (Pierini 1986)

This phenomenon, far from trivial (it would be if the waves were linear, but in that case no phase shift would arise) is shown in Fig. 9.19.

It is worth adding that, regardless of the resolution of Eq. (9.49) by numerical methods (such as, for example, the finite difference one adopted in the simulations

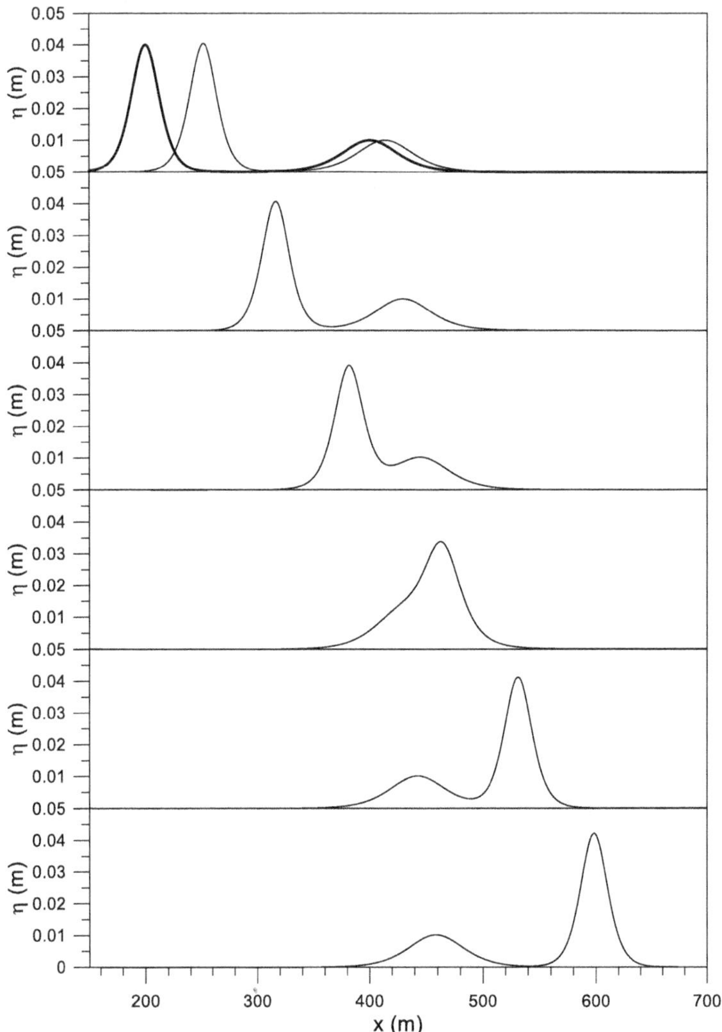

Fig. 9.19 Interaction between solitons. In the first graph, the thick line indicates the initial condition composed of two solitons of amplitudes respectively $a = 0.04$ m and $a = 0.01$ m. The six thin lines show the evolution of the wave at regular intervals determined by KdV Eq. (9.49) in a shallow sea with $D = 2$ m. The simulation was carried out in the reference frame moving with speed c_0 using an implicit finite difference numerical scheme (Pierini 1986)

reported here), the KdV equation can be solved analytically and exactly (an exceptional case for nonlinear evolution equations) using the method of *inverse scattering transform* by Gardner et al. (1967). The same method can be generalised to solve a wide range of nonlinear partial differential evolution equations, which are therefore called *completely integrable*. This aspect, of great mathematical and applicative interest, is beyond the scope of this discussion (the interested reader can refer to Osborne 2010).

We conclude by considering an instructive scaling exercise that will reveal the weight of the terms that appear in the KdV equation. Let us switch to the dimensionless variables defined as follows:

$$\eta' = \frac{\eta}{A}; x' = \frac{x}{L}; t' = t\frac{c_0}{L}.$$

If the parameters A and L are chosen in such a way as to well represent the length scales respectively of the height and width of the wave, we will have:

$$O(\eta') \sim O\left(\frac{\partial \eta'}{\partial t'}\right) \sim O\left(\frac{\partial \eta'}{\partial x'}\right) \sim O\left(\frac{\partial^3 \eta'}{\partial x'^3}\right) \sim O(1). \tag{9.51}$$

Consequently, the dimensionless version of the KdV Eq. (9.49) is

$$\eta'_{t'} + \eta'_{x'} + \xi \eta' \eta'_{x'} + \vartheta \eta'_{x'x'x'} = 0 \tag{9.52}$$

where

$$\xi = \frac{3}{2}\frac{A}{D} \ll 1; \vartheta = \frac{D^2}{6L^2} \ll 1.$$

It is evident that the first order terms are represented by the wave translation with speed c_0 while the nonlinear and dispersive terms, weighted respectively by the dimensionless parameters ξ and ϑ, represent small corrections that, however, asymptotically cause the fundamental effect of producing solitons. At this point it is obvious to introduce the Ursell number for KdV waves:

$$U_r = \frac{\xi}{\vartheta} = 9\frac{AL^2}{D^3}. \tag{9.53}$$

U_r represents the relative weight of the nonlinear term compared to the dispersive one. As stated above, it is obvious that $U_r = U_r(t) \rightarrow O(1)$, which is nothing more than the order of magnitude that competes with a soliton. It can be shown that the number of solitons n that emerge from a compact positive initial condition characterised by a certain U_{r0} is given by $n = O(U_{r0}) + 1$. Therefore, if initially the wave is substantially linear ($O(U_{r0}) \sim 0$) a single soliton will asymptotically emerge.

Bibliography

Dean, R.G., Dalrymple, R.A.: Water wave mechanics for engineers and scientists. Prentice-Hall, Englewood Cliffs, New Jersey (1984)

Gardner, C.S., Greene, J.M., Kruskal, M.D., Miura, R.M.: Method for solving the Korteweg-deVries equation. Phys. Rev. Lett. **19**, 1095–1097 (1967)

Hasselmann, K.: On the non-linear energy transfer in a gravity wave spectrum: Part 1. General theory. J. Fluid Mech. **12**, 481–500 (1962)

Hasselmann, K.: On the non-linear energy transfer in a gravity wave spectrum: Part 2. Conservation theorems; wave-particle analogy; irreversibility. J. Fluid Mech. **15**, 273–281 (1963a)

Hasselmann, K.: On the non-linear energy transfer in a gravity wave spectrum: Part 3. Evaluation of the energy flux and swell-sea interaction for a Neumann spectrum. J. Fluid Mech. **15**, 385–398 (1963b)

Korteweg, D.J., de Vries, G.: On the change of form of long waves advancing in a rectangular canal, and on a new type of long stationary waves. The London, Edinburgh, Dublin Phil. Mag. J. Sci. **39**, 422–443 (1895)

Le Blond, P.H., Mysak, L.A.: Waves in the Oceans. Elsevier, New York (1978)

Osborne, A.: Nonlinear Ocean Waves and the Inverse Scattering Transform. Elsevier, International Geophysics Series (2010)

Pierini, S.: Solitons in a channel emerging from a three-dimensional initial wave. Il Nuovo Cimento **C9**, 1045–1061 (1986). https://doi.org/10.1007/BF02507422

Whitham, G.B.: Linear and Nonlinear Waves. John Wiley & Sons, New York (1974)

Further Recommended Reading

Defant, A.: Physical Oceanography, vol. II. Pergamon Press, New York (1960)

Gill, A.E.: Atmosphere-Ocean Dynamics. Academic Press, New York (1982)

Kinsman, B.: Wind Waves. Prentice-Hall, Englewood Cliffs, New Jersey (1965)

Neumann, G., Pïerson, W.J.J.: Principles of Physical Oceanography. Prentice-Hall, Englewood Cliffs, New Jersey (1966)

Phillips, O.M.: The Dynamics of the Upper Oceans. Cambridge University Press, Cambridge (1966)

Shapiro, A.H.: Illustrated Experiments in Fluid Mechanics. The MIT Press, Cambridge, Massachusetts (1972)

Stoker, J.J.: Water Waves, the Mathematical Theory with Applications. Interscience Publishers, New York (1957)

Chapter 10
Surface Gravity Waves: Energy and Statistics

To complete what was presented in Chap. 9, this chapter deals with the main aspects concerning the energy of surface gravity waves. The energy of a sinusoidal wave and the power spectral density of an irregular wave are analysed, the complex problem of the generation of surface waves by the atmosphere is briefly examined, and finally, the main aspects of the statistics of sea level and wave heights are illustrated.

10.1 Wave Energy, Power Spectral Density

We start by considering the energy of a sinusoidal wave. Each oscillation of the sea surface is associated with energy. The fluid elements move vertically from their equilibrium position and this involves a change in their gravitational potential energy; moreover, each displacement involves a certain kinetic energy. Given the continuous nature of the phenomenon, the evaluation of the total energy of a wave must necessarily refer to a certain horizontal extension.

Referring to the sinusoidal wave $\eta_s = a\cos(kx - \sigma t)$ (also called *regular wave*, now indicated with η_s to distinguish it from the composite wave, also called *irregular*), for which, without loss of generality, propagation along x is assumed, we now proceed to calculate its total energy extended to a wavelength λ and to a unit length in the transverse direction. The result will be independent of time; therefore, for convenience, $t = 0$ is considered, so in the following $\eta_s = a\cos(kx)$.

The potential energy density V (referred to that in the absence of the wave) of a sinusoidal wave per unit surface averaged over a wavelength is given by:

$$V = \frac{1}{\lambda}\int_0^{\lambda}\left(\rho g\int_0^{\eta} z dz\right)dx = \frac{\rho g}{\lambda}\frac{1}{2}\int_0^{\lambda}\eta_s^2 dx = \frac{1}{4}\rho g a^2. \tag{10.1}$$

S. Pierini, *Oceanic and Atmospheric Fluid Dynamics*, UNITEXT for Physics,
https://doi.org/10.1007/978-3-031-77991-6_10

In the last step, considering the indefinite integral

$$\int \cos^2 \xi d\xi = \frac{\xi}{2} + \frac{1}{4} \sin(2\xi) + c$$

we get

$$\int_0^\lambda a^2 \cos^2(kx)dx = \frac{\lambda a^2}{2\pi} \int_0^{2\pi} \cos^2 x' dx' = \frac{\lambda a^2}{2},$$

from which Eq. (10.1) follows.

The kinetic energy density K of a sinusoidal wave per unit surface averaged over a wavelength is given by:

$$K = \frac{1}{\lambda} \int_0^\lambda \left(\frac{1}{2}\rho \int_{-D}^0 |\mathbf{u}|^2 dz \right) dx = \frac{\rho \sigma^2 a^2}{4k} \coth(kD) = \frac{1}{4}\rho g a^2, \tag{10.2}$$

where (i) the upper limit of the integral on z can be assumed $= 0$ instead of $= \eta$ thanks to the linearity of the wave, (ii) in the first step the velocity field obtainable from Eq. (9.19) was considered and (iii) in the second step the dispersion relation (Eq. 9.20) was exploited.

In conclusion, the total energy density $E_s = K + V$ of a sinusoidal wave per unit surface averaged over a wavelength is given by:

$$E_s = \frac{1}{2}\rho g a^2. \tag{10.3}$$

It is interesting to note that the equality between kinetic and potential energy (valid exclusively for linear waves) is the same as that which occurs in the harmonic oscillator, as well as in any other conservative dynamical system in the presence of small oscillations. Moreover, the energy density does not depend on the depth of the fluid D but only on the amplitude of the wave at the surface.

Now imagine being at $x = 0$ (any other point is equivalent) and asking what the variance $\overline{\eta_s^2}$ of the sinusoidal signal $\eta_s = a\cos(\sigma t)$ calculated over a period is. The following calculation exploits again the indefinite integral reported above:

$$\overline{\eta_s^2} = \frac{1}{T_p} \int_0^{T_p} \eta_s^2 dt = \frac{1}{T_p} \int_0^{T_p} a^2 \cos^2(\sigma t) dt = \frac{a^2}{2}$$

(the period here is indicated with the symbol T_p to distinguish it from the parameter T that will be introduced below) from which, thanks to Eq. (10.3), the following relationship between energy and the variance of the sinusoidal signal results

$$E_s = \rho g \overline{\eta_s^2}. \tag{10.4}$$

This result will prove fundamental in the subsequent discussion of the spectral power density of an irregular wave.

In Sect. 9.4 it was seen that an idealised wave group propagates with the group velocity c_g that competes with the k of the local oscillation (but note that this happens for groups of any shape and structure). In light of Eq. (10.3) it is now evident that c_g is the speed with which the *energy* of a wave group propagates. The same speed is also the one with which energy is transferred in a sinusoidal wave, as we will now demonstrate. Let $\mathcal{F}$ be the energy flux given by the work done per unit time by the pressure forces acting on a surface arranged vertically and extended from the bottom to the surface, aligned with the lines $\eta = cost$ and of unit length in the transverse direction. Consider $\mathcal{F}$ averaged over a period for a wave that propagates along x (as in Eq. (10.2), in the integral on z the upper limit η can be approximated to 0 thanks to the linearity of the wave):

$$\overline{\mathcal{F}} = \frac{1}{T_p} \int_t^{t+T_p} \int_{-D}^{0} p\, u\, dz\, dt,$$

where the pressure is given by Eq. (9.10),

$$p = -\rho g z - \rho \frac{\partial \varphi}{\partial t},$$

with φ given by Eq. (9.19) and $u = \partial\varphi/\partial x$. Laborious but simple calculations lead to the following result:

$$\overline{\mathcal{F}} = E_s c_g \tag{10.5}$$

where c_g is given by Eq. (9.32). Therefore, in a sinusoidal wave there is an energy transfer (equal to the average energy of the wave) along the direction of propagation with speed c_g. This result is perfectly consistent with the interpretation of the group velocity given in Sect. 9.4.

We now move on to analyse the power spectral density by extending Eqs. (10.3 and 10.4) to irregular waves. Considering the complexity of the generic irregular wave field, the approach to the problem cannot be based on a deterministic description but will rather require a statistical description that considers the state of the sea surface as a stochastic process to which to apply the Fourier transform method. In the present treatment, we will resort to a simplified—yet rigorous—analysis that does not make direct reference to the theory of stochastic processes. For a more in-depth treatment, one can refer to more specialist texts, such as those by Kinsman (1965), Phillips (1966) and Le Blond and Mysak (1978).

Imagine a realistic situation in which the sea surface is described by a combination of a large number of regular waves with different **k** (and therefore $\sigma(k)$) and phases ϕ. It will therefore be necessary to refer to the Fourier integral representation already introduced in Eq. (9.28), but now written in the frequency domain:

$$\eta(t) = \frac{1}{2\pi} \int_{-\infty}^{+\infty} \tilde{\eta}(\sigma) e^{i\sigma t} d\sigma \tag{10.6}$$

where the Fourier transform is given by

$$\tilde{\eta}(\sigma) = \int_{-\infty}^{+\infty} \eta(t) e^{-i\sigma t} dt. \tag{10.7}$$

To extend Eq. (10.4) to the realistic case, we proceed to evaluate the variance of the composite signal (Eq. 10.6). For this purpose, consider the fundamental Parseval's theorem,

$$\int_{-\infty}^{+\infty} |\eta(t)|^2 dt = \frac{1}{2\pi} \int_{-\infty}^{+\infty} |\tilde{\eta}(\sigma)|^2 d\sigma \tag{10.8}$$

and introduce the truncated Fourier transform $\tilde{\eta}_T$

$$\tilde{\eta}_T(\sigma) = \int_{-T/2}^{T/2} \eta(t) e^{-i\sigma t} dt. \tag{10.9}$$

Combining Eqs. (10.8 and 10.9), we obtain the expression of the variance $\overline{\eta^2}$ of the (real) signal η:

$$\overline{\eta^2} = \lim_{T\to\infty} \frac{1}{T} \int_{-T/2}^{T/2} \eta^2 dt = \lim_{T\to\infty} \frac{1}{2\pi T} \int_{-T/2}^{T/2} |\tilde{\eta}_T(\sigma)|^2 d\sigma. \tag{10.10}$$

If we now introduce the unidirectional *power spectral density* (also known as *power spectrum*) of the waves,

$$\Phi(\sigma) = \lim_{T\to\infty} \frac{1}{2\pi T} |\tilde{\eta}_T(\sigma)|^2 \tag{10.11}$$

we obtain the fundamental relation:

$$\overline{\eta^2} = \int_{-\infty}^{+\infty} \Phi(\sigma) d\sigma. \tag{10.12}$$

Given that $\tilde{\eta}_T$ represents the coefficient of the Fourier component of η at frequency σ, that Φ is proportional to its square and that according to Eq. (10.4) this is in turn proportional to the energy E_s of the sinusoidal wave, it follows that the power spectrum Φ provides the *wave energy per unit area and per unit frequency interval.* Consequently, we obtain the following generalisation of Eq. (10.4) for an irregular wave:

$$E = \rho g \overline{\eta^2} = \rho g \int_{-\infty}^{+\infty} \Phi(\sigma) d\sigma. \tag{10.13}$$

In other words, $\Phi(\sigma)d\sigma$ represents the energy per unit surface area (divided by ρg) of the wave relative to the Fourier components with angular frequencies between σ and $\sigma + d\sigma$. Note that, for Eqs. (10.9 and 10.11), Φ is a real, positive and even function ($\Phi(-\sigma) = \Phi(\sigma)$); for this last property, it is therefore sufficient to refer to positive frequencies. Figure 10.1 shows a typical form of power spectrum at statistical equilibrium (see Sect. 10.2).

It should be noted that thanks to the existence of the dispersion relation, and therefore a one-to-one correspondence between σ and k, it is possible to express the power spectrum $\Phi(\sigma)$ also in terms of the wave number by a function $\Psi(k)$. It is also possible to distinguish between the directions θ of propagation of the individual waves considering the *directional energy spectrum* $\Phi(\sigma, \theta)$. Finally, note that the power spectra do not provide any information on the phases of the Fourier components, whose knowledge is otherwise considered superfluous in a statistical analysis of the wave field.

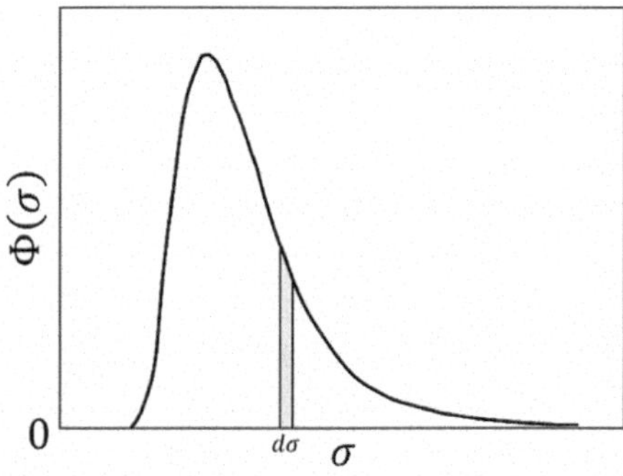

Fig. 10.1 Schematic representation of a typical power spectrum at statistical equilibrium for surface gravity waves. The area of the grey sector represents the wave energy contained in the range of angular frequencies from σ to $\sigma + d\sigma$

10.2 Phillips and Miles Generation Mechanisms

We now briefly consider the very complex mathematical problem of the generation of surface gravity waves by the atmosphere (for a detailed treatment see, for example, Kinsman 1965; Phillips 1966; Le Blond and Mysak 1978). It is possible to formulate the evolution of wave energy using the power spectrum $\Phi(\sigma, t)$, or alternatively $\Psi(k, t)$, which will now depend on time. In general, the evolution of $\Psi(k, t)$ will be described by the equation

$$\frac{\partial \Psi(k, t)}{\partial t} = S(k, t), \qquad (10.14)$$

where S, called the *source function*, will contain various terms, each of which will correspond to a different generation mechanism.

The first mechanism is due to Phillips (1957). Imagine starting from a perfectly flat sea surface with vanishing wind velocity. The establishment of a wind will produce a weak wave field due to the normal stresses caused by the turbulent surface atmospheric pressure fluctuations associated with the wind itself. We are in the presence of a resonance mechanism for which the wave with wave number k will be excited by pressure fluctuations characterised by k and σ corresponding to those of the wave. In this case we will have

$$S_{Phillips} \propto E_p[k, \sigma(k)], \qquad (10.15)$$

where $E_p(k, \sigma)$ (which is assumed to be independent of time in a statistical equilibrium situation) is the spectrum of the turbulent field of surface atmospheric pressure (note that, unlike a wave field, in turbulence there is no specific relationship between k and σ, so E_p can cover a wide spectrum of frequencies and wave numbers that are uncorrelated with each other). This mechanism will lead to a very slow growth of the wave field energy, as, for Eq. (10.14), Ψ will grow *linearly* with time:

$$\Psi \propto t.$$

The second mechanism, due to Miles (1957), assumes a wave field already developed as a consequence of the Phillips mechanism. In this case the waves will be able to interact with the wind field inducing a perturbation in the overlying atmospheric boundary layer, which, in turn, will interact with the wave field itself in a constructive way giving rise to a *positive feedback* mechanism that refers to the so-called *critical layer*. Essentially Miles showed that the source function has the following form:

$$S_{Miles} \propto \Psi(k, t) \frac{\left.\frac{d^2 U}{dz^2}\right|_{z=z_c}}{\left.\frac{dU}{dz}\right|_{z=z_c}}. \qquad (10.16)$$

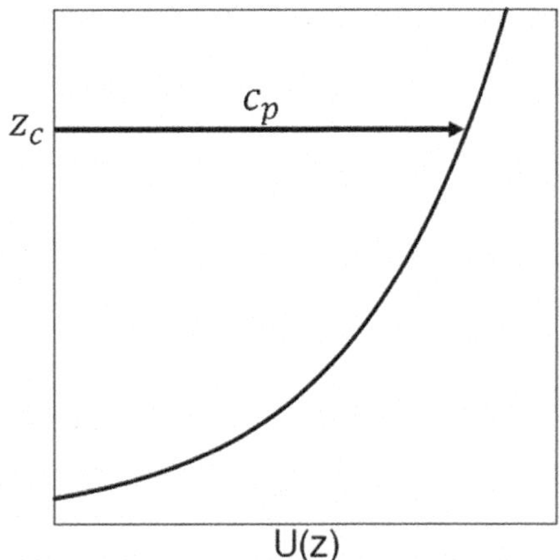

Fig. 10.2 Logarithmic wind profile in the atmospheric boundary layer and critical height z_c

(with a negative proportionality factor). In Eq. (10.16), $U(z)$ is the average wind speed profile which, in the surface atmospheric layer (the lowest part of the planetary boundary layer limited to $z \lesssim 100\,\text{m}$) is logarithmic (Fig. 10.2),

$$U(z) = \frac{u_*}{\kappa} \ln \frac{z}{z_0}$$

For the definitions of the friction velocity u_*, the Von Karman constant κ and the roughness parameter z_0 see Sect. 15.4; moreover, z_c is the critical height at which the wind speed equals the phase velocity of the wave. For this profile, from Eq. (10.16) it follows that S is inversely proportional to the critical height:

$$S_{Miles} \propto \Psi(k,t)\frac{1}{z_c(k)} \tag{10.17}$$

(with a positive proportionality factor). From this expression two important properties are derived.

The first property is provided by the solution of Eq. (10.14). Given that, due to the feedback mechanism, the energy input into the wave field is proportional to its own energy ($S_{Miles} \propto \Psi(k,t)$), it will be

$$\Psi \propto e^{\frac{t}{\tau}}$$

(the expression of τ is irrelevant in this discussion). Therefore, in the Miles phase, the wave grows much faster than in the Phillips phase, even exponentially. Note that this mechanism can only take place in the presence of a vertical wind shear; hence it is referred to as the *shear instability mechanism*. Furthermore, the existence of such positive feedback makes, in this context, the air-sea system coupled; it is therefore

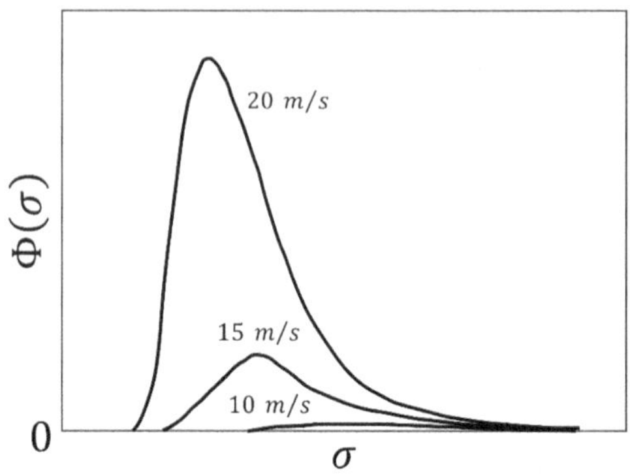

Fig. 10.3 Schematic representation of a typical power spectrum at statistical equilibrium for surface gravity waves for three developed seas corresponding to different wind intensities

possible to define the Miles mechanism as a phenomenon of instability of the air-sea system.

The second property in turn explains two important characteristics of the spectral energy density as a function of frequency, as shown in Fig. 10.3. First of all, as the wind speed increases, S_{Miles} increases, and therefore the total energy of the wave field at equilibrium (the area under each curve). In fact, the height z_c at which, for example, the wind speed of 15 $m\ s^{-1}$ equals the wave speed will be lower than that for a wind of 10 $m\ s^{-1}$ and, since $S \propto z_c^{-1}$ the energy input will be greater in the first case. Moreover, considering that the Miles mechanism involves the wind profile up to a limited height, the stronger the wind the higher will be the maximum speed of the waves it generates. But for the wind waves (in deep water) we have $\sigma = g/c_p$, so as the wind increases the minimum frequency of the waves it generates decreases, which explains the shift towards ever lower frequencies of the spectrum as the wind intensity increases, as shown in Fig. 10.3. It is also worth noting that, since $k = \sigma^2/g$, as σ decreases the wavelength increases.

In summary, the stronger the wind, the lower the minimum frequency, the higher the maximum period and the greater the maximum length of the waves it generates. These characteristics are all easily deducible from a careful qualitative observation of the sea surface.

The mechanisms of Phillips and Miles (and others not considered here, primarily those pertaining to weakly nonlinear interactions between the various Fourier components, cf. Hasselmann 1962, 1963a,b) continue to act in the presence of a wind, assumed constant, until a statistical equilibrium is reached (Fig. 10.3) at which as much energy is input by the wind as is dissipated (by the breaking of the waves, by the turbulence thus generated and through mechanisms of different nature that it is not possible to delve into here).

It is emphasised that the generation of waves does not depend only on the local wind but also on the area over which the wind acts, the fetch. The larger the fetch, the more intense the wave field generated, as a wave generated in one sea area propagates, contributing to increase the energy of the wave field even at a considerable distance.

The discussion concludes by mentioning that there are various standard wave spectra determined empirically and valid in different situations. Generally, a standard normalised and unidirectional energy spectral density can be expressed as

$$\Phi(\sigma) = H_{1/3}^2 \varphi(\sigma, \overline{T_I}), \tag{10.18}$$

where $H_{1/3}$ is the *significant wave height*, $\overline{T_I}$ is the average period of the irregular wave (both defined in the following paragraph) and φ is a certain function of frequency. The spectrum can be mentioned of Pierson-Moskowitz, that of Bretschneider, the JONSWAP, etc.. A discussion of this aspect is beyond the scope of this text; for further information, one can refer to specialist texts.

10.3 Statistics of Sea Level and Wave Heights

A complementary approach to spectral analysis for the description of surface gravity wave field is to resort to a statistical study of some fundamental parameters, with the relative definition of their respective probability distributions. In this paragraph, the two main parameters in this context will be considered: the sea level η and the wave height H.

We start by considering the time series of the surface elevation $\eta(t)$ measured at a certain point (Fig. 10.4), from which any trends have been eliminated to make the variable zero-mean. As already extensively discussed, this signal results from the superposition of a large number of gravity waves, also considered here of small amplitude.

The analysis of a time series of η of temporal length T_L measured with a sampling time Δt will produce a sequence of $N = T_L/\Delta t$ values η_i of the surface elevation. For sufficiently large N, it will be possible to group these values into M ($\ll N$) classes, thus obtaining a probability distribution P_k for the average values of each class $\eta_k (k = 1, \ldots, M)$, with $\sum_{k=1}^{M} P_k = 1$. The transition to a continuous distribution (easier to handle analytically) will lead to the definition of a probability density

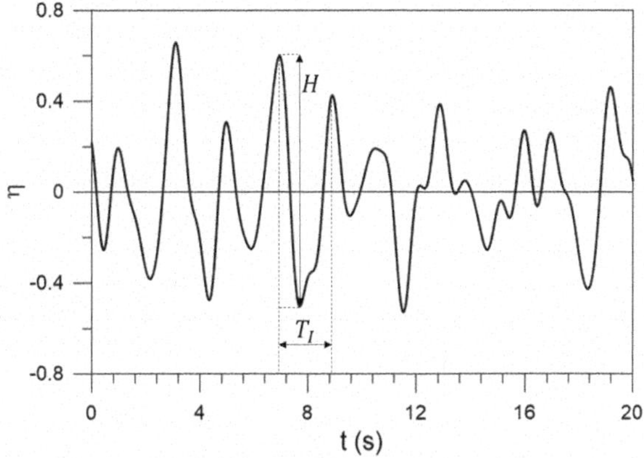

Fig. 10.4 Example of a time series of surface elevation with the definition of wave height H and the period of the irregular wave T_I

function $P(\eta)$. $P(\eta)d\eta$ is the probability that a measurement of the surface elevation will yield a value between η and $\eta + d\eta$ (obviously, $\int_{-\infty}^{+\infty} P(\eta)d\eta = 1$).

In this case, it is hypothesised that the *central limit theorem* applies, which states that the random variable given by the sum of a large number of independent random variables is normally distributed (i.e., according to a Gaussian), and this holds regardless of the probability distribution of the addend variables. This hypothesis, experimentally verified with good approximation for small amplitude waves, is based on the independence of the various Fourier components of the wave field as, as already mentioned, their nonlinear interactions are weak, albeit not entirely negligible. In conclusion, it is possible to hypothesise that the sea state represented by η is a Gaussian stationary random process, for which, therefore, the following probability density function applies

$$P_\eta(\eta) = \frac{1}{\sigma_\eta\sqrt{2\pi}} e^{-\frac{\eta^2}{2\sigma_\eta^2}} \tag{10.19}$$

where σ_η is the corresponding standard deviation. Figure 10.5 shows the distribution for $\sigma_\eta = 0.1\ m$.

Another fundamental parameter for the description of an irregular wave is the wave height H (Fig. 10.4), defined as the vertical distance between an absolute maximum (wave crest) and the subsequent absolute minimum (wave trough); naturally, in a regular wave, H is simply double the amplitude: $H = 2a$. This parameter has not only a significant scientific value but also a practical utility, as it is the one that is best suited to be evaluated qualitatively by an observer. Along with H, it is also possible to define the period T_I as the temporal distance between one crest and the next (Fig. 10.4; sometimes T_I is instead defined—substantially equivalently—as the time that elapses between a passage of η from negative to positive values and the next, *zero up-crossing*). Therefore, in a time series of η of an irregular wave, it is possible to identify a certain number of "waves", understood as single oscillations each characterised by a specific value of H and T_I.

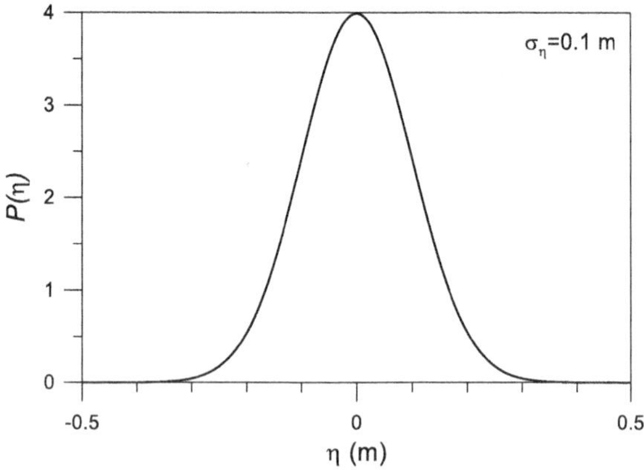

Fig. 10.5 Gaussian probability density function of the surface elevation η of a linear irregular wave with standard deviation $\sigma_\eta = 0.1$ m

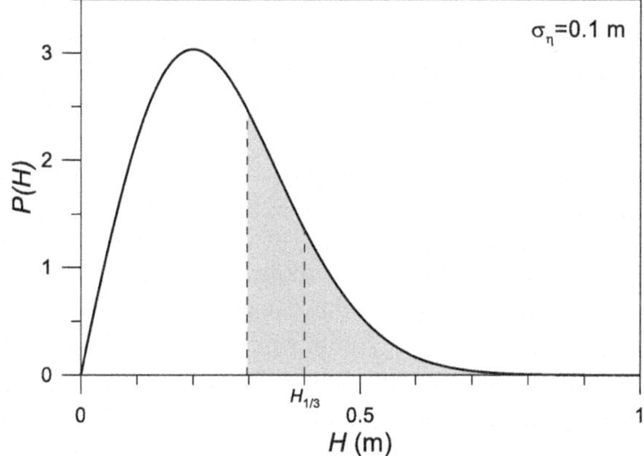

Fig. 10.6 Rayleigh probability density function of the wave height H for a standard deviation of the surface elevation $\sigma_\eta = 0.1$ m (cf. Fig. 10.5). The area of the white (grey) zone of the graph is equal to 2/3 (1/3). The significant wave height $H_{1/3} = 4\sigma_\eta$ is the average height of the waves included in the grey zone

It is obvious to ask whether there is a theoretical probability distribution that can statistically describe the wave height. Longuet-Higgins (1952) showed that if η is normally distributed with a *narrow-band frequency spectrum* and the wave components have uncorrelated phases, then H follows the so-called *Rayleigh distribution*

$$P_H(H) = \frac{H}{4\sigma_\eta^2} e^{-\frac{H^2}{8\sigma_\eta^2}} \tag{10.20}$$

(figure 10.6), where σ_η is the standard deviation of η (Eq. 10.19). The narrow band implies that $\Phi(\sigma)$ covers a relatively limited range of frequencies, as often happens for small amplitude surface waves, especially for the swell (Sect. 9.5). Experimental data confirm that Eq. (10.20) indeed provides a good description of the statistics of wave heights.

Note that the only parameter that appears in $P_H(H)$ is the variance of η: $\sigma_\eta^2 = \overline{\eta^2}$. Therefore, the statistics of the sea level automatically defines that of the wave heights (provided the above conditions are met). Moreover, Eq. (10.13) links σ_η^2 to the energy of the irregular wave:

$$\sigma_\eta^2 = E/\rho g. \tag{10.21}$$

Therefore, Eq. (10.21) expresses the link between the spectral approach (Eqs. 10.12 and 10.13), the statistics of the sea level (Eq. 10.19) and that of the wave heights (Eq. 10.20).

We conclude by introducing an important parameter in the context of wave heights: the so-called *significant wave height* H_s (already mentioned at the end of Sect. 10.2), defined as the average of the highest wave heights to which $1/3$ of the total probability of H is attributed (for this reason H_s is also referred to as $H_{1/3}$). In Fig. 10.6, the range of $P_H(H)$ over which the average is calculated is indicated by the grey area; in the specific case $H_{1/3} = 0.4$ m, i.e. 4 times the standard deviation of η. In fact, it can be shown that the following relationship is generally valid:

$$H_{1/3} = 4\sigma_\eta. \tag{10.22}$$

The relevance of the significant wave height, generally considered as the standard measure of the height of surface waves, lies in the fact that, in a visual estimate of the average wave height, the observer tends to underestimate the lower H and instead relies mainly on the higher H. Such a qualitative justification, with even psychological implications, of a mathematical parameter actually has a historical origin. In fact, $H_{1/3}$ was introduced by Munk (1944) in order to characterise in a simple and effective way the state of the sea as a trained observer could do.

The reader interested in delving into the more practical aspects of the theory of surface waves, will have acquired from the introductory Chaps. 9 and 10 the basis to proceed smoothly with more specialised texts.

Bibliography

Hasselmann, K.: On the non-linear energy transfer in a gravity wave spectrum: Part 1. General theory. J. Fluid Mech. **12**, 481–500 (1962)

Hasselmann, K.: On the non-linear energy transfer in a gravity wave spectrum: Part 2. Conservation theorems; wave-particle analogy; irreversibility. J. Fluid Mech. **15**, 273–281 (1963a)

Hasselmann, K.: On the non-linear energy transfer in a gravity wave spectrum: Part 3. Evaluation of the energy flux and swell-sea interaction for a Neumann spectrum. J. Fluid Mech. **15**, 385–398 (1963b)

Kinsman, B.: Wind Waves. Prentice-Hall, Englewood Cliffs, New Jersey (1965)

Le Blond, P.H., Mysak, L.A.: Waves in the Oceans. Elsevier, New York (1978)

Longuet-Higgins, M.S.: On the statistical distribution of the heights of sea waves. J. Mar. Res. **11**, 245–266 (1952)

Miles, J.: On the generation of surface waves by shear flows. J. Fluid Mech. **3**, 185–204 (1957)

Munk, W.H: Proposed Uniform Procedure for Observing Waves and Interpreting Instrument Records. Wave Project at the Scripps Institution of Oceanography, La Jolla, California (1944)

Phillips, O.: On the generation of waves by turbulent wind. J. Fluid Mech. **2**, 417–445 (1957)

Phillips, O.M.: The Dynamics of the Upper Oceans. Cambridge University Press, Cambridge (1966)

Further Recommended Readings

Dean, R.G., Dalrymple, R.A.: Water Wave Mechanics for Engineers and Scientists. Prentice-Hall, Englewood Cliffs, New Jersey (1984)

Defant, A.: Physical Oceanography, vol. II. Pergamon Press, New York (1960)

Neumann, G., Pierson, W.J.J.: Principles of Physical Oceanography. Prentice-Hall, Englewood Cliffs, New Jersey (1966)

Whitham, G.B.: Linear and Nonlinear Waves. John Wiley & Sons, New York (1974)

Chapter 11
Internal Gravity Waves

This chapter introduces the main aspects of internal gravity waves, initially considering linear waves and subsequently weakly nonlinear waves. We first study the case of a continuously stratified fluid; we then move on to the simpler, but very significant, case of a two-layer fluid. We touch on the phenomenology and generation mechanisms of internal waves and finally consider nonlinear effects, with particular reference to internal solitons.

11.1 Linear Internal Gravity Waves

We have seen how surface gravity waves owe their existence to the gravitational restoring force that arises as a result of the displacement of fluid elements from their equilibrium position at the air-sea interface. On the other hand, in Sect. 3.3 we saw how within a stratified fluid there is another restoring force much weaker than gravity, called reduced gravity, which acts in a similar way to the first. This gives rise to *internal waves*, both oceanic and atmospheric, which consist in the propagation of disturbances within the fluid. There are both small-scale internal waves, whose speed has a significant vertical component, and larger-scale ones, whose disturbances propagate essentially horizontally. In the following discussion, we will limit ourselves to dealing with this second category. We start with the analysis of internal waves in a continuously stratified fluid. We will subsequently consider the case of internal waves in a two-layer fluid.

Consider an incompressible fluid with density

$$\rho = \rho_0(z) + \rho'(x, y, z, t), \tag{11.1}$$

where ρ_0 represents a known stable stratification, to which a density field ρ' associated with the motion is superimposed. Similarly, the pressure is decomposed into

S. Pierini, *Oceanic and Atmospheric Fluid Dynamics*, UNITEXT for Physics,
https://doi.org/10.1007/978-3-031-77991-6_11

the atmospheric pressure p_a, assumed constant, the hydrostatic pressure p_0 and the anomaly p' associated with the motion:

$$p = p_a + p_0(z) + p'(x, y, z, t)\cdot \tag{11.2}$$

The complete set of linearised equations of motion for an incompressible fluid given by Eq. (5.12) will, in this case, take the following form

$$\begin{cases} (a)\frac{\partial \mathbf{u}_H}{\partial t} &= -\frac{1}{\overline{\rho}}\nabla_H p' \\ (b)\frac{\partial w}{\partial t} &= -\frac{1}{\overline{\rho}}\frac{\partial p'}{\partial z} - g\rho'/\overline{\rho} \\ (c)\nabla_H \cdot \mathbf{u}_H + \frac{\partial w}{\partial z} &= 0 \\ (d)\frac{\partial \rho'}{\partial t} + w\frac{\partial \rho_0}{\partial z} &= 0 \end{cases} \tag{11.3}$$

where $\mathbf{u}_H = (u, v)$ and $\nabla_H = (\partial/\partial x, \partial/\partial x)$. Equations (11.3a, b) represent the Euler equations decomposed, respectively, into the horizontal and vertical components. Considering the small density variations at play, in Eq. (11.3a) the density appearing in the denominator in the pressure gradient force is given by a constant reference density $\overline{\rho}$, which can be seen as the value of ρ_0 averaged over the depth: this is known as the *Boussinesq approximation*. The equation for vertical acceleration (Eq. 11.3b) is obtained from Eqs. (11.1 and 11.2) considering the hydrostatic equilibrium $0 = -\partial p_0/\partial z - \rho_0 g$ and also imposing the Boussinesq approximation here. The term $-g\rho'/\overline{\rho} = -g'$ is called *buoyancy* and represents the effect of the reduced gravity g' introduced in Sect. 3.3. Equation (11.3c) is the continuity equation, while Eq. (11.3d) represents the linearised incompressibility condition.

By deriving Eq. (11.3a) with respect to z, applying the operator ∇_H to Eq. (11.3b) and subtracting, we obtain:

$$\frac{\partial}{\partial t}\left(\frac{\partial \mathbf{u}}{\partial z} - \nabla_H w\right) = \frac{g}{\overline{\rho}}\nabla_H \rho'.$$

Now we differentiate with respect to time and exploit Eq. (11.3d):

$$\frac{\partial^2}{\partial t^2}\left(\frac{\partial \mathbf{u}}{\partial z} - \nabla_H w\right) = -\frac{g}{\overline{\rho}}\frac{\partial \rho_0}{\partial z}\nabla_H w.$$

Here we take the horizontal divergence of both members:

$$\frac{\partial^2}{\partial t^2}\left(\nabla_H \cdot \frac{\partial \mathbf{u}}{\partial z} - \nabla_H^2 w\right) = -\frac{g}{\overline{\rho}}\frac{\partial \rho_0}{\partial z}\nabla_H^2 w.$$

Here we eliminate the term $\nabla_H \cdot \mathbf{u}_z$ using the derivative with respect to z of Eq. (11.3c), finally obtaining the following equation for the sole vertical component of the velocity:

$$\frac{\partial^2}{\partial t^2}\nabla^2 w + N^2(z)\nabla_H^2 w = 0, \tag{11.4}$$

where

$$N^2(z) = -\frac{g}{\overline{\rho}}\frac{\partial \rho_0}{\partial z}. \tag{11.5}$$

N is the so-called (angular) *Brunt-Väisälä frequency*. This is the natural frequency of oscillation of a fluid element displaced vertically from its equilibrium position in the case of stable stratification ($N^2 > 0$, N, real), therefore with a ρ_0 that, for example in the oceans, increases with depth. N can have values of $10^{-2} - 10^{-3}$ rad s^{-1} in the thermocline, decreasing sharply above and below it where the fluid is almost homogeneous. In terms of periods, for example, $T = 2\pi/N \cong 10$ min for a typical value in the thermocline $N = 10^{-2}$ rad s^{-1}, up to reaching values of hours outside of it.

The evolution equation for w (Eq. 11.4) allows studying all possible forms of linear internal waves (in the absence of velocity shear). For example, it is possible to study very short waves that propagate in all directions within the thermocline. As already pointed out, here we will focus on longer waves that propagate horizontally. It will therefore be appropriate to adopt the same nomenclature of Chap. 9 for spatial coordinates,

$$\mathbf{x} = (x, y)$$

and we will also here look for solutions in the form of sinusoidal waves of the type

$$w(\mathbf{x}, z, t) = W(z)\cos(\mathbf{k}\cdot\mathbf{x} - \sigma t + \gamma), \tag{11.6}$$

where W provides the vertical structure of the wave. For convenience of calculation, we will resort to Euler's formula (Sect. 9.4), which allows to write

$$w(\mathbf{x}, z, t) = \Re\text{e}\left[W(z)e^{i(\mathbf{k}\cdot\mathbf{x}-\sigma t)}\right] \tag{11.7}$$

where W can be a complex function, thus also including a phase factor. This simplifies calculations because it allows to avoid resorting to trigonometry formulas, but it is worth emphasising that this procedure can only be adopted for linear problems, like the one in question.

Substituting Eq. (11.7) into Eq. (11.4), after simple calculations, we obtain the ordinary differential equation for the vertical structure of w:

$$W'' + \left[\frac{N^2(z)}{\sigma^2} - 1\right]k^2 W = 0. \tag{11.8}$$

The boundary conditions to be imposed on the surface (assuming the existence of a rigid surface, or *rigid lid*, which eliminates the surface waves without practically affecting the internal dynamics) and on the bottom are:

$$\begin{cases} W = 0;\ z = 0 \\ W = 0;\ z = -D \end{cases} \tag{11.9}$$

Equation (11.8), together with Eq. (11.9), constitutes a so-called *eigenvalue problem* belonging to the category of *Sturm–Liouville problems*. This differential problem is not solvable for every choice of σ and k: for a given σ there is an infinite and discrete set of wave numbers k_n, each of which corresponds to a certain W_n:

$$\begin{cases} W_n(\sigma, z) \\ k_n(\sigma) \end{cases} n = 1, 2, 3, \ldots$$

The functions W_n and the wave numbers k_n are respectively the eigenfunctions (*normal modes*) and the eigenvalues of the problem.

Figure 11.1 schematically describes the structure of the first two normal modes. In Fig. 11.1a a typical profile of N and a particular frequency σ are shown. In the two grey background areas $N^2 < \sigma^2$, so Eq. (11.8) implies that locally the solution has an exponential character; conversely, in the white background area $N^2 > \sigma^2$, and there the solution has a sinusoidal character. Naturally, the three solutions must match at the two depths that mark the change of regime.

In Fig. 11.1b the first normal mode W_1 is shown, corresponding to a half-cycle in the area where $N^2 > \sigma^2$: this corresponds to an oscillation of the thermocline that essentially preserves its thickness. In Fig. 11.1c the second normal mode W_2 is shown: this instead corresponds to an oscillation of the thickness of the thermocline that essentially leaves its average depth unchanged. Such oscillations propagate horizontally with phase velocities respectively equal to $c_p = \sigma/k_1$ e $c_p = \sigma/k_2$. Note

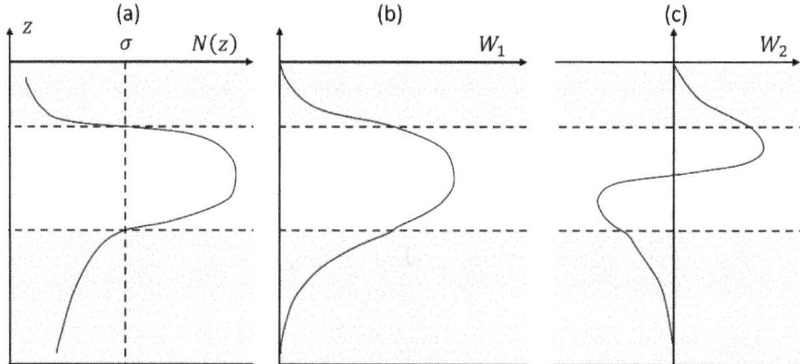

Fig. 11.1 (**a**) Typical profile of the Brunt-Väisälä frequency. (**b, c**) profiles of $W(z)$ respectively of the first and second normal mode

that the waves exist provided that $\sigma \leq N_{max}$, so the maximum value of the Brunt-Väisälä frequency represents a threshold value for the existence of internal waves; alternatively, internal waves have a period that is necessarily greater than $2\pi/N_{max}$.

11.2 Internal Gravity Waves in a Two-Layer Fluid

In the case of an arbitrary profile of $N(z)$, the resolution of the eigenvalue problem must necessarily make use of numerical methods, but in very particular cases it is possible to analytically solve the problem. In the following, the case of two homogeneous fluids separated by an interface, which is equivalent to an extremely thin thermocline, will be analytically solved. This case, known as a *two-layer fluid*, is particularly interesting because it constitutes the simplest system that supports internal waves while preserving the main dynamic characteristics of the problem.

Consider two layers of immiscible fluid of thickness D_1 and D_2 ($D = D_1 + D_2$) and densities ρ_1 and $\rho_2 > \rho_1$ separated by an interface (Fig. 11.2a). Naturally, in this case the only mode of oscillation corresponds to the first normal mode W_1 of the case with continuous stratification. In both layers, Eq. (11.8) reduces to $W'' - k^2 W = 0$, for which:

$$W(z) = \begin{cases} A\sinh(kz); & -D_1 < z \leq 0 \\ B\sinh[k(z+D)]; & -D \leq z < -D_1 \end{cases}$$

Imposing the conditions of Eq. (11.9), we obtain:

$$\frac{A}{B} = -\frac{\sinh(kD_2)}{\sinh(kD_1)} = -C.$$

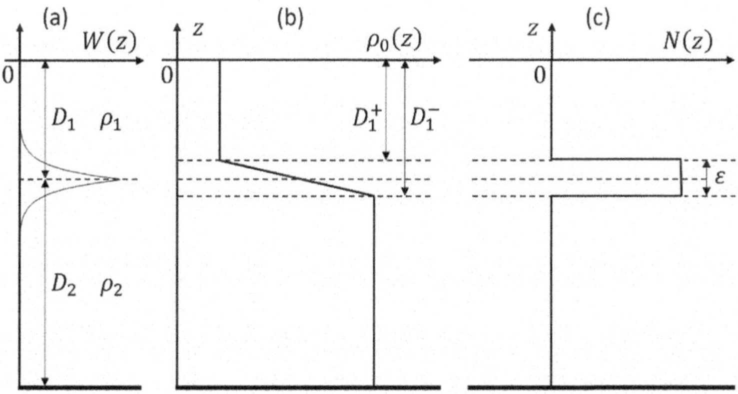

Fig. 11.2 **(a)** Structure of the two-layer system with the cusp-shaped solution given by Eq. (11.10) with arbitrary C. Variation of ρ_0 **(b)** and the corresponding N **(c)** in a system with a thin thermocline

Hence:

$$W(z) = \begin{cases} C\sinh(kz); & -D_1 < z \le 0 \\ -C\frac{\sinh(kD_1)}{\sinh(kD_2)}\sinh[k(z+D)]; & -D \le z < -D_1 \end{cases} \tag{11.10}$$

This solution presents a cusp at the interface (Fig. 11.2a), the width C of which depends on σ (and on k through σ). Here we omit the derivation of C and the horizontal velocity field and instead move on to the derivation of the dispersion relation $\sigma(k)$.

For this purpose, consider a stratification composed of two layers of fluid of different density separated by a thin transition zone of thickness ε (Fig. 11.2b; the corresponding N is shown in Fig. 11.2c). If Eq. (11.8) is integrated through the thermocline, taking into account that W is practically constant within it and that $N^2 = -g\delta\rho/\varepsilon\overline{\rho} = -g'/\varepsilon$ with $\delta\rho = (\rho_2 - \rho_1)$, we obtain:

$$\int_{-D_1^-}^{-D_1^+} W''dz = W'(-D_1^+) - W'(-D_1^-) = \left(\varepsilon + \frac{g'}{\sigma^2}\right)k^2 W(-D_1) \tag{11.11}$$

Combining Eqs. (11.10) and (11.11) we get:

$$\begin{aligned}\varepsilon + \frac{g'}{\sigma^2} &= \frac{W'(-D_1^+) - W'(-D_1^-)}{k^2 W(-D_1)} \\ &= \frac{Ck\cosh(kD_1) + Ck\sinh(kD_1)\cosh(kD_2)/\sinh(kD_2)}{Ck^2\sinh(kD_1)} \\ &= \frac{1}{k}[\coth(kD_1) + \coth(kD_2)].\end{aligned}$$

In the limit $\varepsilon \to 0$ we finally obtain the following dispersion relation for internal gravity waves in a two-layer fluid (we consider waves that propagate along the positive x-direction):

$$\sigma = \sqrt{\frac{g'k}{\coth(kD_1) + \coth(kD_2)}}. \tag{11.12}$$

From this formula the following expression for the phase velocity is derived:

$$c_p = \sqrt{\frac{g'}{k[\coth(kD_1) + \coth(kD_2)]}}. \tag{11.13}$$

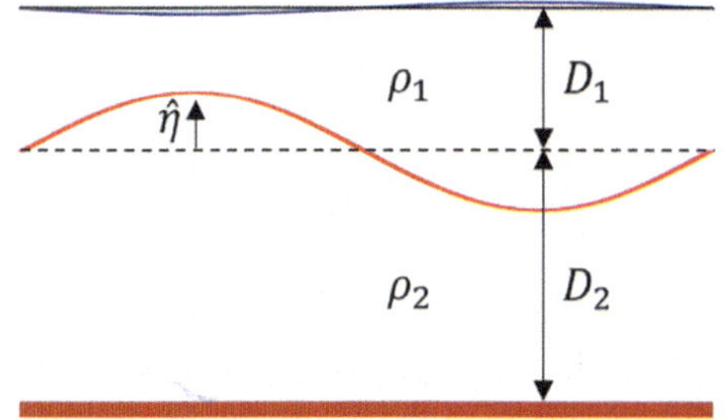

Fig. 11.3 Disturbed interface in a two-layer system. The vertical displacements of η (blue line) and $\hat{\eta}$ (red line) are not to scale

In the treatment of internal waves in a two-layer fluid it is often convenient to refer to the displacement of the interface $\hat{\eta}$ from its equilibrium position (Fig. 11.3) as a dynamic parameter, in analogy with what was done with η for surface waves.

Since $\hat{\eta}(x, t) = w(x, -D_1, t)$, the interface will have the same temporal dependence (with different phase) and, obviously, the same dispersion relation of the sinusoidal wave expressed in terms of w (Eq. 11.6). Therefore, the sinusoidal internal wave expressed in terms of the interface displacement is

$$\hat{\eta}(\mathbf{x}, t) = a\cos[\mathbf{k} \cdot \mathbf{x} - \sigma(k)t + \phi]. \tag{11.14}$$

This expression perfectly corresponds to the displacement of the free surface expressed by Eq. (9.2) for the surface sinusoidal wave, but, of course, for the latter the dispersion relation is Eq. (9.21) while for Eq. (11.14) it is Eq. (11.12).

The analysis of Eqs. (11.12) and (11.13) reveals strong analogies with surface waves. If the wavelength is much less than the thickness of both layers (*short internal waves*, $kD_1, kD_2 \to \infty,\ \coth(kD_1), \coth(kD_2) \to 1$) Eq. (11.13) gives:

$$c_p = \sqrt{\frac{g'}{2k}}. \tag{11.15}$$

In the same limit, Eq. (11.12) reduces to

$$\sigma = \sqrt{\frac{g'k}{2}},$$

from which the group velocity is derived:

$$c_g = \frac{c_p}{2}. \tag{11.16}$$

The expression of the phase velocity is similar to that for gravity surface waves in deep water, with two differences: (i) the 2 in the denominator in the square root of Eq. (11.15) comes from the presence of two layers and (ii) the gravitational acceleration g is replaced by the reduced gravity g' (Sect. 3.3), which, it is recalled, is much smaller than g: for example, in the ocean $g/g' = O(10^{-3})$ so, for the same λ, the internal waves have a speed that can be $\sim 1\% - 5\%$ of that of the surface waves (and their length can be much greater). In general, not only are internal waves slower and longer, but they also have much greater amplitudes than surface ones, reaching even tens of metres. The relationship between c_g and c_p in Eq. (11.16) is instead identical to that for surface waves.

If, on the contrary, the wavelength is much greater than the thickness of both layers (*long internal waves*, $kD_1, kD_2 \cong 0$, $\coth(kD_1) \cong 1/(kD_1)$, $\coth(kD_2) \cong 1/(kD_2)$) Eq. (11.13) gives:

$$c_p = \sqrt{g'\overline{D}};\ \overline{D} = \frac{D_1 D_2}{D_1 + D_2}. \tag{11.17}$$

In the same limit, Eq. (11.12) reduces to

$$\sigma = \sqrt{g'\overline{D}}k,$$

from which the group velocity is derived:

$$c_g = c_p. \tag{11.18}$$

The expression of the phase velocity is similar to that for surface gravity waves in shallow water, with two differences: g is replaced by g' and D is replaced by $\overline{D}$. Also, just like for surface shallow water waves, long internal waves are non-dispersive (Eq. 11.18). An analysis of the velocity field for long internal waves shows that the vertically integrated horizontal transport is instantaneously zero (Fig. 11.4b) while, for the corresponding surface motion, the transport is zero only if averaged over a whole period (Fig. 11.4a, note that in this case the stratification has no appreciable effect on the dynamics). We speak of *barotropic* (or *external*) mode for the surface mode and *baroclinic* mode for the internal one (refer to Sect. 13.1 for a definition of these important concepts). Note that this dual dynamic mode exists for all long waves in oceanic and atmospheric fluid dynamics, such as Rossby waves (Chap. 19) and Kelvin, Poincaré waves, etc., not covered in this text.

Finally, still within the framework of long internal waves, if $\lambda \gg D_1$ but $\lambda \ll D_2$ we have the so-called *reduced gravity fluid*, for which:

$$c_p = c_g = \sqrt{g'D_1}. \tag{11.9}$$

This case schematically represents the ocean surface mixed layer superimposed on a deep quiescent layer (see Sect. 17.4 for further details), but also the atmospheric mixed layer, in which case the quiescent layer is the upper one.

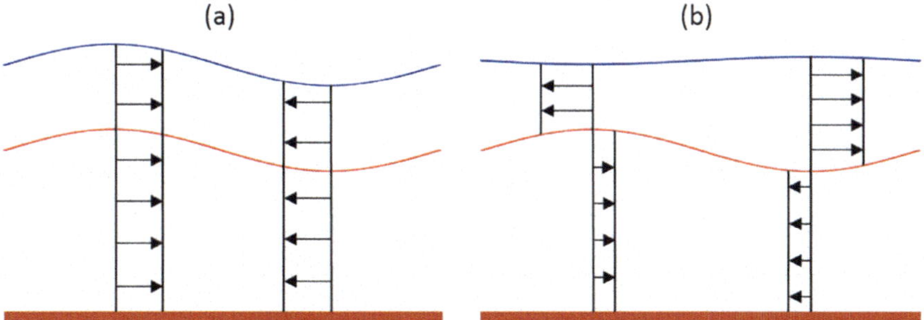

Fig. 11.4 Structure of the velocity field for long surface (a, barotropic case) and internal (b, baroclinic case) waves. The vertical displacements of η (blue line) and $\hat{\eta}$ (red line) are not to scale

11.3 Phenomenology and Generation of Internal Gravity Waves, Internal Solitons

Internal waves represent dynamic structures omnipresent in the oceans that, despite representing a small fraction of the total kinetic energy, play an important role in the transfer of momentum, heat and tracers of various kinds. Moreover, their breaking contributes substantially to the generation of turbulence, with consequent dissipation of energy on small scales, indirectly contributing to larger scale dynamics. The Garrett and Munk spectrum (1972, 1975) provides a description of the spectral distribution of internal waves with characteristics of universality, ranging from the Brunt-Väisälä frequency (Sect. 11.1) to the inertial one (Sect. 12.2).

Oceanic internal waves owe their existence to various generation mechanisms. Several instability mechanisms in the presence of stratification and vertical velocity shear, such as the Kelvin–Helmholtz instability, can give rise to internal waves. The atmosphere can generate oceanic internal waves through a resonant mechanism similar to that of Phillips (Sect. 10.2), as pressure fluctuations extend in depth. However, these will have to contribute with wave numbers and frequencies much smaller than those capable of generating surface waves: for example, pressure fluctuations of $\sim$ 1 mbar associated with an average surface wind with a speed of $\sim 0.5\,\mathrm{m\,s^{-1}}$ and a length scale of $\sim$ 500 m produce an internal wave with a growth of $\sim$ 1 m per day. Another generation mechanism of atmospheric origin is that of Ekman pumping/suction, which is the vertical velocity produced by a wind field with vorticity (Sect. 16.4): this induces a vertical displacement of the isopycnals, whose relaxation gives rise to internal waves.

Another generation mechanism consists in the intrusion of gravity currents into a stratified fluid. Figure 11.5, taken from Bourgault et al. (2016), shows an example of this phenomenon observed in a Canadian fjord. Note the region of entry of the intrusion, the area of generation of internal waves on a thin thermocline and their subsequent propagation. In the figure, the formation of an internal wave with characteristics qualitatively similar to those of the surface soliton discussed in Sect. 9.8 is evident, but here the thermocline shows a depression instead of an elevation, as well as a much greater amplitude. In fact, depression waves like the one highlighted in Fig. 11.5 show a profile accurately described by Eq. (9.50) (Fig. 11.6), but with different parameters (these will be introduced in the next Sect. 11.4). We can therefore speak of *internal solitary waves* (or *internal solitons*).

Internal solitons like those just discussed also originate from the interaction of a tidal current-possibly superimposed on a shear flow-with a strong topographic variation. This happens particularly (but not exclusively) in straits, where the geographical narrowing on the one hand intensifies the currents and on the other hand involves the presence of a topographic saddle associated with a strong and localized decrease in depth.

Figure 11.7, taken from Brandt et al. (1997), schematically shows the mechanism of generation of an internal disturbance as a result of this interaction, whose evolution can subsequently give rise to a train of internal solitons. In Figure 11.7a, a tidal flow coming from the left produces a *hydraulic jump* in the depression of the thermocline

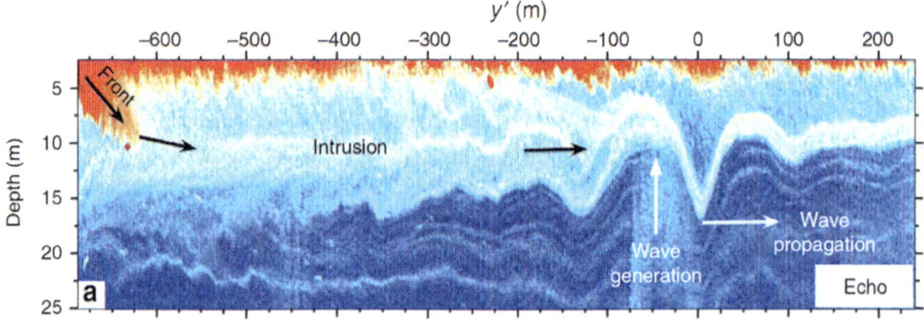

Fig. 11.5 Internal waves generated by the intrusion of a gravity current into a Canadian fjord revealed by an echo sounder. Figure adapted from Bourgault et al. (2016). Work distributed under a Creative Commons Attribution 4.0 International License

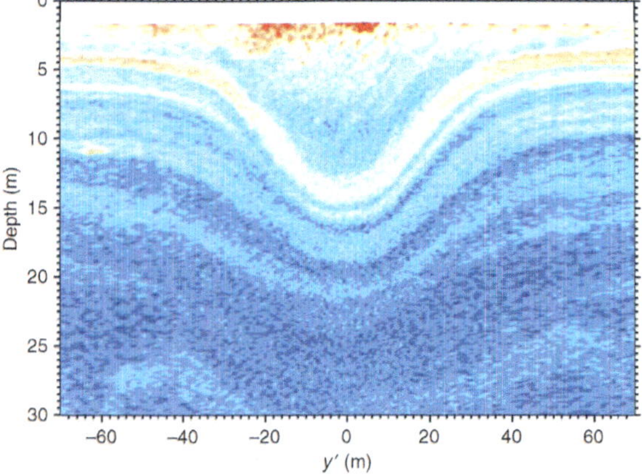

Fig. 11.6 Echogram of an internal solitary wave generated by the intrusion of a gravity current into a Canadian fjord. Figure adapted from Bourgault et al. (2016). Work distributed under a Creative Commons Attribution 4.0 International License

downwind of the obstacle: this sinking of the thermocline remains blocked and this implies that its phase speed (towards the left) is balanced by the speed of the flow in the opposite direction. When the tide decreases in intensity, the disturbance begins to move to the left, as its speed (in magnitude) now exceeds that of the flow. When the tide changes sign (Fig. 11.7c, d), the disturbance is carried by the flow itself, overcoming the saddle and moving away indefinitely. This generation mechanism is typically nonlinear and essentially non-dispersive, but phase dispersion becomes important in subsequent evolution. It is therefore not surprising that the latter often shows trains of internal solitons similar to what is illustrated, for example, in Fig. 9.17

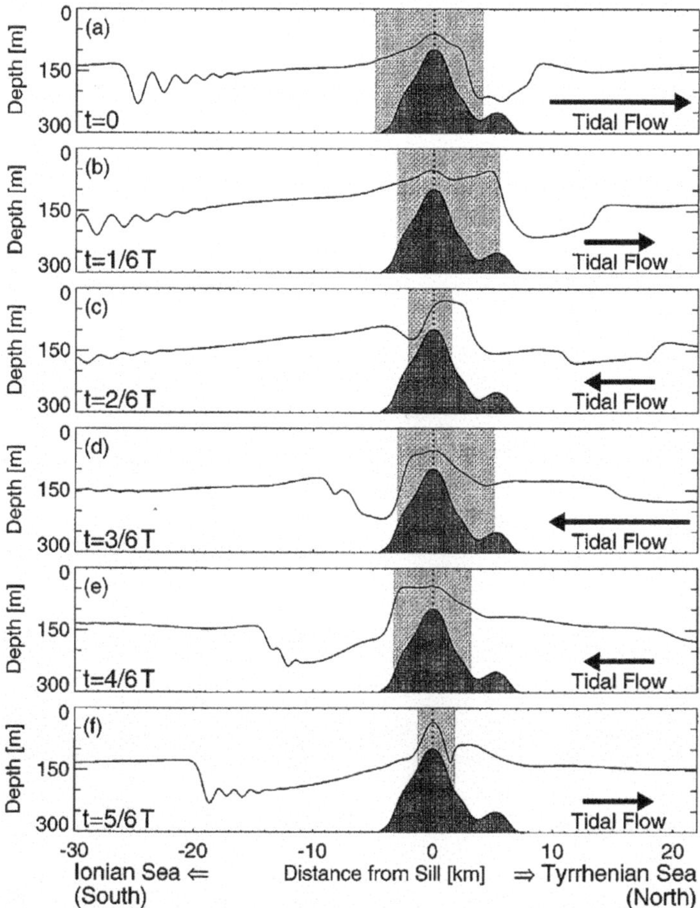

Fig. 11.7 Spatio-temporal evolution of the interface of a fluid composed of a surface layer of Tyrrhenian Surface Water (TSW) and a lower layer of Levantine Intermediate Water (LIW) within the Strait of Messina. The dynamics, caused by the interaction of the tidal current with the local topographic saddle, was simulated using a mathematical model by Brandt et al. (1997). Figure taken from Brandt et al. (1997), © American Meteorological Society. Used with permission

for surface waves starting from an initial elevation disturbance; however, for internal waves, the sign is generally opposite (again, see the next Sect. 11.4 for the explanation of this difference).

Figure 11.8 shows two very significant examples of this behaviour relating to the Straits of Gibraltar and Messina (for the Strait of Messina see also Alpers and Salusti 1983), where the tidal currents are very strong and the underwater obstacle is represented by a very pronounced saddle. The propagation of the depression of the thermocline generated in the strait by the tide (predominantly semi-diurnal) through the mechanism just described, highlights the formation of a train of solitons, often already within the strait itself. The subsequent evolution in the open sea reveals the phenomenon of diffraction, which involves the formation of two-dimensional waves, as clearly shown in Fig. 11.8. Many other similar cases are reported in the rich atlas of solitary internal waves by Jackson (2004).

These images were obtained from active microwave sensors (on board low polar orbit satellites) called *Synthetic Aperture Radar* (SAR), which detect particular surface manifestations of internal waves, as explained by Alpers' theory (1985): the front of a soliton is indeed associated with a high surface roughness which corresponds to a bright band in the radar image.

While there are no surface gravity waves in the atmosphere, there are very similar internal gravity waves. In the upper atmosphere (particularly in the mesosphere), long

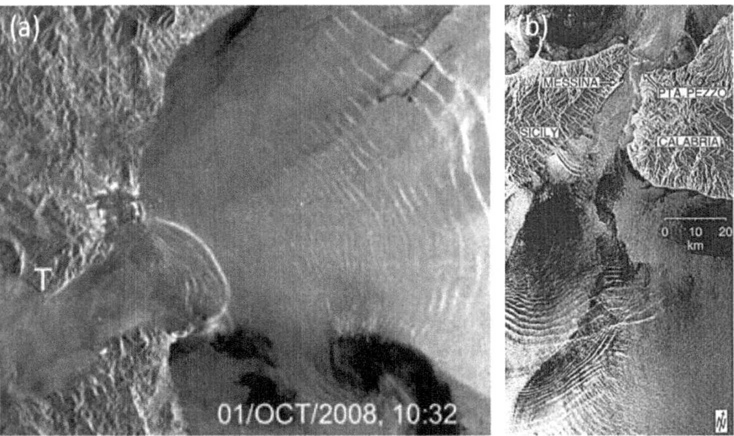

Fig. 11.8 **(a)** Image obtained from the Advanced Synthetic Aperture Radar (ASAR) of the European Space Agency's (ESA) Envisat satellite describing internal solitary waves emerging from the Strait of Gibraltar (figure adapted from Vazquez et al. 2009. American Geophysical Union—AGU, John Wiley & Sons. Used with permission). **(b)** image obtained from the Synthetic Aperture Radar (SAR) of the European Remote Sensing Satellite-1 (ERS-1) of the European Space Agency (ESA) describing internal solitary waves emerging from the Strait of Messina (figure taken from Brandt et al. 1997, © American Meteorological Society. Used with permission)

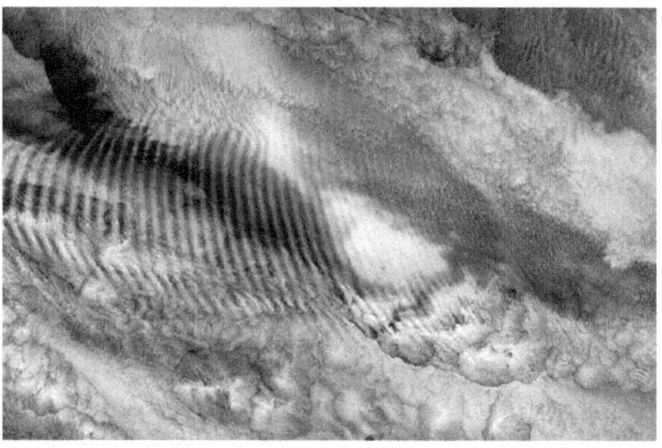

Fig. 11.9 Internal gravity waves in the atmosphere highlighted by a sequence of stratocumulus clouds. Natural colour image obtained from the multi-angle imaging spectroradiometer (NASA/ GSFC/JPL/LaRC MISR Science Team; https://misr.jpl.nasa.gov/)

internal waves can propagate with very large amplitudes due to the low air density. In the troposphere, the main internal waves are those generated by a mechanism similar to the oceanic one of a topographic nature: they indeed derive from the interaction of winds with hills or mountains in a situation of stable stratification. Downwind of the obstacle, instead of hydraulic jumps, almost linear internal waves, called *lee waves*, are generated. These are highlighted by tubular clouds aligned along the crests of the waves, where the vertical displacement of air masses reaches the lifting condensation level. Figure 11.9 shows an example of this phenomenon. Note that a nonlinear regime can also be reached in the atmosphere, with the consequent formation of internal solitons with characteristics similar to those observed in the oceans.

11.4 Nonlinear Effects: Kadomtsev–Petviashvili Equation

As just seen, solitonic dynamics is also present in stratified fluids in the form of internal solitons. This can be easily explained as the mathematical form that describes a weak phase dispersion and a weak nonlinearity for long internal waves is the same discussed in Sect. 9.8 for long surface waves. In fact, for a two-layer fluid, Eq. (9.49) KdV is rewritten for the interface $\hat{\eta}$ as follows:

$$\hat{\eta}_t + c_{0i}\hat{\eta}_x + \delta_i\hat{\eta}\hat{\eta}_x + \gamma_i\hat{\eta}_{xxx} = 0, \tag{11.20}$$

where c_{0i} is the speed c_p defined by Eq. (11.17) and

$$\delta_i = \frac{3}{2}c_{oi}\frac{D_1 - D_2}{D_1 D_2};\ \gamma_i = \frac{c_{oi}}{6}D_1 D_2.$$

It is important to note that, while for surface waves δ is positive, δ_i can take both signs depending on the predominance of the thickness of one layer over the other. A dynamics similar to the surface one occurs for $D_1 > D_2$; however, in the oceans the opposite case is by far the most common, since the depth of the thermocline is generally much less than the depth of the sea. In this case, the steepening of the wave front for solitons to asymptotically form requires negative amplitudes, so that $c_{0i} + \delta_i \hat{\eta} > c_{0i}$. This explains why internal solitons (like, for example, those shown in Figs. 11.5, 11.6, and 11.8) are *waves of depression* deriving from initial conditions also of depression, like those in Fig. 11.7.

This discussion concludes by noting that the assumption of unidimensionality inherent in the KdV equation is not verified by the trains of internal solitons in Fig. 11.8; the same applies to most of the internal solitons observed in the oceans (Jackson 2004). As for the Straits of Gibraltar and Messina, the phenomenon of diffraction has already been invoked, which leads to a two-dimensional wave field when the internal solitons begin to propagate in the open sea. In this regard, the obvious question is whether it is possible to generalise the KdV dynamics to take into account two-dimensional effects.

This is possible under the assumption of *weak bidimensionality*. By this, it is meant that the characteristic length scale along the prevailing propagation direction (assumed along x in accordance with the nomenclature used so far) is much less than that along the transverse direction (an inspection of Fig. 11.8 shows that this assumption is verified with good approximation). Referring to a sinusoidal wave, this condition is expressed in terms of wave numbers as follows (Fig. 11.10):

$$\frac{k_2}{k_1} \ll 1. \tag{11.21}$$

At this point, considering the simple linear and non-dispersive dynamics (Eq. 9.46), the following approximate dispersion relation is obtained for waves propagating along the **k** direction slightly inclined relative to x:

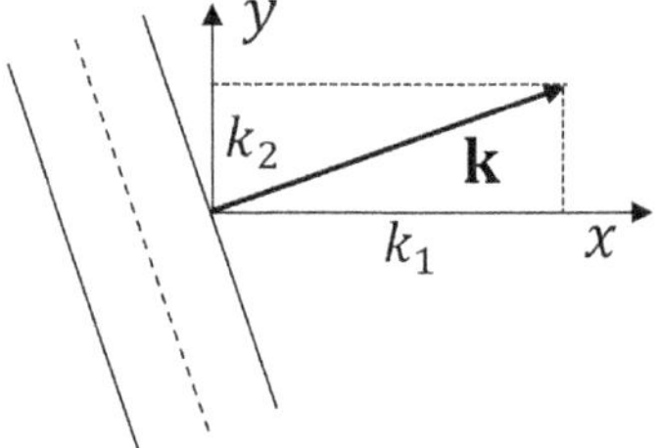

Fig. 11.10 Propagation of a sinusoidal wave along a slightly inclined direction relative to x

$$\sigma = c_{0i}k = c_{0i}\sqrt{k_1^2 + k_2^2} = c_{0i}k_1\sqrt{1 + \frac{k_2^2}{k_1^2}} \cong c_{0i}k_1\left(1 + \frac{k_2^2}{2k_1^2}\right). \tag{11.22}$$

Just as was done to derive Eq. (9.46), the corresponding linear evolution equation can be obtained by considering a sinusoidal wave (Eq. 9.45):

$$\hat{\eta} = e^{i(k_1x + k_2y - \sigma t)}. \tag{11.23}$$

It is straightforward to verify that the dispersion relation Eq. (11.22) satisfies the following evolution equation:

$$\hat{\eta}_t + c_{0i}\hat{\eta}_x - i\frac{c_{0i}}{2k_1}\hat{\eta}_{yy} = 0 \tag{11.24}$$

The imaginary unit can be eliminated by deriving with respect to x. In this way, the evolution equation for linear and non-dispersive waves is obtained, which generalises Eq. (9.46) to the weakly two-dimensional case for internal waves:

$$(\hat{\eta}_t + c_{0i}\hat{\eta}_x)_x + \frac{c_{0i}}{2}\hat{\eta}_{yy} = 0. \tag{11.25}$$

Now, proceeding heuristically as was done in Sect. 9.8, the dispersive term and the nonlinear one are added within the bracket, finally obtaining, for internal waves (the same equation also applies to surface waves, $\hat{\eta} \to \eta$, provided the corresponding parameters are replaced):

$$(\hat{\eta}_t + c_{0i}\hat{\eta}_x + \delta_i\hat{\eta}\hat{\eta}_x + \gamma_i\hat{\eta}_{xxx})_x + \frac{c_{0i}}{2}\hat{\eta}_{yy} = 0. \tag{11.26}$$

This is called the Kadomtsev–Petviashvili (1970, KP) equation and represents the generalisation of Eq. (11.20) KdV to the case of weakly two-dimensional waves.

As an example, a modelling study will now be briefly illustrated in which, for the first time, internal solitons were simulated using the KP equation (Pierini 1986, 1989). The case concerns the propagation in the Alboran Sea of the internal solitons generated in the Strait of Gibraltar, whose phenomenology is well described by Fig. 11.8a.

Figure 11.11 shows the schematic domain adopted in the simulation. Along the Z segment, a time-dependent boundary condition is imposed that describes a depression of the thermocline based on experimental data and already equipped with a steep wave front resulting from an initial nonlinear evolution. Up to $x = 0$ the evolution is one-dimensional while for $x > 0$ the diffraction of the wave in the Alboran Sea produces significant two-dimensional effects.

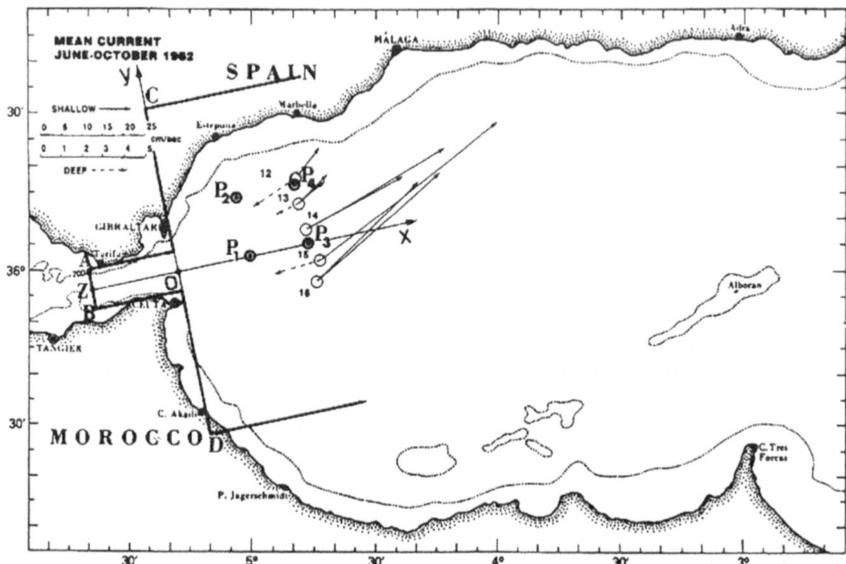

Fig. 11.11 Integration domain of the KP equation adopted in the simulation of the internal solitons observed in the Alboran Sea. Figure taken from Pierini (1989), © American Meteorological Society. Used with permission

Figure 11.12 shows the internal solitons (with opposite sign for graphic clarity) simulated numerically at two subsequent time instants. The complex evolution resulting from the combined effects of nonlinearity, phase dispersion and wave diffraction is evident. It is clear that the formation of solitons (which in this case would be more appropriate to define as *solitary waves*, as their shape is not invariant due to the two-dimensionality of the wave) depends essentially on the possibility that the wave has to spread laterally under the effect of transverse pressure gradients. Consequently, the use of the KdV equation in cases like this would be inappropriate. Figure 11.13 clearly shows the limits of a modelling approach based on a one-dimensional equation like the KdV in cases like the one considered here.

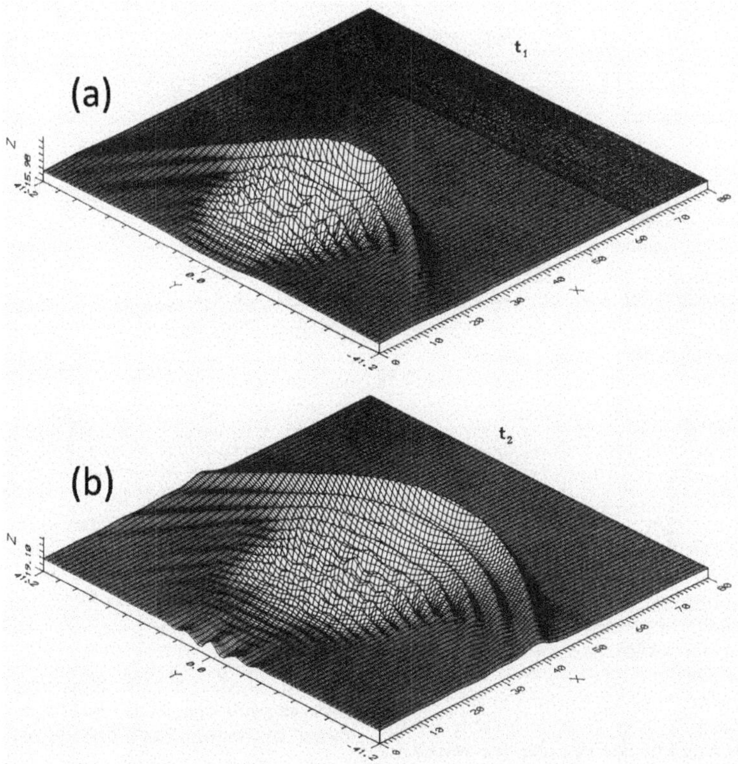

Fig. 11.12 Internal solitons simulated in the Alboran Sea using the KP equation at two successive time instants (**a**, **b**). The simulation was carried out using an implicit finite difference numerical scheme (Pierini 1986). Figures adapted from Pierini (1989), © American Meteorological Society. Used with permission

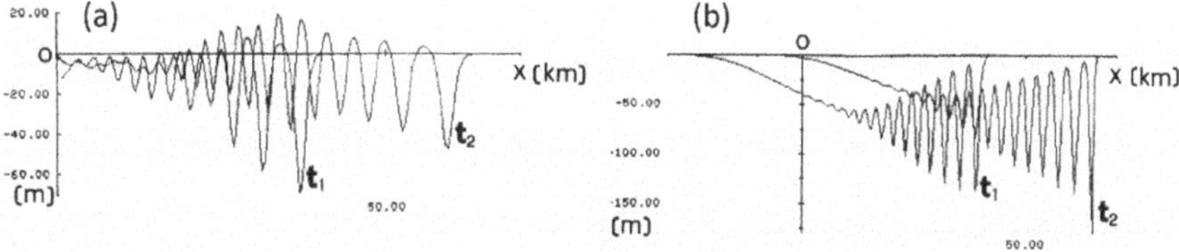

Fig. 11.13 (**a**) Profiles of $\widehat{\eta}$ along x for $y = 0$ at times t_1 and t_2 corresponding to the two-dimensional maps shown in Fig. 11.12. (**b**) same profiles obtained using the KdV equation. Figures adapted from Pierini (1989), © American Meteorological Society. Used with permission

Bibliography

Alpers, W.: Theory of radar imaging of internal waves. Nature **314**, 245–247 (1985)
Alpers, W., Salusti, E.: Scylla and Charybdis observed from space. J. Geophys. Res. Oceans **88**, 1800–1808 (1983)
Brandt, P., Rubino, A., Alpers, W., Backhaus, J.O.: Internal waves in the Strait of Messina studied by a numerical model and Synthetic Aperture Radar images from the ERS 1/2 Satellites. J. Phys. Oceanogr.Oceanogr. **27**, 648–663 (1997)
Bourgault, D., Galbraith, P., Chavanne, C.: Generation of internal solitary waves by frontally forced intrusions in geophysical flows. Nat. Commun **7**, 13606 (2016)
Garrett, C., Munk, W.: Space-time scales of internal waves. Geophys. Fluid Dyn. **3**, 225–264 (1972)
Garrett, C., Munk, W.: Space-time scales of internal waves: a progress report. J. Geophys. Res.Geophys. Res. **80**, 291–297 (1975)
Jackson, C.R.: An Atlas of Internal Solitary-Like Waves and Their Properties, 2nd ed. Office of Naval Research Code 322PO (2004). https://www.internalwaveatlas.com/Atlas2_index.html
Kadomtsev, B.B., Petviashvili, V.I.: On the stability of solitary waves in weakly dispersive media. Soviet Phys. Dokl. **15**, 539–541 (1970)
Pierini, S.: Solitons in a channel emerging from a three-dimensional initial wave. Il Nuovo Cimento C **9**, 1045–1061 (1986). https://doi.org/10.1007/BF02507422
Pierini, S.: A model for the Alboran Sea internal solitary waves. J. Phys. Oceanogr. **19**, 755–772 (1989)
Vazquez, A., Flecha, S., Bruno, M., Macıas, D., Navarro, G.: Internal waves and short-scale distribution patterns of chlorophyll in the Strait of Gibraltar and Alboran Sea. Geophys. Res. Lett. **36**, L23601 (2009)

Further Recommended Reading

Le Blond, P.H., Mysak, L.A.: Waves in the Oceans. Elsevier, New York (1978)
Gill, A.E.: Atmosphere-Ocean Dynamics. Academic Press, New York (1982)
Phillips, O.M.: The Dynamics of the Upper Oceans. Cambridge University Press, Cambridge (1966)
Pond, S., Pickard, G.L.: Introductory Dynamical Oceanography. Pergamon Press, Oxford (1983)
Wallace, J.M., Hobbs, P.V.: Atmospheric Science, An Introductory Survey. Elsevier, Amsterdam (2006)

Part II
Fluid Dynamics in Rotating Reference Frames

Chapter 12
Coriolis Force, Geostrophic Motions

In this chapter, we derive the Coriolis force and determine the Coriolis parameter, which expresses the effect of the rotation of the reference frame on a plane tangent to the Earth's surface. After introducing the Rossby and Ekman dimensionless numbers, of great significance in fluid dynamics in rotating reference frames, we analyse the motions governed by the geostrophic balance.

12.1 Coriolis Force

In Sect. 1.3 it was seen that, in the case the fluid is in a non-inertial reference frame, Newton's equations of motion must include, in the balance of forces, appropriate apparent forces. In the case of geophysical fluids, the apparent force to consider for motions with sufficiently large time scale, so that the rotation affects the dynamics, is the Coriolis force. In this paragraph, this force will be derived for a rotating reference frame in which the rotation manifests in the direction perpendicular to the plane on which the fluid is contained. In the next paragraph, the force will be adapted to a plane tangent to the Earth's surface.

Consider a rotating reference frame with constant angular velocity Ω, which is assumed to be aligned along the z-axis: $\mathbf{\Omega} = (0, 0, \Omega)$. We identify with I the inertial reference frame and with R the rotating one. To fix the ideas one can imagine an experiment (Fig. 12.1) in which, in a laboratory (reference frame I) a cylindrical tank containing a liquid, rotating with angular velocity Ω (reference frame R) is placed. If Ω is much greater than the angular velocity of the Earth, with respect to R, I will effectively constitute a substantially inertial reference frame.

Let $\mathbf{A}$ be a constant vector in R (Fig. 12.2): we ask what is the time derivative of $\mathbf{A}$ with respect to I.

From Fig. 12.2 it is evident that:

S. Pierini, *Oceanic and Atmospheric Fluid Dynamics*, UNITEXT for Physics,
https://doi.org/10.1007/978-3-031-77991-6_12

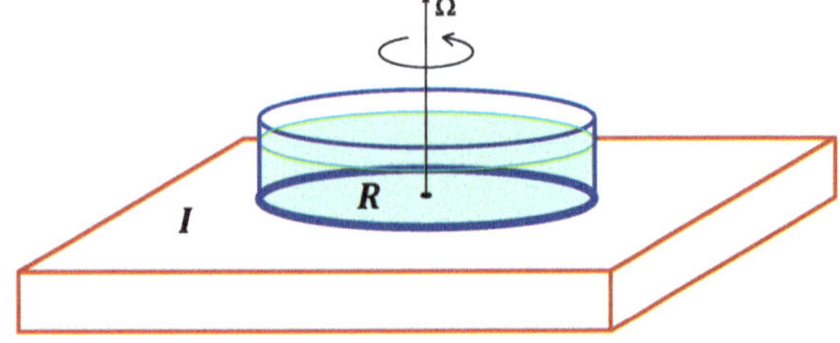

Fig. 12.1 A liquid is contained in a rotating circular tank (rotating reference frame R) placed in a laboratory (inertial reference frame I)

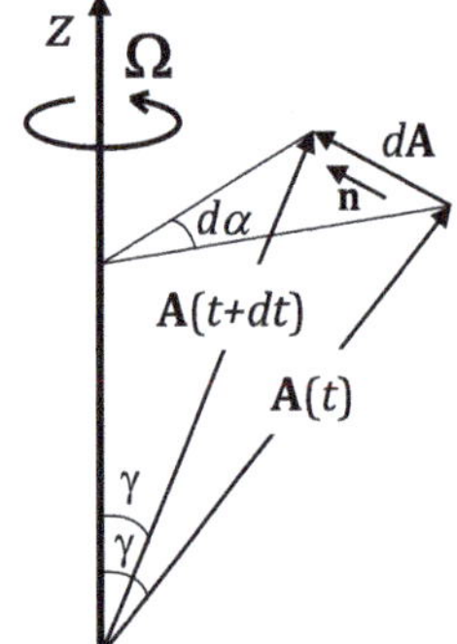

Fig. 12.2 Rotation with respect to I of a constant vector **A** in R

$$d\mathbf{A} = \mathbf{A}(t + dt) - \mathbf{A}(t) \cong \mathbf{n} A d\alpha \sin \gamma.$$

But $d\mathbf{A}$ is perpendicular to both **A** and **Ω**, hence:

$$\mathbf{n} = \boldsymbol{\Omega} \times \mathbf{A} / |\boldsymbol{\Omega} \times \mathbf{A}|.$$

Considering then that $\Omega = d\alpha/dt$ we will get:

$$\left(\frac{d\mathbf{A}}{dt}\right)_I = \boldsymbol{\Omega} \times \mathbf{A}, \tag{12.1}$$

where the subscript I indicates that the derivative is taken with respect to the inertial reference frame. If, on the other hand, the vector -which now is convenient to call **B**—also varies with respect to R we will have:

$$\left(\frac{d\mathbf{B}}{dt}\right)_I = \left(\frac{d\mathbf{B}}{dt}\right)_R + \boldsymbol{\Omega} \times \mathbf{B}, \tag{12.2}$$

where the subscript R indicates that the derivative is taken with respect to the rotating reference frame.

If we now apply Eq. (12.2) to $\mathbf{B} = \mathbf{r}$, where **r** is the position vector, we get:

$$\mathbf{u}_I = \mathbf{u}_R + \boldsymbol{\Omega} \times \mathbf{r}. \tag{12.3}$$

Then applying Eq. (12.2) to $\mathbf{B} = \mathbf{u}_I$ (the velocity of the fluid element with respect to the inertial reference frame) and taking into account Eq. (12.3), we get:

$$\left(\frac{d\mathbf{u}_I}{dt}\right)_I = \left(\frac{d\mathbf{u}_R}{dt}\right)_R + 2\boldsymbol{\Omega} \times \mathbf{u}_R + \boldsymbol{\Omega} \times (\boldsymbol{\Omega} \times \mathbf{r}). \tag{12.4}$$

Note that in the Navier–Stokes equation in an inertial reference frame the acceleration, which until now has been expressed as $d\mathbf{u}/dt$, is nothing more than

$$\frac{d\mathbf{u}}{dt} \equiv \left(\frac{d\mathbf{u}_I}{dt}\right)_I. \tag{12.5}$$

Therefore, substituting Eq. (12.5) into Eq. (5.8) yields the (vector) Navier–Stokes equation expressed in terms of acceleration $(d\mathbf{u}_R/dt)_R$ and *relative velocity* $\mathbf{u}_R$ evaluated with respect to the rotating reference frame:

$$\left(\frac{d\mathbf{u}_R}{dt}\right)_R + 2\boldsymbol{\Omega} \times \mathbf{u}_R + \boldsymbol{\Omega} \times (\boldsymbol{\Omega} \times \mathbf{r}) = \mathbf{g} - \frac{1}{\rho}\nabla p + \nu\nabla^2\mathbf{u}_R. \tag{12.6}$$

Equation (12.6) needs to be reformulated considering the following aspects and conventions.

First of all, for convenience of notation, the subscript R will be omitted, so in this second part of the text with $\mathbf{u}$ and $d\mathbf{u}/dt$ we will always refer to the quantities related to the rotating reference frame. Moreover, in the meteorological and oceanographic applications of interest for this text, the fluid is always in a state of turbulent motion (Chap. 6); consequently all dependent variables ($\mathbf{u}, p, \rho$ in Eq. (12.6), but other variables appear in more general equations, such as temperature T, salinity S in the ocean, the mixing ratio q in the atmosphere, etc.) will be understood as the average fields defined in Sect. 6.3, however implying the average symbol $\langle\cdot\rangle$ for convenience of notation. The turbulent nature of the flow also means that in the balance of forces the terms on the right-hand side of Eq. (6.7) containing the Reynolds stresses appear. In Sect. 6.4 it was shown how the effect of turbulence on the average flow can be parameterised, for example, as in Eq. (6.13),

$$\mathbf{F}_V = K_H\nabla_H^2\mathbf{u} + K_v\frac{\partial^2\mathbf{u}}{\partial z^2} \tag{12.7}$$

compared to which the effect of molecular viscosity is entirely negligible (of course except in the thin boundary layer in contact with solid boundaries). Other parameterisations are possible, but Eq. (12.7) is the one that will be used in the subsequent discussion. Finally, the term $\boldsymbol{\Omega} \times (\boldsymbol{\Omega} \times \mathbf{r})$ in Eq. (12.6) represents the centripetal acceleration, which can be expressed as the gradient of a potential function. Therefore, the effect of this force can be incorporated into a total potential energy $V_{tot} = V_g + V_c$ in

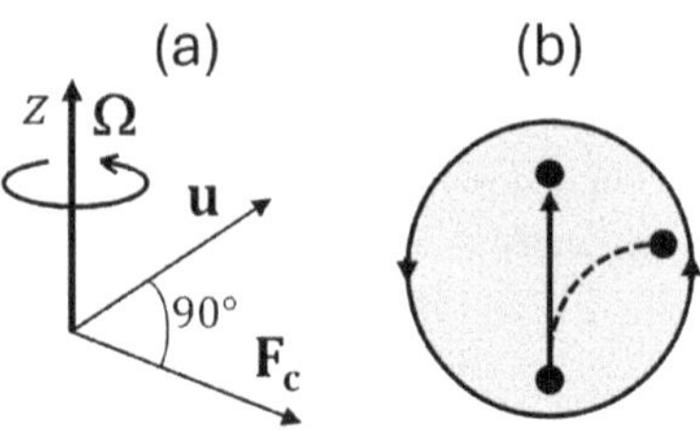

Fig. 12.3 The Coriolis force (**a**) accounts for the deviation of the trajectory in a rotating reference frame (**b**) of a body moving in a straight line in an inertial reference frame

the expression of the volume force per unit mass (Eq. 2.1). However, this difference is entirely negligible in the meteorological and oceanographic context.

Consequently, considering that the gradient of a scalar (the pressure) is invariant with respect to rotation, the dynamical equations for a fluid in a state of turbulent motion in a rotating reference frame are given by

$$\frac{d\mathbf{u}}{dt} = \mathbf{F}_C + \mathbf{g} - \frac{1}{\rho}\nabla p + \mathbf{F}_V, \tag{12.8}$$

where $\mathbf{F}_V$ is expressed by Eq. (12.7) and

$$\mathbf{F}_C = -2\boldsymbol{\Omega} \times \mathbf{u} \tag{12.9}$$

is the *Coriolis force* per unit mass (from the French mathematician and engineer Gaspard-Gustave de Coriolis, 1792–1843, who determined its expression). Naturally, given the apparent nature of this force, its opposite will represent the *Coriolis acceleration*

$$\boldsymbol{a}_c = 2\boldsymbol{\Omega} \times \mathbf{u}. \tag{12.10}$$

Note that $\mathbf{F}_C$ is a non-conservative force that, moreover, does no work as it is perpendicular to the displacement of the fluid element (this is analogous to the Lorentz force $\mathbf{F}_L = q\mathbf{u} \times \mathbf{B}$ to which a point electric charge q in motion with velocity $\mathbf{u}$ immersed in a magnetic induction field $\mathbf{B}$ is subject). Figure 12.3a shows the orientation of the three vectors in play: for a counterclockwise rotation of R, $\mathbf{F}_C$ is oriented to the right of the velocity vector. In Fig. 12.3b the dashed line shows the curved trajectory that appears in a rotating reference frame, of a body moving in a straight line in an inertial reference frame.

12.2 Coriolis Parameter

Consider, at a certain latitude φ_0, a marine or atmospheric area of limited meridional extent so that it can be considered approximately flat. For the purposes of the Coriolis force, the generic vector $\boldsymbol{\Omega}$ in Eq. (12.9) will be replaced with the angular velocity

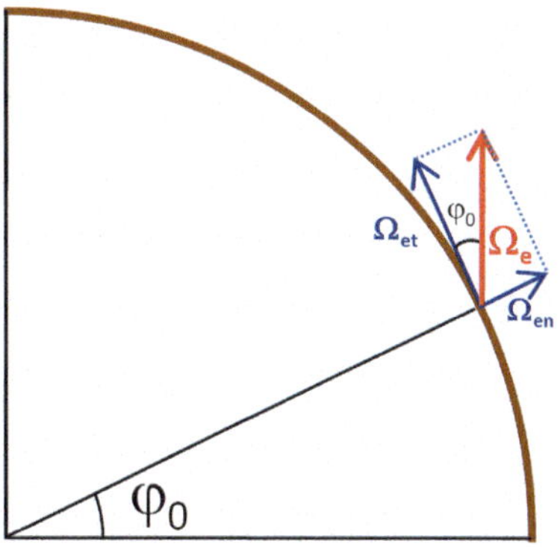

Fig. 12.4 Decomposition of $\boldsymbol{\Omega}_e$ into two components that are, respectively, normal and tangent to the plane in turn tangent to the Earth at the latitude φ_0

of the Earth $\boldsymbol{\Omega}_e$ (Fig. 12.4). Now this vector, except at the poles, is not perpendicular to the horizontal plane as happens for the rotating tank considered in the previous paragraph. It is therefore convenient to decompose $\boldsymbol{\Omega}_e$ into a vector normal to the surface $\boldsymbol{\Omega}_{en}$ and one tangent to it $\boldsymbol{\Omega}_{et}$ (Fig. 12.4).

Agreeing that the x and y axes are aligned respectively with the parallel and the meridian passing through the central point of the area considered, in the decomposition $2\boldsymbol{\Omega}_e = 2\boldsymbol{\Omega}_{en} + 2\boldsymbol{\Omega}_{et}$ we have $2\boldsymbol{\Omega}_{et} = \left(0, \hat{f}_0, 0\right)$ and $2\boldsymbol{\Omega}_{en} = (0, 0, f_0)$, with $\hat{f}_0 = 2\Omega_e \cos\varphi_0$ and

$$f_0 = 2\Omega_e \sin\varphi_0, \tag{12.11}$$

called *Coriolis parameter*. The Coriolis force (Eq. 12.9) on a plane tangent to the Earth will therefore be given by

$$\mathbf{F}_C = \left(f_0 v - \hat{f}_0 w, -f_0 u, \hat{f}_0 u\right). \tag{12.12}$$

It is possible to make a useful simplification. In motions for which the Coriolis force is relevant (see the next paragraph and Chap. 17), the horizontal length scale of the motion is typically much greater than the vertical one and the same goes for the corresponding velocity scales. This implies that at mid-latitudes, in the component along x of $\mathbf{F}_C$ the term $-\hat{f}_0 w$ is completely negligible compared to $f_0 v$. Furthermore, under the same assumptions, the component along z is $\hat{f}_0 u \sim 10^{-6} - 10^{-4}\ m\ s^{-2}$ (depending on whether the ocean or the atmosphere is considered), therefore it is several orders of magnitude smaller than the gravitational acceleration. Consequently, the following expression for the Coriolis force in large-scale motions can be assumed with excellent approximation:

$$\mathbf{F}_C \cong -2\boldsymbol{\Omega}_{en} \times \mathbf{u} = -f_0 \mathbf{k} \times \mathbf{u} \tag{12.13}$$

where $\mathbf{k} = (0, 0, 1)$. Equation (12.13) expresses the so-called *f-plane approximation*, which will be useful in a series of subsequent analyses.

It is essential to note that, while in the Northern Hemisphere $f_0 > 0$, for which the orientation of the vectors shown in Fig. 12.3a applies (thus with the Coriolis force acting to the right of the velocity), in the Southern Hemisphere the Coriolis force acts to the left of the velocity, as $f_0 < 0$.

In an approximation that also takes into account the meridional extension of the domain on the tangent plane, it is convenient to develop $2\Omega_{en}$ in a Taylor series around $\varphi = \varphi_0$ (which is made to correspond to $y = 0$) stopping at the first order:

$$f = f_0 + \beta y \tag{12.14}$$

where

$$\beta = \left.\frac{df}{dy}\right|_{y=0} = \left.\frac{1}{r_e}\frac{df}{d\varphi}\right|_{\varphi=\varphi_0} = \frac{2\Omega_e}{r_e}\cos\varphi_0$$

and where $r_e \cong 6370$ km is the Earth's radius. This is called the *beta-plane approximation*:

$$\mathbf{F}_C \cong -f\mathbf{k} \times \mathbf{u}. \tag{12.15}$$

The term βy introduces an important dynamic effect at the base of the very existence of planetary Rossby waves (Chap. 19) and the westward intensification of wind-driven ocean currents (Sect. 20.4). Regarding the typical values of these parameters, at mid latitudes, for example for $\varphi = 45°$, $f_0 \cong 10^{-4}$ $rad\ s^{-1}$ and $\beta \cong 1.6 \times 10^{-11} rad\ m^{-1} s^{-1}$. Around the equator $f = \beta y$ (with $\beta = 2\Omega_e/r_e$); here the equatorial beta effect supports the existence of equatorial Rossby and Kelvin waves which are at the base of the oceanic variability in the equatorial belt and regulate, among other things, the dynamics of El Niño. These phenomena will not be treated in this introductory text.

12.3 Rossby and Ekman Numbers

It is obvious that the Earth, despite being a rotating reference frame, in many cases behaves like an inertial reference frame. For example, for motions with very small time scales during which the Earth rotates by a very small angle, $\mathbf{F}_C$ will be negligible in the force balance: this case includes all the phenomena treated in Part I of the text. How can we estimate a priori whether, for a given motion, the Coriolis force is important or not?

Just like for the Reynolds number, which estimates the ratio between the weight of the inertial terms and that of the viscous terms in the Navier–Stokes equation, it is possible to introduce a dimensionless number that estimates the ratio between the weight of the inertial acceleration and that of the Coriolis acceleration, called the *Rossby number* ε. Considering the typical scales of speed and horizontal length U

and L and the typical time scale $T = L/U$ we get:

$$O\left(\left|\frac{d\mathbf{u}}{dt}\right|\right) = O\left(\frac{U^2}{L}\right); \quad O(|f\mathbf{k} \times \mathbf{u}|) = O(fU).$$

The Rossby number can therefore be defined as:

$$\varepsilon_1 = \frac{U}{f\mathrm{L}} \tag{12.16}$$

(the "order of magnitude" symbol is omitted).

Let us now define the *inertial period* T_i as the one corresponding to the Coriolis parameter (which is an angular frequency):

$$T_i = \frac{2\pi}{f}. \tag{12.17}$$

Some values of T_i are, for example: $T_i(\varphi = 90°) = 12\ h$, $T_i(\varphi = 45°) \cong 17.5\ h$, $T_i(\varphi = 30°) \cong 24\ h$ (as we approach the equator T_i tends to infinity, but this simply implies that the f-plane approximation there is no longer valid; in this regard, see the mention of the equatorial beta-plane approximation made in the previous paragraph). In Sect. 16.2 inertial currents will be described, which allow to attribute an immediate dynamic value to the inertial period. Therefore, ε_1 can also be expressed as the ratio between two time scales: $\varepsilon_1 = T_i/(2\pi T)$. Alternatively, the following definition of ε, essentially equivalent to ε_1, is also used:

$$\varepsilon_2 = \frac{T_i}{T}. \tag{12.18}$$

In conclusion, if $\varepsilon << 1$, $\mathbf{F}_C$ prevails over the inertial acceleration and the corresponding motions are called *subinertial*; conversely, if $\varepsilon >> 1$, $\mathbf{F}_C$ can be neglected and the corresponding motions are called *superinertial*.

An example of superinertial motions is given by surface gravity waves: it was seen in Sect. 9.5 that their periods are of the order of seconds; in that span of time the earth is essentially non-rotating so $\mathbf{F}_C$ plays no role. In that case, for $\varphi = 45°$ and choosing $T = 5\ s$ we have $\varepsilon_2 \cong 17.5\ h/0.0014\ h \cong 1.25 \times 10^4 >> 1$.

Examples of subinertial motions are provided by large-scale atmospheric and oceanic circulation. In an atmospheric circulation in the mid-troposphere, for example at 500 *mbar*, one can choose $U \cong 10\ ms^{-1}$ and $L \cong 1000\ km$, in which case for $\varphi = 45°$, $\varepsilon_1 \cong 0.1$. The same value is obtained for a western boundary current, such as the Gulf Stream (Sects. 20.2 and 20.4), for which $U \cong 1\ ms^{-1}$ and $L \cong 100\ km$. In these cases, $\mathbf{F}_C$ plays a fundamental role in the dynamics.

As for the forces of turbulent viscosity $\mathbf{F}_V$, it is possible to proceed similarly. Noting that the two terms in Eq. (12.7) scale as

$$O\left(\left|K_H \nabla_H^2 \mathbf{u}\right|\right) = O\left(K_H \frac{U}{L^2}\right);\ O\left(\left|K_v \frac{\partial^2 \mathbf{u}}{\partial z^2}\right|\right) = O\left(K_v \frac{U}{D^2}\right)$$

(where D is the typical scale of vertical motions), two dimensionless numbers, called *Ekman numbers*, are introduced, which estimate the ratio between the weight, respectively, of the horizontal and vertical turbulent viscosity terms and that of the Coriolis acceleration:

$$E_H = \frac{K_H}{fL^2};\ E_v = \frac{K_v}{fD^2}. \tag{12.19}$$

If $E_H << 1$ and $E_v << 1$ the viscous terms $\mathbf{F}_V$ are negligible compared to the Coriolis force $\mathbf{F}_C$ in Eq. (12.8).

12.4 Geostrophic Motions

For subinertial motions ($\varepsilon << 1$) with negligible turbulent viscosity ($E_H, E_v << 1$), the equations of motion (Eq. 12.8) for a homogeneous and incompressible fluid on the f-plane with the Coriolis acceleration expressed by Eq. (12.13) are reduced to:

$$f_0 \mathbf{k} \times \mathbf{u} \cong \mathbf{g} - \frac{1}{\rho} \nabla p. \tag{12.20}$$

The sign $\cong$ is to underline that in this approximation isobars with large radius of curvature are allowed, in which case, therefore, the inertial acceleration (the nonlinear terms $(\mathbf{u} \cdot \nabla)\mathbf{u}$ for stationary flows) is not completely negligible (for example, see the subsequent Fig. 12.6). However, for convenience, $\cong$ will be replaced with $=$.

Along z the Eq. (12.20) satisfies the hydrostatic balance while the horizontal components provide the so-called *geostrophic balance*

$$f_0 \mathbf{k} \times \mathbf{u}_g = -\frac{1}{\rho} \nabla_H p, \tag{12.21}$$

where $\nabla_H = \left(\partial/\partial_x, \partial/\partial_y\right)$ and $\mathbf{u}_g = \left(u_g, v_g\right)$ is the so-called *geostrophic velocity*. In explicit terms we have:

$$\begin{cases} -f_0 v_g = -\frac{1}{\rho}\frac{\partial p}{\partial x} \\ f_0 u_g = -\frac{1}{\rho}\frac{\partial p}{\partial y} \end{cases} \tag{12.22}$$

or, alternatively,

$$\mathbf{u}_g = \frac{1}{\rho f_0} \mathbf{k} \times \nabla p. \tag{12.23}$$

If we define

$$\psi = p/(\rho f_0) \tag{12.24}$$

it is immediate to verify that the geostrophic balance is nothing more than the expression of the horizontal velocity field in terms of the streamfunction ψ as defined in Sect. 4.3, for which the incompressibility condition is automatically verified. Therefore, the winds and oceanic geostrophic currents flow along the isolines $\psi = const$, that is, along the isobars.

Figure 12.5 helps to intuitively understand why this happens and how such a balance is achieved. Imagine that a fluid element initially at rest (thus with $\mathbf{F}_C = 0$) be subjected to a pressure gradient force $\mathbf{F}_{gp}$ indicated by the red arrow in Fig. 12.5a. The acceleration imparted to it will initially produce a small velocity $\mathbf{u}$ parallel to $\mathbf{F}_{gp}$: at this point the Coriolis force, normal to $\mathbf{u}$, will intervene to deviate to the right (in the Northern hemisphere) the trajectory of the fluid element. This deviation will continue until $\mathbf{u}$ aligns with the local isobar: at this point $\mathbf{F}_{gp}$ and $\mathbf{F}_C$ will balance exactly (Fig. 12.5b), establishing the geostrophic balance.

It should be specified that the process of achieving the geostrophic balance as schematised neglects a series of important factors that instead intervene in the real process, called *geostrophic adjustment*, such as the disturbance induced by the system from the displacement of fluid masses and the presence of a transient phase characterised by the activation of gravity waves, possibly modified by the effect of the Coriolis force (Poincaré waves, see Sect. 16.2 for a brief description of their characteristics). However, the simple scheme of Fig. 12.5 helps to understand the essence of the process that leads to the alignment of the flow to the isobars.

As for the scales involved in the winds and geostrophic currents, in the atmosphere the typical slope of the isobars is $\tan \alpha = O(10^{-4})$ (1 m/10 km), corresponding

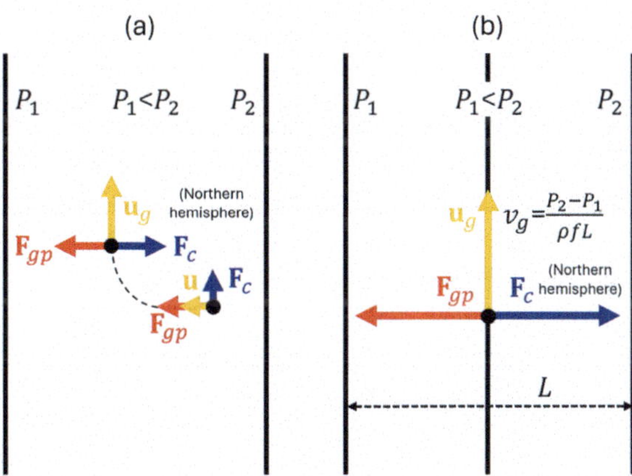

Fig. 12.5 Schematic representation of the geostrophic adjustment process

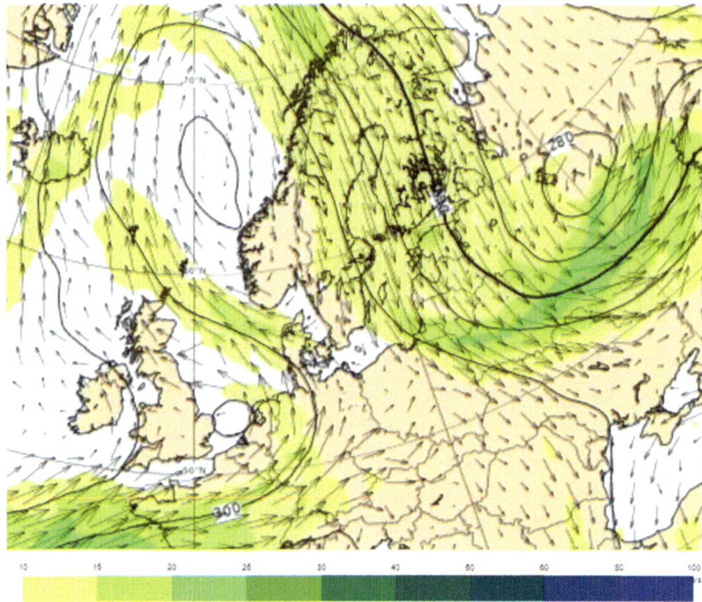

Fig. 12.6 Winds (arrows) and geopotential heights (isolines) at 700 mbar ($z \cong 3000$ m) in Northern Europe from high-resolution forecast for 5 May 2024, 06 UTC (the colour indicates the wind amplitude in ms^{-1}). The corresponding map on the ground is shown in Fig. 15.1. Image obtained from © European Centre for Medium-Range Weather Forecasts (ECMWF, https://www.ecmwf.int/, published under a Creative Commons Attribution 4.0 International, CC BY 4.0)

to geostrophic winds in the free atmosphere (Sect. 14.2) at mid-latitudes of a few tens of m/s. Figure 12.6 shows an example of winds at 700 mbar ($z \cong 3000$ m) superimposed on the geopotential height isolines (these are almost coincident with the isobars at a given geopotential height and can be obtained by a conversion factor calculable from the hypsometric equation). The substantial alignment of the winds along the isobars is evident. In Chap. 14 it will be seen how geostrophic winds vary with altitude due to the pressure field associated with the inclination of the isopycnals. In the oceans the typical slope of the isobars is $\tan\alpha = O(10^{-6})$ (1 mm/1 km), corresponding to geostrophic currents at mid-latitudes of the order of a few tens of cm/s.

Recommended Readings

Gill, A.E.: Atmosphere-Ocean Dynamics. Academic Press, New York (1982)
Greenspan, H.P.: The Theory of Rotating Fluids. Cambridge University Press, Cambridge (1968)
Holton, J.R.: An Introduction to Dynamic Meteorology. Elsevier, Amsterdam (2004)
Marshall, J., Plumb, R.A.: Atmosphere, Ocean, and Climate Dynamics. Elsevier, Amsterdam (2008)

Monin, A.S.: Theoretical Geophysical Fluid Dynamics. Kluwer Academic Publishers, Dordrecht (1990)
Pedlosky, J.: Geophysical Fluid Dynamics. Springer-Verlag, New York (1987)
Vallis, G.K.: Atmospheric and Oceanic Fluid Dynamics. Cambridge University Press, Cambridge (2006)
Wallace, J.M., Hobbs, P.V.: Atmospheric Science, an Introductory Survey. Elsevier, Amsterdam (2006)

Chapter 13
Two-Dimensionality in Rotating Fluids

In subinertial and barotropic flows, the Coriolis force imparts a horizontal two-dimensionality to the motion; this entails, among other things, a peculiar mode of mixing of fluid masses. This chapter deals with this particular dynamics, which will be considered the reference point for subsequent approximations.

13.1 Taylor-Proudman Theorem, Taylor Columns

It has already been observed that in motions for which the Coriolis force is important ($\varepsilon \ll 1$) the horizontal scales of motion are generally much larger than the vertical ones. In this context, the (horizontal) geostrophic balance already highlights one of the characteristics of the two-dimensional flows discussed in Sect. 4.3. It will now be shown that in a barotropic geostrophic flow (whose definition will now be given) the Coriolis force imparts a horizontal two-dimensionality to the motion in the sense of Eq. (4.14).

Consider Eq. (12.8) under the geostrophic assumption ($\varepsilon, E_H, E_v \ll 1$):

$$2\mathbf{\Omega} \times \mathbf{u} = \mathbf{g} - \frac{1}{\rho}\nabla p \tag{13.1}$$

(on a plane tangent to the Earth, $2\mathbf{\Omega} = 2\mathbf{\Omega}_e \cong f_0\mathbf{k}$, this reduces to Eq. 12.20). Taking the curl ($\nabla\times$) of both members yields:

$$\nabla \times (2\mathbf{\Omega} \times \mathbf{u}) = -\left(\frac{1}{\rho}\nabla \times \nabla p - \frac{1}{\rho^2}\nabla\rho \times \nabla p\right).$$

The first term on the right-hand side is zero as the curl of a gradient is identically zero. As for the first member, use the identity Eq. (5.29) with $\mathbf{A} = 2\mathbf{\Omega}$, $\mathbf{B} = \mathbf{u}$ and

S. Pierini, *Oceanic and Atmospheric Fluid Dynamics*, UNITEXT for Physics,
https://doi.org/10.1007/978-3-031-77991-6_13

consider that $\nabla \cdot \mathbf{u} = 0$ (we limit ourselves, as usual, to incompressible fluids) and that $\boldsymbol{\Omega} = const$: from this it follows $\nabla \times (2\boldsymbol{\Omega} \times \mathbf{u}) = -(2\boldsymbol{\Omega} \cdot \nabla)\mathbf{u}$. Hence:

$$(2\boldsymbol{\Omega} \cdot \nabla)\mathbf{u} = \frac{1}{\rho^2}\nabla p \times \nabla \rho. \tag{13.2}$$

In Sect. 14.1 we will see that this important relationship allows us to explain the variation of the geostrophic flow along the vertical. The second member term, called *baroclinic*, appears in the vorticity equation (Eq. 5.8) and, as discussed in Sect. 5.31, induces vorticity due to the misalignment of the isobaric surfaces with the isopycnic ones.

We now introduce the further assumption of a *barotropic state of motion*, for which the density is constant along the isobaric surfaces: $\rho = \rho(p)$ (naturally a homogeneous and incompressible fluid falls into this category). In this case $\nabla \rho \times \nabla p = \rho' \nabla p \times \nabla p = 0$, so Eq. (13.2) reduces to:

$$(\boldsymbol{\Omega} \cdot \nabla)\mathbf{u} = 0. \tag{13.3}$$

Finally, considering that $\boldsymbol{\Omega}$ is aligned along the vertical, $\boldsymbol{\Omega} \cdot \nabla = (0, 0, \Omega\partial/\partial z)$, we get:

$$\frac{\partial \mathbf{u}}{\partial z} = 0. \tag{13.4}$$

This simple but significant relationship, known as the *Taylor-Proudman theorem*, implies that a barotropic geostrophic flow is two-dimensional in the sense defined by Eq. (4.14), which is here rewritten for convenience:

$$\mathbf{u} = \left[u(x, y, t), v(x, y, t), 0\right]. \tag{13.5}$$

Indeed, the horizontal velocities do not depend on z and $w(x, y, t) = 0$ since, given that $w = 0$ on the lower boundary, which here is assumed to be flat, this must hold throughout the fluid column.

Equation (13.5) implies that the flow organises itself into vertical columns, known as *Taylor columns* (from the English physicist and mathematician G.I. Taylor, 1886–1975, who made many fundamental contributions to fluid dynamics), that therefore cannot tilt or expand or contract. Ultimately, under the aforementioned assumptions, the fluid columns move horizontally as if they were rigid. In this regard, it is worth remembering a classic laboratory experiment (Taylor 1922) that clearly highlights this phenomenon. The following description will illustrate the technique followed by Marshall and Plumb (2008) in their GFD VII laboratory experiment.

A disc of height less than the thickness of the fluid (orange rectangle in Fig. 13.1a) is placed at the bottom of a rotating cylindrical tank containing a liquid. The fluid is brought to a state of rigid-body rotation ($\mathbf{u}_R = 0$) and coloured floating crystals are released, used as passive surface tracers (blue dots). Then the angular rotation speed

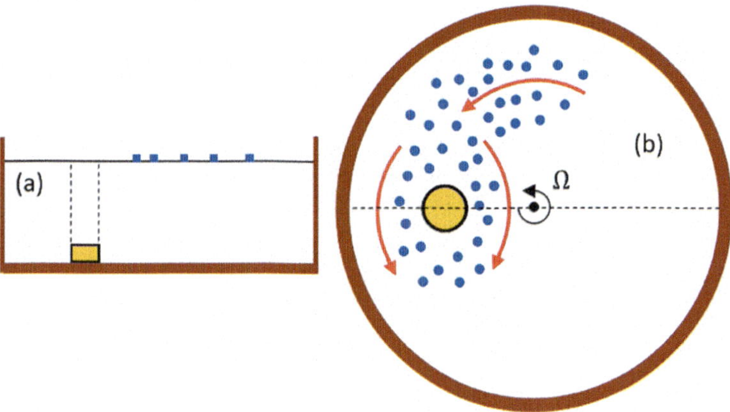

Fig. 13.1 Diagram of the experiment aimed at describing the dynamics of Taylor columns

is slightly slowed down in order to induce a circular motion. During the transient phase, it is observed (Fig. 13.1b) how the surface tracers, flowing towards the obstacle, bypass it along its vertical projection as if the fluid columns that transport them were rigid. Ultimately, in the extension of the theorem to variable bottom, a fluid column must follow the line of constant depth (see Eq. 19.7): in fact, passing at different depths, for the conservation of mass the height and width of the column would necessarily have to vary, but this would contradict what is implied by Eq. (13.5).

13.2 Mixing in Two-Dimensional Flows

An important consequence of the two-dimensionality of the flow in the conditions of Sect. 13.1 concerns the mixing of tracers transported by the fluid motion. Marshall and Plumb (2008) present the results of two laboratory experiments (GFD 0) that highlight very well the features of the mixing in a two-dimensional flow.

In the first experiment, the tank is at rest (so the reference frame is inertial). A turbulent motion is induced by inserting a hand into the fluid and moving it gently. Subsequently, passive tracers of different colours are introduced to highlight the flow and the mixing. The turbulence produces an efficient three-dimensional mixing, also evident on a vertical profile. The structure of the mixing is similar to that shown in Fig. 6.3.

In the second experiment, the tank rotates in a condition that satisfies the Taylor-Proudman theorem. The fluid is brought to a state of rigid-body rotation, then a motion is induced in the manner already described and finally the dyes are introduced. Now the situation appears completely different: veins of dye that precipitate vertically are highlighted while the horizontal motions form vertical tendrils that stretch and wrap around each other in an increasingly intricate manner over time (this process

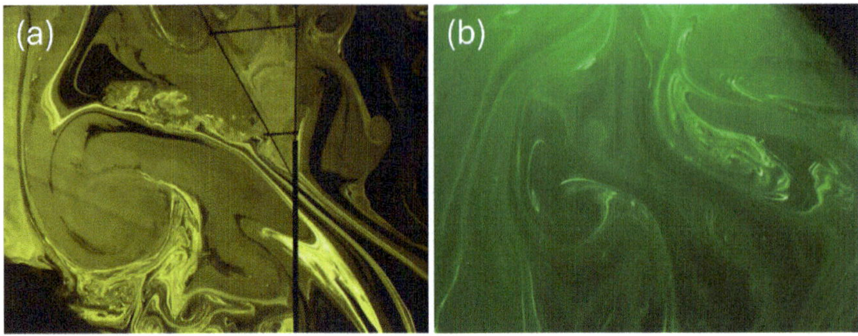

Fig. 13.2 (**a**) Example of two-dimensional mixing highlighted in one of the laboratory experiments conducted with the 13-m diameter "Coriolis" rotating platform of LEGI-CNRS (Grenoble), for the study of western boundary currents in the presence of a lateral gap (Pierini et al. 2022). The image is obtained thanks to the illumination of a fluorescent dye (rhodamine) by a horizontal laser beam. (**b**) Photo taken during another laboratory experiment. Figure (a) is adapted from Pierini et al. (2022, Sci. Rep. 12, 1375; Springer Nature. Work distributed under a Creative Commons Attribution 4.0 International License)

is usually described as the *stirring and folding* of fluid columns). This is due to the two-dimensionality of the motion that causes the dye traces, transported by Taylor columns, to be forced to evolve on a horizontal plane.

Figure 13.2 shows examples of two-dimensional mixing taken from a laboratory experiment conducted with a rotating platform for the study of a particular oceanographic phenomenon. The stretching and folding of the dye traces in the presence of a substantially two-dimensional motion are evident.

13.3 Vorticity in Rotating Reference Frames

In the case of rotating reference frames, there are three forms of vorticity:

- the vorticity observed from the rotating reference frame, called *relative vorticity* $\boldsymbol{\omega}_R$,
- the vorticity -observed from an inertial reference frame- of a fluid at rest in a reference frame rotating with angular velocity $\boldsymbol{\Omega}$, called *planetary vorticity* $\boldsymbol{\omega}_P$ (by the very definition of vorticity, $\boldsymbol{\omega}_P = 2\boldsymbol{\Omega}$),
- the vorticity -observed from an inertial reference frame- of a fluid in motion in a reference frame rotating with angular velocity $\boldsymbol{\Omega}$, called *absolute vorticity* $\boldsymbol{\omega}_a$.

These three forms of vorticity are linked by the obvious relationship:

$$\boldsymbol{\omega}_a = \boldsymbol{\omega}_R + 2\boldsymbol{\Omega}. \tag{13.6}$$

For example, referring to Fig. 12.1, $\boldsymbol{\omega}_R$ is the vorticity observed in the rotating frame R. If $\boldsymbol{\omega}_R = 0$, the vorticity observed by I is simply $2\boldsymbol{\Omega}$. In the more general case ($\boldsymbol{\omega}_R \neq 0$) the (absolute $\boldsymbol{\omega}_a$) vorticity observed from I will be the sum of the relative vorticity $\boldsymbol{\omega}_R$ plus the planetary one $2\boldsymbol{\Omega}$.

In a substantially two-dimensional flow on a plane tangent to the Earth, $\boldsymbol{\omega}_R \cong (0, 0, \zeta)$ (where $\zeta = v_x - u_y$). Therefore, the vertical component of the absolute vorticity (the only non-zero one) will be given by

$$\omega_{a3} = \zeta + f. \tag{13.7}$$

In Chap. 18, we will see how absolute vorticity forms the basis for the definition of the important concept of potential vorticity.

Finally, noting that $O(|\zeta|) = U/L$, it is immediate to verify that the Rossby number ε_1 defined by Eq. (12.16) also represents the ratio between the weight of the relative vorticity and that of the planetary vorticity:

$$\varepsilon_1 = \frac{O(|\boldsymbol{\omega}_R|)}{O(|\boldsymbol{\omega}_P|)}. \tag{13.8}$$

Therefore, for subinertial motions, the vorticity of a fluid element seen from an inertial reference frame is mainly due to Earth's rotation (f). On the other hand, it is precisely on that small relative vorticity observed on Earth (ζ) that our interest will focus.

13.4 The Various Approximations

In Sect. 12.4, the geostrophic balance was introduced, which well represents a wide and relevant class of meteorological and oceanographic motions. In this chapter, the condition of barotropicity has also been examined, which describes z-independent, two-dimensional geostrophic motions; this further condition leads to some realistic flows. However, meteorological and oceanographic dynamics is very varied and several relevant flows present in the oceans and in the atmosphere do not fall into the case considered here.

In the following chapters, the phenomenology of these flows will be described and the related theoretical treatments will be developed. For this purpose, it is very useful to highlight which effects differentiate one case from the other, because this allows to obtain a clear and organic overall view of the varied case studies of meteorological and oceanographic flows.

In this context, Fig. 13.3 schematically summarises the approximations that will be considered in the following chapters. The case (a) dealt with in this chapter will serve as a useful dynamic point of reference from which we will deviate, from time to time, due to the different approximations.

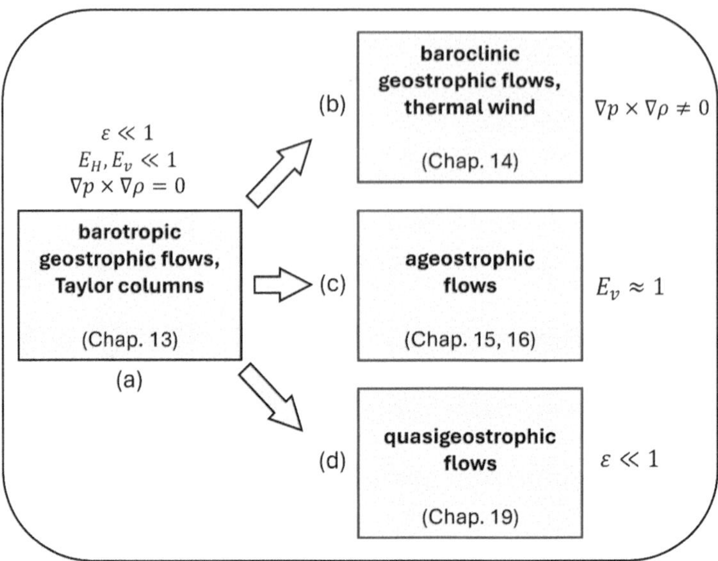

Fig. 13.3 Diagram of the transition from barotropic geostrophic flows (**a**) to baroclinic geostrophic flows (**b**), ageostrophic flows (**c**) and quasi-geostrophic ones (**d**). In each box, the chapter in which that dynamics is treated is indicated and next to it is reported the main condition that characterises it compared to (**a**)

By removing the assumption of barotropic flow ($\nabla p \times \nabla \rho \neq 0$), baroclinic flows will be obtained, which are associated with a vertical variation of the geostrophic flow observed both in the atmosphere and in the ocean (the so-called thermal wind). This case will be considered in Chap. 14.

In the atmospheric boundary layer and in the oceanic boundary layers on the surface and on the seabed, vertical eddy viscosity cannot be ignored. In this case, we will talk about ageostrophic flows, for which $E_v \approx 1$. This case will be considered in Chap. 15 for the atmosphere and in Chap. 16 for the oceans.

Finally, in Chap. 19, to obtain an evolution equation for the streamfunction that expresses the geostrophic balance, the flows will no longer be considered stationary (but must still be subinertial) and the nonlinear terms and horizontal turbulent viscosity will no longer be ignored. This dynamics, called quasi-geostrophic, will allow us to easily study the delicate but very relevant problem of atmospheric and oceanic Rossby waves.

Bibliography

Marshall, J., Plumb, R.A.: Atmosphere, ocean, and climate dynamics. Elsevier, Amsterdam (2008)

Pierini S., De Ruggiero, P., Negretti, M.E., Schiller-Weiss, I., Weiffenbach, J., Viboud, S., Valran, T., Dijkstra, H.A., Sommeria, J.: Laboratory experiments reveal intrinsic self-sustained oscillations in ocean relevant rotating fluid flows. Sci. Rep. **12**, 1375 (2022)
Taylor, G.I.: The motion of a sphere in a rotating liquid. Proc. R. Soc. Lond. A **102**, 180–189 (1922)

Further Recommended Reading

Cushman-Roisin, B.: Introduction to Geophysical Fluid Dynamics. Prentice-Hall, Englewood Cliffs, New Jersey (1994)
Gill, A.E.: Atmosphere-Ocean Dynamics. Academic Press, New York (1982)
Holton, J.R.: An Introduction to Dynamic Meteorology. Elsevier, Amsterdam (2004)
Monin, A.S.: Theoretical Geophysical Fluid Dynamics. Kluwer Academic Publishers, Dordrecht (1990)
Pedlosky, J.: Geophysical Fluid Dynamics. Springer-Verlag, New York (1987)
Vallis, G.K.: Atmospheric and Oceanic Fluid Dynamics. Cambridge University Press, Cambridge (2006)

Chapter 14
The Thermal Wind

The barotropic hypothesis implies the independence of geostrophic flows from the vertical coordinate. However, geostrophic winds at mid latitudes exhibit a strong dependence on altitude and, similarly, geostrophic currents in the oceans generally depend significantly on depth. To describe this behaviour, in this chapter we consider baroclinic geostrophic flows. The resulting thermal wind relationship explains the experimental observations.

14.1 The Thermal Wind Relationship

In the previous chapter, we saw how the assumptions ε, E_H, $E_v \ll 1$ and $\nabla p \times \nabla \rho = 0$ correspond to barotropic geostrophic motions, i.e. to horizontal flows satisfying the geostrophic balance and independent of z. However, the geostrophic motions observed in the atmosphere and in the oceans often show significant vertical variations in the velocity field.

As for the atmosphere, Fig. 14.1 shows a map of the zonal component of the winds averaged zonally (i.e., over all longitudes) and annually as a function of latitude and pressure, and therefore of altitude (it is well known that this zonal component prevails over the meridional one). As we will see in Chap. 15, above the planetary boundary layer the geostrophic balance is effectively verified. Yet, except for the low latitudes (cyan band in Fig. 14.1), the average—westerly—zonal geostrophic winds depend substantially on altitude. In fact, at low and mid latitudes we go from winds of a few m/s just above the planetary boundary layer ($p \lesssim 900 - 850$ *mbar* in daytime hours) to reach average values of $\sim\ 25 - 30$ m/s (but with instantaneous values that can be much higher) corresponding to the *tropospheric subtropical jet stream*, located just below the tropopause ($p \cong 200$ mbar) around the latitude of 30°, between the Hadley and Ferrel cells.

S. Pierini, *Oceanic and Atmospheric Fluid Dynamics*, UNITEXT for Physics,
https://doi.org/10.1007/978-3-031-77991-6_14

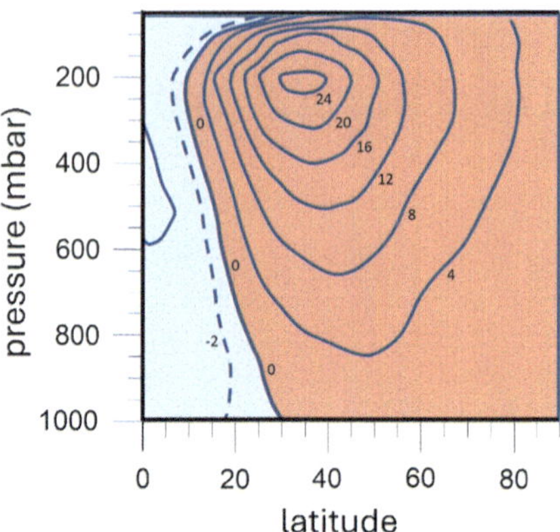

Fig. 14.1 Map of the zonally and annually averaged tropospheric zonal winds as a function of latitude and pressure in the northern hemisphere (units in m s^{-1})

As for the ocean, Fig. 14.2 shows a typical vertical profile of the geostrophic velocity within the Gulf Stream. Here too, the current goes from $\sim$ 1 m s^{-1} at the surface to a velocity practically zero below 2000 m.

How can such a blatant contradiction between these observational data and the Taylor-Proudman theorem be explained? The explanation lies in the assumption, not verified in these cases, of fluid in a state of barotropic motion: here the fluid is instead in a baroclinic state, in which, therefore, the isobaric and isopycnic surfaces do not coincide (obviously this is possible only in a stratified fluid, a situation entirely normal in oceanic and atmospheric fluids).

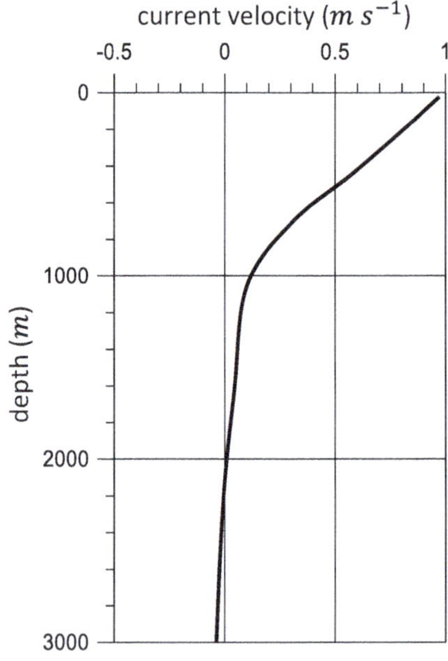

Fig. 14.2 Geostrophic velocity of the Gulf Stream as a function of depth

Therefore, to theoretically interpret the observations, we start from Eq. (13.2), which is reported here again for convenience:

$$(2\boldsymbol{\Omega} \cdot \nabla)\mathbf{u} = \frac{1}{\rho^2}\nabla p \times \nabla \rho.$$

Under the assumption of barotropic flow ($\rho = \rho(p)$) the right-hand side is null and we obtain the Taylor-Proudman theorem. Conversely, in the cases just considered $\rho \neq \rho(p)$ and, therefore, $\nabla p \times \nabla \rho \neq 0$ (remember that this term induces vorticity through the baroclinic torque mechanism, see Sect. 5.8): in this case we speak of fluid in a state of *baroclinic motion*. This occurs when ρ also depends on the temperature T ($\rho = \rho(p, T)$), so that isobaric and isopycnic surfaces do not coincide. The above equation is called the *thermal wind relationship*.

Considering that the fluid verifies with excellent approximation the hydrostatic balance $\nabla p/\rho = -g\mathbf{k}$, remembering that $\boldsymbol{\Omega} \cdot \nabla = (0, 0, \Omega\partial/\partial z)$ and substituting $2\boldsymbol{\Omega}$ with $f\mathbf{k}$, the following thermal wind balance for the geostrophic velocity $\mathbf{u}_g = (u_g, v_g)$ is obtained:

$$\frac{\partial \mathbf{u}_g}{\partial z} = -\frac{g}{\rho f}\mathbf{k} \times \nabla \rho. \tag{14.1}$$

It is worth noting that the assumption $\nabla p \propto \mathbf{k}$ (horizontal isobars) is not in contradiction with the small inclination of the isobars necessary for the very existence of geostrophic winds. Explicitly, the components are:

$$\left(\frac{\partial u_g}{\partial z}, \frac{\partial v_g}{\partial z}\right) = \frac{g}{\rho f}\left(\frac{\partial \rho}{\partial y}, -\frac{\partial \rho}{\partial x}\right). \tag{14.2}$$

Therefore, the presence of a density variation in the generic direction $\boldsymbol{\ell}$ on the horizontal plane involves the presence of a vertical shear in the velocity component along the direction perpendicular to $\boldsymbol{\ell}$. This happens because the inclination of the isopycnals introduces a relative pressure field dependent on z, an inclination of the isobaric surfaces and, therefore, a geostrophic velocity also dependent on z.

A clarifying example is presented in the next paragraph, where geostrophic zonal winds will be considered for which Eq. (14.2) is reduced to:

$$\frac{\partial u_g}{\partial z} = \frac{g}{\rho f}\frac{\partial \rho}{\partial y}. \tag{14.3}$$

We conclude by noting that the phenomenon of the thermal wind shows how the rotation of the reference frame, and therefore the Coriolis force, allows a density field with inclined isopycnals to maintain itself in balance in the presence of a well-defined vertical velocity shear, preventing the natural tendency to level those surfaces without energy expenditure.

14.2 The Thermal Wind in the Atmosphere

As a relevant example of thermal wind in the atmosphere, we now consider the case of geostrophic zonal winds in the northern hemisphere; in this case Eq. (14.3) applies, in which the meridional density gradient can be converted into that of temperature using the equation of state of perfect gas $\rho = p/(RT)$:

$$\frac{1}{\rho}\frac{\partial \rho}{\partial y} = \frac{RT}{p}\frac{p}{R}\left(-\frac{1}{T^2}\right)\left(\frac{\partial T}{\partial y}\right)_p = -\frac{1}{T}\left(\frac{\partial T}{\partial y}\right)_p$$

where the subscript p indicates that the derivative is calculated along an isobaric surface (but the slope of the latter is so small that this derivative practically coincides with the partial derivative at constant z). From this follows:

$$\frac{\partial u_g}{\partial z} = -\frac{g}{Tf}\left(\frac{\partial T}{\partial y}\right)_p . \tag{14.4}$$

To verify this relationship with experimental data, the map of the temperature averaged zonally and annually as a function of latitude and pressure shown in Fig. 14.3 can be used.

Between 0° and $\sim$ 20°N the temperature does not present appreciable meridional gradients in the troposphere; consequently, for Eq. (14.4) the westerly geostrophic winds u_g are barotropic, that is, they do not vary with the altitude; this is in agreement with what is shown in Fig. 14.1. This situation is summarised schematically in Fig. 14.4.

On the contrary, at mid-latitudes the westerly geostrophic winds are baroclinic as T records strong meridional variations, which cause a strong variation in the intensity of the geostrophic wind along z. This situation is schematically summarised in Fig. 14.5.

Note that in an area of about $\pm 15°$ around $\lambda = 40°N$ and up to an altitude of $\sim$ 500 mbar the meridional temperature gradient is almost constant. Therefore,

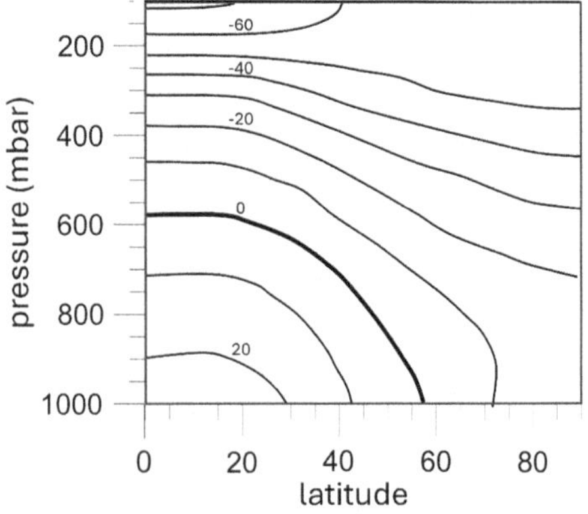

Fig. 14.3 Map of tropospheric temperatures averaged zonally and annually as a function of latitude and pressure in the northern hemisphere (units in °C)

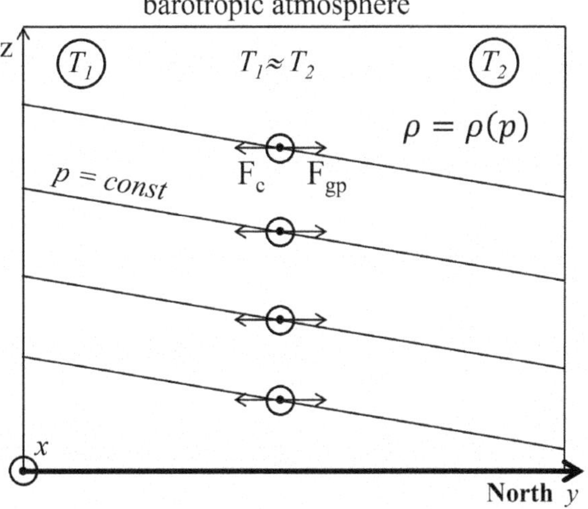

Fig. 14.4 Schematic of a barotropic atmosphere in the northern hemisphere (the inclination of the isobars is greatly amplified)

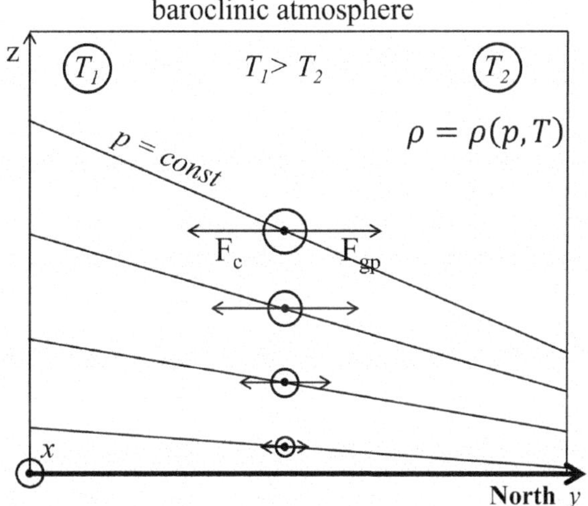

Fig. 14.5 Schematic of a baroclinic atmosphere in the northern hemisphere (the inclination of the isobars is greatly amplified)

it is useful to refer to this zonal band and to these heights for a simple quantitative verification of Eq. (14.4). At $\lambda = 40°N$ the isotherm $T = 0°C$ is located at $p \cong 720\,\text{mbar}$ ($z \cong 2800\,\text{m}$); at the same pressure, the isotherm $T = -10°C$ corresponds to $\lambda \cong 56°N$, that is, $\cong$1780 km further North. The corresponding meridional temperature gradient will therefore be:

$$\left(\frac{\partial T}{\partial y}\right)_p \cong \frac{-10}{1.78 \times 10^6} \cong -5.6 \times 10^{-6}\,°K\ \text{m}^{-1}.$$

With this value Eq. (14.4) provides the following geostrophic wind shear:

$$\frac{\partial u_g}{\partial z} \cong \frac{9.8 \times 10^4}{273 - 5} 5.6 \times 10^{-6} \cong 2 \times 10^{-3}\,\mathrm{s}^{-1}. \tag{14.5}$$

On the other hand, in the map of Fig. 14.1, at the intermediate latitude considered above between 40°N and 56°N, the isoline $u_g = 4\,\mathrm{m\,s}^{-1}$ corresponds, for example at $\lambda = 45°N$, to $p \cong 850\,\mathrm{mbar}$ ($z \cong 1500\,\mathrm{m}$) while the isoline $u_g = 8\,\mathrm{m\,s}^{-1}$ corresponds to $p \cong 650\,\mathrm{mbar}$ ($z \cong 3500\,\mathrm{m}$). Therefore, for the shear in Fig. 14.1 we have

$$\frac{\partial u_g}{\partial z} \cong \frac{4}{2 \times 10^3} = 2 \times 10^{-3}\,\mathrm{s}^{-1}, \tag{14.6}$$

in perfect agreement with the estimate provided by Eq. (14.5) (referring to the same zonal and altitude band) starting from the temperature field. In essence, in the area of the (φ, p) plane considered here, the westerly winds increase by $\Delta u_g = 1\,\mathrm{m\,s}^{-1}$ every $\Delta z \cong 500\,\mathrm{m}$.

14.3 The Thermal Wind in the Ocean: The Relative Currents

The situation in the ocean is similar to that of the atmosphere, but there are obviously also significant differences. First of all, it should be noted that there are no prevailing "zonal" currents in the oceans due to the geometric constraints imposed by the coastlines, so the oceanic geostrophic currents do not have a prevailing direction. However, the y-axis can be locally aligned along $\nabla\rho$ so that, for convenience of treatment, we can continue to refer to the thermal wind in the form given by Eq. (14.3) (otherwise Eqs. (14.1–14.2) will have to be used).

Integrating Eq. (14.3) vertically we get:

$$u_g(z) = u_g(z_0) + \frac{g}{f}\int_{z_0}^{z} \frac{1}{\rho}\frac{\partial \rho}{\partial y} dz'. \tag{14.7}$$

This relationship provides the geostrophic velocity at each depth if the density field ρ and $u_g(z_0)$ are known. Eq. (14.7) is obviously valid for both the ocean and the atmosphere but is of particular importance in dynamical oceanography, as it provides the only method for estimating, albeit indirectly, the large-scale oceanic circulation at depth, the knowledge of which is necessary to evaluate the transport of heat and mass by the oceans with fundamental implications for understanding global climate. In fact, direct current measurements at depth on a synoptic scale are simply impossible to carry out with the required frequency and spatial resolution.

On the contrary, a huge amount of *hydrological data* is available (i.e., temperature and salinity as a function of pressure, from which density is derived using the equation of state of seawater) obtained from oceanographic ships (but also from so-called *ships of opportunity*) over the years with in-situ measurements made with a CTD device (*Conductivity, Temperature and Depth*) and/or with the launch of XBTs (*expendable bathythermograph*). The result is the knowledge of a climatological (average) mass field from which it is possible to trace—under the geostrophic hypothesis—the shear of the average current speed through the thermal wind relationship and, therefore, to the current as a function of depth using Eq. (14.7). It should be added that the same method is also applied to the more limited Mediterranean seas and their sub-basins, in which case it is possible to carry out oceanographic measurement campaigns also using complementary experimental methods, such as gliders, ADCPs, current chains, Lagrangian measurements, etc. (methods that can obviously be used locally even in large oceans).

The methodology summarised by Eq. (14.7), known as the *classical method of dynamic computation of relative currents*, however, suffers from a series of limitations. The geostrophic hypothesis may not be entirely consistent with the nature of the measured mass field. For example, this may reflect the "instantaneous" image of a time-varying field rather than a quasi-stationary situation. The horizontal curvature of the isopycnals might contradict the hypothesis of quasi-straight currents. Ultimately, the hypothesis of subinertial motion at the base of geostrophy may not be adequately verified.

Furthermore, another fundamental limitation lies in the need to know the geostrophic velocity at a certain depth $u_g(z_0)$ so that one can pass from the velocity shear to the current velocity itself as a function of z; Fig. 14.6 illustrates this problem. From the mass field obtained from in-situ hydrological measurements within the Gulf Stream the vertical shear of the geostrophic velocity is obtained: this is represented, for example, by three speed profiles corresponding exactly to the same shear but differing by an arbitrary offset. Which of these profiles (or of infinite others obtained with different offsets) provides the real geostrophic velocity? The knowledge of u_g at a generic depth z_0 solves the problem, but such information is far from simple to obtain.

Traditionally, a *level of no motion*, also known as the *reference level*, is assumed (i.e., at a given depth it is assumed that $u_g = 0$). For example, in Fig. 14.6 if the reference level is $z = -2000$ m the profile to select is the one given by the solid line.

The sub-surface geostrophic current ($z = 0$) is due to an inclination of the free surface η, corresponding to the inclination of the surface isobar (in Sect. 16.3 it will be seen how this inclination can be generated). If the isopycnals were parallel to the surface isobar at every depth (for example—but not necessarily—if the density were constant), the current velocity would be in turn constant, verifying at every depth the same geostrophic balance between the *absolute pressure field* associated with the inclination of the free surface and the Coriolis force (barotropic motion), until it vanishes on the bottom due to the no-slip condition. Conversely, the velocity shear shown in Fig. 14.6 is the result of the sum of the absolute pressure field (independent of z) and the *relative pressure field* (dependent on z, hence the term *relative currents*); the latter is associated with the inclination of the isopycnals (in the specific case of opposite sign to that of the free surface) and therefore with a baroclinic flow.

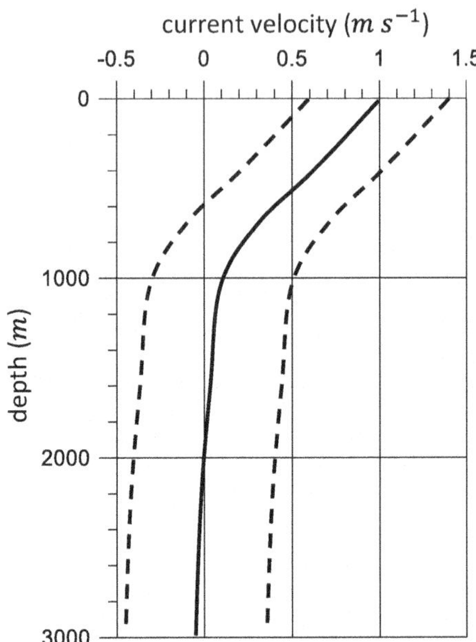

Fig. 14.6 Schematic representation of the geostrophic velocity profile of the Gulf Stream as a function of depth. Each of the three profiles is consistent with the method of dynamic computation of relative currents (thermal wind); the profile in solid line is the one to select assuming the level of no motion at 2000 dbar (Fig. 14.2)

The decrease of geostrophic current with depth is denoted *baroclinic compensation*. In the example of Fig. 14.6, the compensation is complete at 2000 *m*, but in other cases, the level of no motion does not exist if the relative pressure field never exactly compensates the absolute one, resulting in a non-zero geostrophic current at every depth. It can also happen that the level of no motion exists but the current changes direction at greater depths: consider the presence of overlapping water masses flowing in different directions, as happens, for example, in the North Atlantic Ocean or—on a smaller scale—in the Sicily Channel in the Mediterranean Sea.

The assumption of a certain level of no motion in large oceanic areas can be just a conjecture or can instead be based on sophisticated inverse methods (e.g., Defant 1961; Veronis 1987), but it is nevertheless useful for obtaining maps of average geostrophic currents on the surface or at different depths. The reference levels used in the Atlantic and Pacific oceans range from 1000 to 2000 m. For example, a map of the dynamic topography of the sea surface relative to 1000 dbar can be obtained (it is recalled that 1 dbar is almost exactly equivalent to 1 m of seawater, see Sect. 3.1) by calculating the geostrophic current at $z = 0$ from hydrological data using the thermal wind relationship imposing the level of no motion at $z = -1000$ m.

Another possibility is to assume a *level of known motion*. For example, if locally there is a current meter mooring that provides a long time series of the current at a depth z_0, it is possible to filter non-geostrophic components from the signal (such as tides, ageostrophic flows, etc.) and extract an average value that can reasonably be assumed to be in geostrophic balance. Substituting it into Eq. (14.7) gives the geostrophic velocity at every depth. This is possible, for example, in a situation like the one represented in Fig. 14.6. If there are more current meters at different depths, it is also possible to verify the correspondence of the profile obtained from hydrological data with the measured velocities; checks of this type highlight in many cases the validity of the adopted method.

Recently, *satellite altimetry* (e.g., Stammer and Cazenave 2019) allows us to assume the level of known motion at $z = 0$. This fascinating and relatively recent remote sensing technique allows us to obtain with centimetre precision the distance between the satellite hosting the radar altimeter and the sea surface. The sea level η is obtained by subtracting the height of the geoid (available with sufficient precision from 2011 thanks to the GOCE mission) from the altimetric height referred to the reference ellipsoid. Finally, after subtracting with sophisticated techniques the non-geostrophic components of the signal (the tides, the isostatic response to pressure fluctuations—the so-called inverse barometer—, the steric anomaly, ageostrophic oscillations of various nature) a surface topography η_g is obtained, called *Absolute Dynamic Topography* (ADT), from which it is possible to obtain, through the geostrophic balance, the sub-surface geostrophic current: this allows us to move from the velocity shear to the vertical velocity profile on a synoptic scale.

We conclude by reporting a classic formula particularly used in dynamical oceanography that adapts the thermal wind relationship to a two-layer fluid. It is easy to recognise that

$$\tan \gamma = \frac{\partial \rho}{\partial y} / \frac{\partial \rho}{\partial z},$$

where γ is the angle formed by the local isopycnal with respect to the horizontal plane. Combining this relationship with Eq. (14.3) gives:

$$\tan \gamma = \frac{\rho f}{g} \frac{\partial u_g}{\partial z} / \frac{\partial \rho}{\partial z}. \tag{14.8}$$

If we now consider a fluid composed of two layers of immiscible fluid (Fig. 14.7) of densities ρ_1 and ρ_2 ($\rho_1 < \rho_2$, $\overline{\rho}=(\rho_1 + \rho_2)/2$) with geostrophic velocities u_{g1} and u_{g2} (normal to the figure) respectively above and below the interface, Eq. (14.8) provides

$$\tan \gamma = \frac{\overline{\rho} f}{g} \left(\frac{u_{g1} - u_{g2}}{\rho_1 - \rho_2} \right), \tag{14.9}$$

known as *Margules' formula*. Generally, the unknown is represented by one of the two velocities: in this way, the jump in geostrophic velocity produced by the inclination of the interface can be evaluated. This formula can also be used in a multilayer ocean by applying it to each pair of layers until the vertical profile of the geostrophic velocity is reconstructed with a discrete approach (which is equivalent to solving the thermal wind equation with the finite difference method).

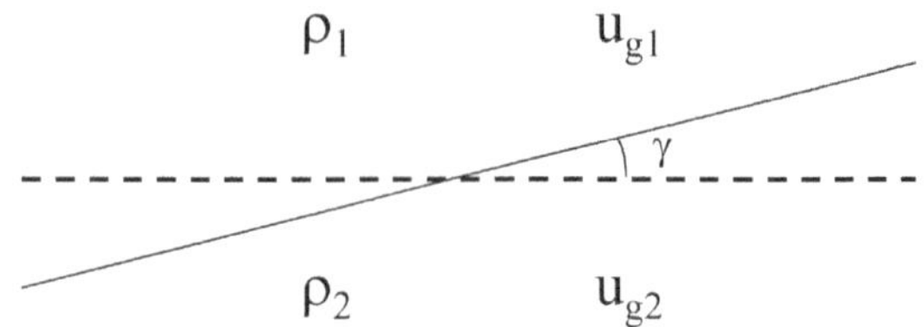

Fig. 14.7 The situation described by Margules' formula (the geostrophic velocities are normal to the figure)

Refer to Sect. 16.4 for the analysis of the mechanisms through which geostrophic currents, both barotropic and baroclinic, are indirectly generated by the surface wind.

Bibliography

Defant, A.: Physical Oceanography, vol. I. Pergamon Press, New York (1961)
Stammer, D., Cazenave, A. (eds.): Satellite Altimetry Over Oceans and Land Surfaces. CRC Press, Boca Raton, Florida (2019)
Veronis, G.: Inverse methods for ocean circulation. In: Abarbanel, H.D.I., Young, W.R. (eds.): General Circulation of the Ocean. Springer-Verlag, New York (1987)

Further Recommended Reading

Cushman-Roisin, B.: Introduction to Geophysical Fluid Dynamics. Prentice-Hall, Englewood Cliffs, New Jersey (1994)
Gill, A.E.: Atmosphere-Ocean Dynamics. Academic Press, New York (1982)
Holton, J.R.: An Introduction to Dynamic Meteorology. Elsevier, Amsterdam (2004)
Marshall, J., Plumb, R.A.: Atmosphere, Ocean, and Climate Dynamics. Elsevier, Amsterdam (2008)
Neumann, G., Pierson jr., W.J.: Principles of Physical Oceanography. Prentice-Hall, Englewood Cliffs, New Jersey (1966)
Pedlosky, J.: Geophysical Fluid Dynamics. Springer-Verlag, New York (1987)
Pond, S., Pickard, G.L.: Introductory Dynamical Oceanography. Pergamon Press, Oxford (1983)
Sverdrup, H.U., Johnson, M.W., Fleming, R.H.: The Oceans, Their Physics, Chemistry and General Biology. Prentice-Hall, New York (1942)
Vallis, G.K.: Atmospheric and Oceanic Fluid Dynamics. Cambridge University Press, Cambridge (2006)
Warren, B.A., Wunsch, C. (ed.): Evolution of Physical Oceanography; Scientific Surveys in Honour of Henry Stommel. The Massachusetts Institute of Technology, Cambridge, Massachusetts (1981)

Chapter 15
Ageostrophic Winds in Atmospheric Boundary Layers

In the planetary boundary layer, due to the interaction of winds with the Earth's surface, turbulent viscosity comes into play; as a result, the winds show a significant deviation from the geostrophic balance. This chapter analyses ageostrophic winds in the planetary boundary layer, both convective and stably stratified, and in the thin surface layer.

15.1 The Planetary Boundary Layer

In the previous chapter, we saw how the failure of one of the conditions required by the Taylor-Proudman theorem (the state of barotropic motion) explains the variation of the geostrophic flow with z. Now we aim to explain the noticeable deviation of the flow from the geostrophic condition (*ageostrophy*) observed in atmospheric and oceanic boundary layers. Here the effect excluded by that theorem, and which now comes into play, is that associated with turbulent viscosity, with particular reference to the diffusive processes that occur vertically. Therefore, in this chapter and the next, the hypothesis $E_v \ll 1$ will be relaxed, considering rather $E_v \approx 1$ (Fig. 13.3c). This chapter discusses ageostrophic flows in the atmospheric boundary layer, also known as the *planetary boundary layer* (PBL).

The PBL is defined as that part of the atmosphere in which the motion is influenced by interaction with the Earth's surface and within which the winds, in geostrophic equilibrium at higher altitudes, change until they cancel out at the surface. This boundary layer is highly turbulent, so eddy viscosity is very important, while molecular viscosity plays a fundamental role only in a very thin viscous layer in contact with the surface.

The structure of the PBL varies significantly during the day. After sunrise and until sunset, especially in the absence of clouds, the solar radiative heat flux warms the Earth's surface which, in turn, heats the overlying air. This produces strong vertical

S. Pierini, *Oceanic and Atmospheric Fluid Dynamics*, UNITEXT for Physics,
https://doi.org/10.1007/978-3-031-77991-6_15

convective motions which, in the presence of unstable or weakly stable stratification, generate vigorous turbulent mixing with the consequent development of a convective mixed layer (also known as the *convective planetary boundary layer*, CPBL), within which the average wind speed and temperature are practically independent of altitude. This layer is bounded above by a thermal inversion zone within which the *entrainment* process takes place. This situation is found at mid and low latitudes especially over land, but also over oceans -particularly at low latitudes- when the sea surface is warmer than the overlying air. A boundary layer of this type can have a vertical extent of $1 - 3\,\text{km}$; on average, at mid-latitudes its thickness is on the order of 1 km. As already pointed out, at higher altitudes there is the *free atmosphere*, where turbulent mixing is negligible and the winds are essentially geostrophic.

On the other hand, at night the outgoing radiative flux cools the air, especially in the absence of clouds: this results in *convective inhibition* that reduces the intensity of turbulence which now, in the absence of convective motions, is supported only by the average wind shear. Consequently, the PBL is in a state of static stability (or neutral stability) and has a smaller thickness than a CPBL, reducing to a few hundred metres. In this case, we speak of a *stably stratified planetary boundary layer* (SPBL). Unlike the CPBL, in an SPBL the average wind speed varies with altitude and connects more gradually to the geostrophic speed in the free atmosphere. Note that SPBLs can also exist during daylight hours in areas where the temperature of the Earth's surface is lower than that of the overlying air; this can occur particularly at high latitudes.

Finally, at lower altitudes, there is a thin, highly turbulent layer (whose thickness is on the order of $100\,\text{m}$), known as the *surface layer*, which is most affected by interaction with the Earth's surface and where the most intense vertical wind shear is recorded, as the wind must vanish at the ground. This layer is supported by an intense vertical turbulent flux of horizontal momentum and dynamically differs from the rest of the PBL, among other things, because the Coriolis force is negligible.

Before considering simple but significant mathematical models capable of describing the most salient features of the CPBL, the SPBL and the surface layer, it is worth focusing on the locally ageostrophic nature of the winds. Fig. 15.1 shows an example of surface winds and the corresponding pressure field: as can be seen, the direction of the winds deviates significantly from the isobars, with a component directed towards the area of lowest pressure. This typical phenomenon has important meteorological implications. In the free atmosphere, winds flow along the isobars, with cyclonic circulation around a low-pressure centre and anticyclonic circulation around a high-pressure centre (Fig. 12.6). At low altitudes, there is instead a mass convergence towards low pressure, where the resulting upward motions can cause masses of humid air to exceed the condensation level and produce clouds, and possibly precipitation. Conversely, around high pressures, there is a mass divergence which, due to the reverse mechanism, is instead associated with scarce precipitation and a sky free from clouds. The reason for this phenomenon will be analysed in the following two paragraphs.

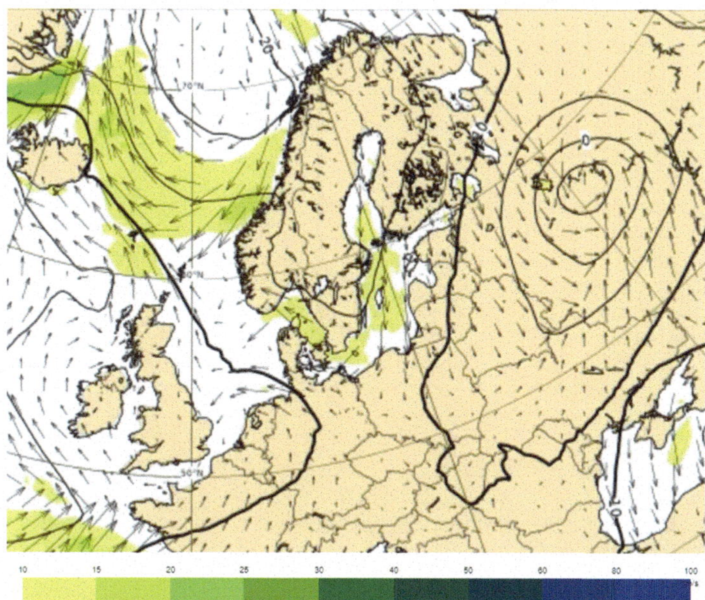

Fig. 15.1 Winds (arrows) and geopotential heights (isolines) at 1000 mbar in Northern Europe from high-resolution forecast for 5 May 2024, 06 UTC (the colour indicates the wind amplitude in m s^{-1}). The corresponding map at 700 mbar is shown in Fig. 12.6. Image obtained from © European Centre for Medium-Range Weather Forecasts (ECMWF, https://www.ecmwf.int/, published under a Creative Commons Attribution 4.0 International, CC BY 4.0)

15.2 Ageostrophic Winds in a Convective Planetary Boundary Layer

In a CPBL, the substantial independence of temperature and wind speed from altitude allows for a particularly simple model to describe the local geostrophic winds. In this case, the boundary layer can be schematised by a fluid slab of thickness d and density ρ_0 in which $\mathbf{u}$ and T are constant while the turbulent flows vary linearly with altitude, cancelling out at $z = d$. In this case, the dynamics can be represented by modifying the geostrophic balance (Eq. 12.21)—to which the flow in the boundary layer is connected above its upper limit—for the presence of a term that represents the effect of turbulent friction:

$$f_0 \mathbf{k} \times \mathbf{u} = -\frac{1}{\rho_0} \nabla_H p + \mathbf{F}_V, \tag{15.1}$$

where ρ_0 is a constant average density and $\mathbf{F}_V$ can be expressed by a *bulk formula* similar to that already encountered in the treatment of the drag exerted on a wing (Eq. 8.7) and another that will be encountered in Sect. 15.4:

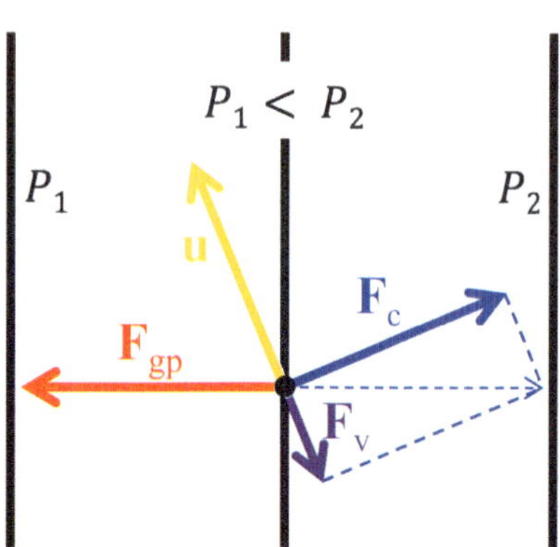

Fig. 15.2 The force balance in a schematic model of a CPBL. $\mathbf{F}_{gp}$, $\mathbf{F}_c$ and $\mathbf{F}_v$ represent, respectively, the pressure gradient force, the Coriolis force (in the northern hemisphere) and the friction force that appear in the dynamic balance (Eq. 15.1). The three black lines indicate isobars; the isobar with lower pressure is the one on the left in the figure

$$\mathbf{F}_V = -\frac{C_D}{d}|\mathbf{u}|\mathbf{u}. \tag{15.2}$$

The drag coefficient C_D has a value on the order of 10^{-3} and can vary whether it is above the sea or above the ground.

Beyond a quantitative analysis, it is useful here to highlight the most salient qualitative aspect. The presence of the friction force introduces a wind component in the direction towards the lowest pressures, as shown by the balance of forces in Fig. 15.2. It is evident that as the drag increases, the angle that the ageostrophic wind forms with the isobars increases.

However, this model is not applicable to a SPBL, in which the wind varies with altitude; therefore, a more sophisticated parameterisation of the effects of turbulence on the average flow, such as that discussed in Sect. 6.4, is required. This analysis will be developed in the next paragraph.

15.3 Ageostrophic Winds in a Stably Stratified Planetary Boundary Layer

In the present discussion, it is assumed that, in the free atmosphere above, the flow is in geostrophic balance with a wind velocity along x: $\mathbf{u}_g = (u_g, 0)$ (it is emphasised that in the present context every direction is equivalent). Therefore, at sufficiently high altitudes the wind velocity will match $\mathbf{u}_g$:

$$\mathbf{u} \to \mathbf{u}_g;\ z \to \infty. \tag{15.3}$$

Furthermore, the Boussinesq approximation can be assumed (Sect. 11.1) as within the PBL the density varies only by $\sim$ 10% compared to that at the surface (cf. Eq. 3.6 and Fig. 3.2). Therefore, for the purposes of the present discussion, in the pressure gradient force ρ can be replaced in the denominator with a constant average value

ρ_0. If a barotropic flow is also assumed, we will have

$$-\frac{1}{\rho_0}\nabla_H p = (0, f_0 u_g), \tag{15.4}$$

where u_g is independent of height ($u_g = const$) in the PBL.

In the case of an SPBL in which horizontal homogeneity is assumed, the equations of motion are given by the geostrophic balance (Eq. 12.21) corrected with the addition of the vertical component of Eq. (12.7),

$$f_0\mathbf{k}\times\mathbf{u} = -\frac{1}{\rho_0}\nabla_H p + K_v\frac{d^2\mathbf{u}}{dz^2}, \tag{15.5}$$

where $\mathbf{u} = (u(z), v(z))$ (the continuity equation for an incompressible fluid is automatically verified), the pressure gradient force is given by Eq. (15.4) and the vertical eddy viscosity coefficient is assumed constant, $K_v = const$. Finally, the velocity must satisfy the no-slip condition on the bottom:

$$\mathbf{u} = 0;\ z = 0. \tag{15.6}$$

Let's rewrite Eq. (15.5) as follows:

$$\frac{d^2u}{dz^2} + bv = 0 \tag{15.7}$$

$$\frac{d^2v}{dz^2} - bu = -bu_g \tag{15.8}$$

where $b = f_0/K_v = const$ (we consider the northern hemisphere, therefore $b > 0$). Thanks to the linearity of the problem, this system of coupled ordinary differential equations can be reduced to a single equation for the complex variable

$$\chi = u + iv,$$

multiplying Eq. (15.8) by the imaginary unit i and adding to Eq. (15.7). In this way we obtain:

$$\frac{d^2\chi}{dz^2} - ib\chi = -ibu_g. \tag{15.9}$$

The solution to this equation is given by the sum of the most general solution of the associated homogeneous equation plus a solution of the non-homogeneous equation:

$$\chi = Ae^{\sqrt{ib}z} + Be^{-\sqrt{ib}z} + u_g, \tag{15.10}$$

where A and B are arbitrary complex constants to be determined imposing the boundary conditions. For the Euler formula, $i = e^{i\pi/2}$, thus $\sqrt{i} = e^{i\pi/4} = (1+i)/\sqrt{2}$; substituting into Eq. (15.10) and defining

$$D^* = \sqrt{2/b} = \sqrt{2K_v/f_0}$$

we get:

$$\chi = Ae^{\frac{z}{D^*}}e^{i\frac{z}{D^*}} + Be^{-\frac{z}{D^*}}e^{-i\frac{z}{D^*}} + u_g. \tag{15.11}$$

For Eq. (15.3), A must be 0 while for Eq. (15.6), $B = -u_g$, from which it follows

$$u + iv = -u_g e^{-\frac{z}{D^*}}\left[\cos\left(\frac{z}{D^*}\right) - i\sin\left(\frac{z}{D^*}\right)\right] + u_g.$$

Equating real and imaginary parts, we finally get the solution of the differential problem given by Eqs. (15.5, 15.3 and 15.6):

$$\begin{cases} u = u_g\left[1 - e^{-\frac{z}{D^*}}\cos\left(\frac{z}{D^*}\right)\right] \\ v = u_g e^{-\frac{z}{D^*}}\sin\left(\frac{z}{D^*}\right) \end{cases} \tag{15.12}$$

This problem was solved in 1905 by the Swedish physicist and oceanographer Vagn Walfrid Ekman (1874-1954); see the following paragraph for an analogous derivation concerning surface ocean currents induced by wind. The wind speed as a function of height is reported in Fig. 15.3 for $u_g = 10\,\mathrm{m\,s^{-1}}$ and $D^* = 500\,\mathrm{m}$ (corresponding to $K_v = 12\,\mathrm{m^2\,s^{-1}}$ at mid-latitudes).

The geostrophic wind exponentially reduces its own amplitude entering the Ekman layer and turning left (in the northern hemisphere, in the opposite direction in the southern hemisphere) and, therefore, towards low pressures. Hence, even for an SPBL, as well as for a CPBL, this fundamental ageostrophic aspect emerges as an effect of turbulent viscosity. Its total contribution can be quantified in terms of the mass transport $\mathbf{Q}$ along y integrated on the vertical,

$$Q_2 = \rho_0 \int_0^\infty v(z)dz, \tag{15.13}$$

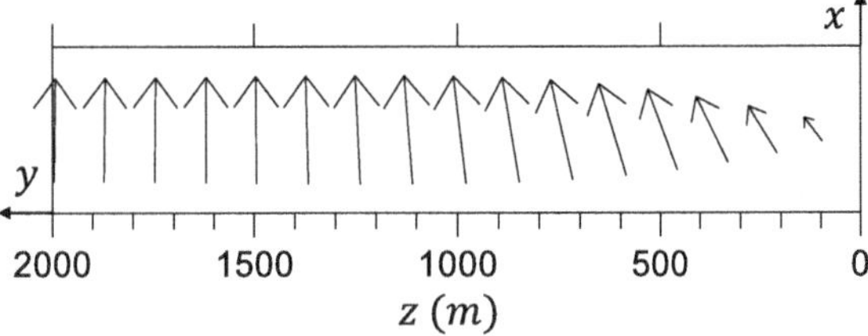

Fig. 15.3 Wind speed in the (x, y) plane as a function of height (m). The map is obtained from Eq. (15.12) with $u_g = 10\,\mathrm{m\,s^{-1}}$ and $D^* = 500\,\mathrm{m}$ ($D_E = 1570\,\mathrm{m}$)

with v given by Eq. (15.12). Defining $\gamma = -z/D^* - \pi/2$ we have:

$$Q_2 = \rho_0 u_g \int_0^\infty e^{-\frac{z}{D^*}} \sin\left(\frac{z}{D^*}\right) dz = \rho_0 u_g D^* e^{\frac{\pi}{2}} \int_{-\pi/2}^{-\infty} e^\gamma \cos\gamma d\gamma.$$

The primitive function $F(\gamma)$ of $f(\gamma) = e^\gamma \cos\gamma$ can be determined by integrating by parts, obtaining $F(\gamma) = e^\gamma(\sin\gamma + \cos\gamma)/2$, from which it follows:

$$Q_2 = \frac{\rho_0 u_g D^*}{2}. \tag{15.14}$$

Note that for $z \to 0$ the angle formed by $\mathbf{u}$ with respect to the direction of the geostrophic wind tends to 45°. Indeed, remembering that for $\varepsilon \ll 1$, $e^\varepsilon \cong 1 + \varepsilon$, in that limit we have: $u \cong u_g(1 - 1 + z/D^*) = u_g z/D^*$ and $v \cong u_g(1 - z/D^*)z/D^* \cong u_g z/D^*$. We also use $D_E^* = \pi D^*$ to denote the thickness of the *bottom Ekman layer*, as at $z = D_E$ the wind is parallel and practically equal to the geostrophic wind.

It is also worth noting that, substituting D^* for D in the expression of the Ekman number in Eq. (12.19) we get $E_v = 0.5 = O(1)$. Therefore, the result of this analysis is (obviously) consistent with the initial hypothesis of turbulent viscosity effects comparable to the Coriolis force. Furthermore, in the example of Fig. 15.3, with $f_0 = 10^{-4}$ rad s^{-1} we have $K_v = 12.5$ m^2 s^{-1}; with this value, Eq. (6.10) allows us to obtain an estimate of the corresponding mixing length averaged over the boundary layer:

$$\langle \ell \rangle \approx \sqrt{\frac{K_v D^*}{u_g}} = 25 \text{ m}.$$

This result is also consistent with the hypothesis (inherent in the very definition of ℓ) that the mixing length is much smaller than the typical vertical scale of motion; in the specific case, in fact, $\ell \ll D^*$.

15.4 Winds in the Surface Layer

We have seen how the Ekman model is able to simulate the ageostrophic veering of winds towards low pressures and the order of magnitude of some parameters involved; however, that model ceases to be valid in the surface layer (Sect. 15.1). Indeed, if it is possible to hypothesise the validity of the flux-gradient relationship ($\mathbf{F}_V = K_v \partial^2 \mathbf{u}/\partial z^2$) even at the lowest levels of the PBL, the hypothesis $K_v = const$ adopted in Ekman's theory is no longer valid, since in the surface layer K_v is found to increase with height.

The parameter that can instead be assumed to be locally constant is the vertical flux of horizontal momentum:

$$\left|\langle u'w'\rangle\right| = u_*^2 \tag{15.15}$$

where we introduce the so-called *friction velocity* u_*; typical values of u_*^2 are of the order of $0.1\,\mathrm{m}^2\,\mathrm{s}^{-2}$. Ultimately, u_*^2 represents the momentum flux that crosses the surface layer towards the earth's surface, where dissipation occurs due to molecular viscosity. Therefore, we will have,

$$u_*^2 = \frac{\tau_{13}}{\rho_0} = K_v \frac{\partial u}{\partial z} = const. \tag{15.16}$$

Another hypothesis based on both measurements and scale analysis suggests that, at these low altitudes, the mixing length increases linearly with z, $\ell = \kappa z$, where κ is called the Von Karman constant and $\kappa \cong 0.4$. With this position, thanks to Eqs. (6.10 and 15.16) provides

$$\frac{\partial u}{\partial z} = \frac{u_*}{\kappa z}, \tag{15.17}$$

from which derives the logarithmic profile of u in the surface layer (Fig. 15.4):

$$u = \frac{u_*}{\kappa} \ln \frac{z}{z_0}. \tag{15.18}$$

In Sect. 10.2 we saw how this profile plays an important role in the generation of gravity waves through the Miles mechanism. Combining Eqs. (15.16 and 15.18) we get

$$K_v = u_* \kappa z = u_* \ell. \tag{15.19}$$

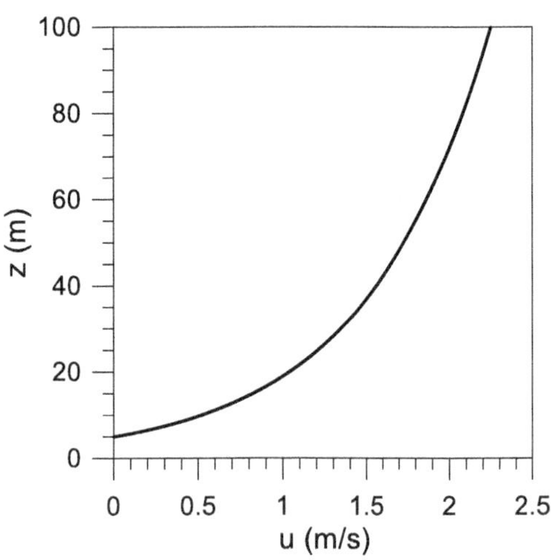

Fig. 15.4 Logarithmic wind profile in the surface layer. In this example, $u_* = 0.3\,\mathrm{m\,s}^{-1}$ and $z_0 = 5\,\mathrm{m}$ in Eq. (15.18)

Therefore, in the surface layer both the mixing length and the vertical eddy viscosity coefficient increase linearly with height.

The arbitrary constant z_0 is called *roughness length* and represents the height below which the formula is not applicable due to the roughness of the surface. This length can vary from mm to a few cm for flat surfaces such as calm sea, icy surfaces, regular grassy surfaces, it can be a few *dm* on cultivated fields, in the presence of small plants, etc., but it can reach values of a few m or even many m in urban centres, forests, etc.. In the case of the sea surface, z_0 obviously depends on the state of the wave field, which in turn depends on u_* (Sect. 10.2). In the theory of air-sea interaction, z_0 is parameterised using the empirical formula (Charnock 1981)

$$z_0 = \alpha_1 u_*^2 / g \tag{15.20}$$

where $\alpha_1 \cong 0.014$.

Finally, following the same empirical approach of Eq. (8.7), which allows the aerodynamic drag to be parameterised by introducing a drag coefficient multiplied by the square of a velocity, in this context the stress τ_{13} (which, remember, is constant in the surface layer) is estimated using the bulk formula

$$\tau_{13}^{(s)} = C_D \rho_0 u_{10}^2 \tag{15.21}$$

((s) stands for *surface* stress) where u_{10} is the wind speed evaluated at *anemometric height*, traditionally chosen at $z = 10\ m$. The dimensionless drag coefficient C_D varies depending on both the degree of static stability and z_0. On the ground, with neutral static stability, it ranges from $C_D \approx 2 \times 10^{-3}$ for z_0 of the order of *mm* to $C_D \approx 2 \times 10^{-2}$ for z_0 of the order of *m*. At sea, C_D itself depends, albeit weakly, on u_{10}, as z_0 depends on u_* (Eq. 15.20) and has values in the range $C_D \approx 1 \div 3 \times 10^{-3}$. Eq. (15.21) is very important as it allows the wind stress to be easily estimated in order to prescribe the momentum flux (thus the effect of the wind) on the sea surface in both idealised models (for example, like the one that will be discussed in Par. 16.1) and in general ocean circulation or wave models; in these cases, u_{10} is obtained from the wind velocity simulated at the lowest vertical level by a meteorological circulation model.

Bibliography

Charnock, A.: Air-sea interaction. In Warren, B.A., Wunsch, C. (eds.): Evolution of Physical Oceanography; Scientific Surveys in Honour of Henry Stommel. The Massachusetts Institute of Technology, Cambridge, Massachusetts (1981)

Further Recommended Reading

Cushman-Roisin, B.: Introduction to Geophysical Fluid Dynamics. Prentice-Hall, Englewood Cliffs, New Jersey (1994)

Dobbins, R.A.: Atmospheric Motion and Air Pollution. John Wiley & Sons, New York (1979)

Gill, A.E.: Atmosphere-Ocean Dynamics. Academic Press, New York (1982)

Holton, J.R.: An Introduction to Dynamic Meteorology. Elsevier, Amsterdam (2004)

Kalnay, E.: Atmospheric Modelling, Data Assimilation and Predictability. Cambridge University Press, Cambridge (2003)

Marshall, J., Plumb, R.A.: Atmosphere, Ocean, and Climate Dynamics. Elsevier, Amsterdam (2008)

Neelin, J.D.: Climate Change and Climate Modelling. Cambridge University Press, Cambridge (2011)

Vallis, G.K.: Atmospheric and Oceanic Fluid Dynamics. Cambridge University Press, Cambridge (2006)

Wallace, J.M., Hobbs, P.V.: Atmospheric Science, an Introductory Survey. Elsevier, Amsterdam (2006)

Chapter 16
Ageostrophic Currents in Oceanic Boundary Layers

Similarly to what happens in the planetary boundary layer, in the surface and bottom oceanic boundary layers, turbulent viscosity plays a fundamental role. This chapter analyses the ageostrophic currents present in these two boundary layers. In addition, the mechanisms through which surface ageostrophic currents indirectly generate geostrophic currents, both in the presence of a coast and in the open sea, are analyzed.

16.1 Oceanic Surface Ekman Layer

During the pioneering polar expedition of the explorer and Norwegian scientist Fridtjof Nansen (1861–1930) carried out with the ship Fram in 1893–96, it was observed that the movement of icebergs was clearly correlated with the prevailing wind and that their direction systematically formed an angle between 20° and 40° to the right of that of the wind. Considering that the submerged part of the icebergs embraces about 90% of their vertical extension and that, moreover, the wind drag is much less than that of the water for the same surface area, it was clear that the direction of the iceberg movement was due to surface marine currents. These had to be consequently generated by the wind and their direction had to deflect to the right of the wind itself.

At the beginning of the 1900s, Nansen presented the problem of this deviation to the physicist and meteorologist Norwegian Vilhelm Bjerknes (1862–1951). He in turn commissioned the Swedish Vagn Walfrid Ekman (already mentioned in Sect. 15.3), then his Ph.D. student, to analyse the effect of the Coriolis deflection force on the generation of surface marine currents by the wind. Ekman (1905) therefore developed an analytical theory of marine currents in the surface and bottom boundary layers. That concerning the bottom has already been treated in the atmospheric context in Sect. 15.3 and will be resumed in Sect. 16.3; in this paragraph the theory of Ekman's wind-generated surface marine currents (also called *drift currents*) will be described.

S. Pierini, *Oceanic and Atmospheric Fluid Dynamics*, UNITEXT for Physics,
https://doi.org/10.1007/978-3-031-77991-6_16

As for the SPBL discussed in Sect. 15.3, here too one can assume $K_v = const$; furthermore, a horizontally homogeneous flow ($\mathbf{u} = (u(z), v(z))$) is considered as a result of the further assumptions that the wind stress is spatially (and also temporally) constant and that the ocean is infinitely extended. In addition, the horizontal pressure gradient force is omitted, since now the Ekman ageostrophic current does not have to connect to a geostrophic current. Therefore, Eq. (15.5) is reduced to

$$f_0 \mathbf{k} \times \mathbf{u} = K_v \frac{d^2\mathbf{u}}{dz^2}. \tag{16.1}$$

The derivation of the solution of this system of two coupled linear ordinary differential equations follows that described in Sect. 15.3. Here Eq. (15.11) is reduced to

$$u + iv = Ae^{\frac{z}{D}}e^{i\frac{z}{D}} + Be^{-\frac{z}{D}}e^{-i\frac{z}{D}} \tag{16.2}$$

where

$$D = \sqrt{2K_v/f_0}.$$

Note that, despite the analytical expression of D being the same as the D^* of Ekman's theory for the bottom boundary layer, two distinct symbols are used to underline the substantial difference in the value that K_v assumes in the two layers.

As for the boundary conditions, it is assumed that the ocean is infinitely deep,

$$\mathbf{u} \rightarrow 0; \quad z \rightarrow -\infty,$$

which implies $B = 0$. By setting $A = Ve^{ic}$, with V and c real constants to be determined based on the boundary condition on the surface, Eq. (16.2). gives:

$$\begin{cases} u = Ve^{\frac{z}{D}} \cos(z/D + c) \\ v = Ve^{\frac{z}{D}} \sin(z/D + c) \end{cases} \tag{16.3}$$

At $z = 0$, the continuity of stress must be imposed: assuming that the wind flows along y, we have:

$$\begin{cases} \left(\rho_w K_v \frac{\partial u}{\partial z}\right)_{z=0} = 0 \\ \left(\rho_w K_v \frac{\partial v}{\partial z}\right)_{z=0} = \tau_{23}^{(s)} \end{cases} \tag{16.4}$$

where in the left-hand sides appears the sea stress just below the surface (ρ_w is the density of sea water, assumed constant) and in the right-hand side along y the wind stress (calculable using Eq. 15.21). Considering the first of the Eq. (16.4), from the first of the Eq. (16.3), we obtain $c = \pi/4$. Then applying the condition along y, we finally get,

$$\begin{cases} u = Ve^{\frac{z}{D}} \cos(z/D + \pi/4) \\ v = Ve^{\frac{z}{D}} \sin(z/D + \pi/4) \end{cases} \tag{16.5}$$

where

$$V = \frac{\tau_{23}^{(s)}}{\sqrt{\rho_w^2 K_v f_0}}. \tag{16.6}$$

The velocity field as a function of z is shown in Figs. 16.1 and 16.2. As the depth increases, $\mathbf{u}$ decreases exponentially in amplitude and rotates clockwise, describing, in projection on the horizontal plane, a curve known as the *Ekman spiral* (Fig. 16.2). On the surface, the speed is in modulus equal to V and forms an angle of 45° with the wind direction to the right in the northern hemisphere (in the southern hemisphere f_0 would appear in modulus in the formulas and the spiral would be flipped by 90°). This result is therefore compatible with the deflection relative to the wind direction observed in the motion of icebergs. At $z = -D_E = -\pi D$, the current speed is opposite to the surface one and has a very small amplitude ($|\mathbf{u}| = Ve^{-\pi} \approx V/20$); D_E is called the thickness of the *surface Ekman layer* (note that D and D_E correspond to the scales D^* e D_E^* of the bottom Ekman layer derived in Sect. 15.3).

Note that V is proportional to the wind stress, that $D_E \propto K_v^{1/2}$ (a more intense turbulence transfers to greater depths the momentum received from the wind stress on the surface), while $V \propto K_v^{-1/2}$ (for the same wind stress, the surface speed decreases as a result of a more extensive distribution of momentum in depth). It is also interesting to note that the thickness of the Ekman layer does not depend on the wind stress.

However, it should be stressed that none of the assumptions on which the theory is based are fully justified. The state of turbulence in the oceanic surface layer is only very approximately described by $K_v = const$. Stratification certainly does not play the crucial role it has in geostrophic currents (Sect. 14.3), however, it can affect K_v as it inhibits vertical motions and therefore the turbulent vertical transfer of momentum.

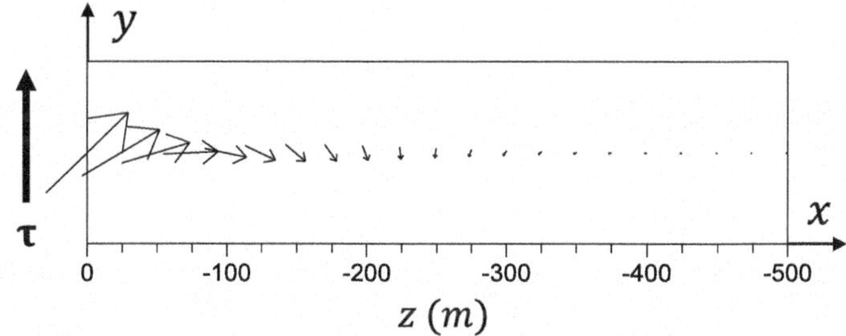

Fig. 16.1 Velocity of the Ekman surface current in the (x, y) plane as a function of depth (in m, northern hemisphere). The surface wind flows along the positive y-direction (thick arrow). The map is obtained from Eq. (16.5) with $V = 0.2\,\mathrm{m\,s^{-1}}$ and $D = 100\,\mathrm{m}$. For the amplitude of the arrows, see Fig. 16.2.

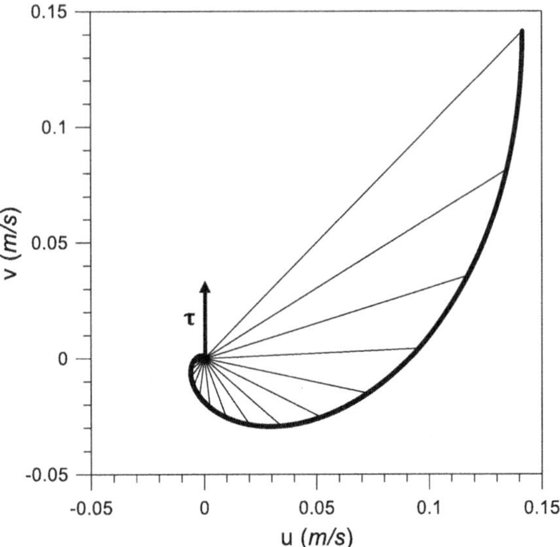

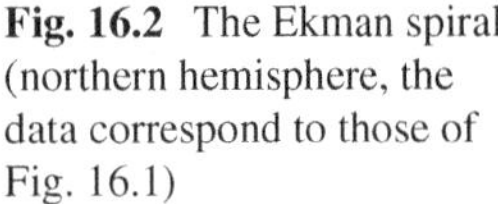
Fig. 16.2 The Ekman spiral (northern hemisphere, the data correspond to those of Fig. 16.1)

Moreover, K_v is likely to depend on the wind -and therefore on the stress τ-while in Ekman's theory K_v and τ are independent of each other. The depth of the sea, if less than D_E, and the presence of coasts have a significant impact on the vertical structure of **u**, Ekman himself extended the analysis presented in this paragraph to take into account these effects (in Sects. 16.4 and 16.5 the important effect of the presence of a coast in an idealised context will be briefly considered).

Despite the simplicity of Ekman's theory, the values of D_E and V derived from it are in reasonable agreement with the observations. As an example, consider the values $K_v = 0.01\,\text{m}^2\,\text{s}^{-1}$, $f_0 = 10^{-4}\,\text{rad}\,\text{s}^{-1}(\varphi \cong 45°)$, $\rho_w = 1028\,\text{kg}\,\text{m}^{-3}$ and, in Eq. (15.21), $C_D = 1.5 \times 10^{-3}$, $\rho_0 = 1.29\,\text{kg}\,\text{m}^{-3}$ and $u_{10} = 10\,\text{m}\,\text{s}^{-1}$. In this way, we obtain $\tau^{(s)} \cong 0.19\,\text{Pa}$, $D_E \cong 45\,\text{m}$ and $V \cong 0.19\,\text{m}\,\text{s}^{-1}$. These values of D_E and V are indeed realistic; in real oceans D_E can vary, depending on latitude, from even lower values up to thicknesses of the order of 100 m or more.

As for the direction of the Ekman current at surface and its deflection with depth expressed by Eq. (16.5), it is very difficult to verify their validity in real oceans. The current theory is valid in deep oceans far from coasts where it is difficult to make direct measurements of currents, even at shallow depths; moreover, the surface current includes the component generated by the wind but also the geostrophic one, and it is problematic to separate the two components. Despite these experimental limitations, there is evidence of a rather significant agreement with the theory.

In this context, the vertically integrated mass transport is of fundamental importance, as it represents a characteristic of the Ekman current more robust than its detailed vertical structure. This mass transport is defined as

$$\mathbf{S} = \rho_w \int_{-\infty}^{0} \mathbf{u}(z)dz. \tag{16.7}$$

With the first of Eq. (16.5) and defining $\gamma = z/D + \pi/4$ we have

$$S_1 = \rho_w V \int_{-\infty}^{0} e^{\frac{z}{D}} \cos\left(\frac{z}{D} + \frac{\pi}{4}\right)dz = \rho_w V D e^{-\frac{\pi}{4}} \int_{-\infty}^{\pi/4} e^{\gamma} \cos\gamma d\gamma.$$

Retracing the calculation of the ageostrophic transport Q_2 (in that case along y) in the SPBL (Eq. 15.14) we obtain

$$S_1 = \frac{\tau_{23}^{(s)}}{f_0}; \; S_2 = 0 \tag{16.8}$$

(the vanishing of **S** along y is obtained with a similar calculation). Therefore, the mass transport of the Ekman current is perpendicular to the wind and is directed to the right (left) in the northern (southern) hemisphere, as shown in Fig. 16.3. Consequently, the total Coriolis force is in turn perpendicular—and therefore antiparallel to the wind stress—thus ensuring the balance of the two forces and a non-accelerated flow.

Equation (16.7) can be written as follows:

$$\mathbf{S} \cong \rho_w D_E \overline{\mathbf{u}}_E^{(s)} \tag{16.9}$$

where

$$\overline{\mathbf{u}}_E^{(s)} = \frac{1}{D_E} \int_{-D_E}^{0} \mathbf{u}(z)dz \tag{16.10}$$

is the surface Ekman current averaged over the thickness of the layer D_E; therefore, $D_E \overline{\mathbf{u}}_E^{(s)}$ represents the net volume transport within the layer. For a wind that flows in

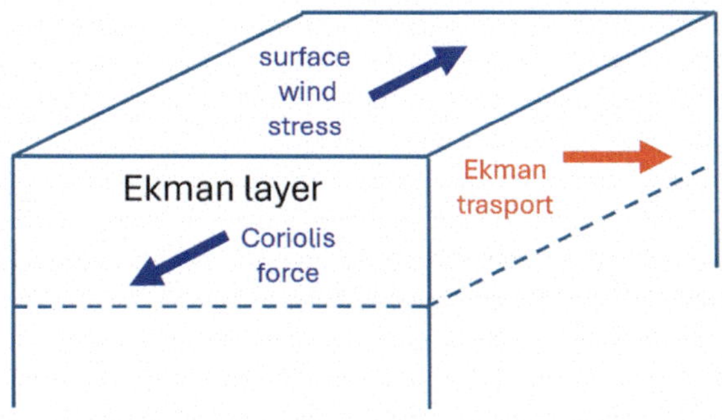

Fig. 16.3 The mass transport of the Ekman current and the balance of forces (northern hemisphere)

a generic direction, Eq. (16.8) generalises to

$$\mathbf{S} = \left(\frac{\tau_{23}^{(s)}}{f_0}, -\frac{\tau_{13}^{(s)}}{f_0} \right). \tag{16.11}$$

This relationship can also be rewritten as

$$f_0 \mathbf{k} \times \overline{\mathbf{u}}_E^{(s)} = \frac{\boldsymbol{\tau}^{(s)}}{\rho_w D_E} \tag{16.12}$$

where $\boldsymbol{\tau}^{(s)} = \left(\tau_{13}^{(s)}, \tau_{23}^{(s)} \right)$.

It is interesting to note that Eq. (16.12) can be easily obtained without going through the explicit derivation of the Ekman currents, but simply by vertically integrating the Ekman balance and imposing the continuity condition of stress. Referring to the divergence of the Reynolds stress tensor that appears in Eq. (6.7) and considering Eq. (6.9), the balance given by Eq. (16.1) can be reformulated as follows (the symbol of partial derivative is reintroduced in view of a dependence of the wind stress, and therefore of $\mathbf{u}_E$, also from the horizontal coordinates, see Sect. 17.1):

$$f_0 \mathbf{k} \times \mathbf{u} = \frac{\partial}{\partial z} \left(K_v \frac{\partial \mathbf{u}}{\partial z} \right). \tag{16.13}$$

Vertically averaging from $z = -D_E$ to $z = 0$ (see Eq. 16.10) and considering that at the base of the Ekman layer the speed is practically null, we obtain:

$$f_0 \mathbf{k} \times \overline{\mathbf{u}}_E^{(s)} = \frac{1}{D_E} \left(K_v \frac{\partial \mathbf{u}}{\partial z} \right)_{z=0}. \tag{16.14}$$

Now imposing the continuity condition of stress (Eq. 16.4) with the wind stress of generic direction, we obtain again Eq. (16.12).

16.2 Inertia Currents

In this paragraph, we will hint at the transient phase that leads to the formation of Ekman currents, without going into the detail of the time-dependent problem. In Sect. 12.4, we referred to the transient process that leads to stationary geostrophic flows (geostrophic adjustment). It is worth asking also for the Ekman currents what is the transient process that leads to the steady flow analysed in the previous paragraph. Imagine a wind stress initially null that grows until it adjusts on a constant value: the asymptotic situation is described by Eq. (16.5), but in the transient phase, the so-called *inertia currents* are excited.

Inertia currents are accelerated currents described by the following equations:

$$\begin{cases} \frac{du}{dt} - f_0 v = 0 \\ \frac{dv}{dt} + f_0 u = 0 \end{cases} \tag{16.15}$$

These currents therefore see a perfect balance between the acceleration of the fluid element and the Coriolis force (the pressure gradient force and viscous effects do not contribute to this dynamics). Multiplying the first equation by u, the second by v and adding, we obtain (for convenience of notation we put $|\mathbf{u}| = V$):

$$\frac{dV^2}{dt} = 0. \tag{16.16}$$

Therefore, such currents have a constant speed. If instead we multiply the first equation by v, the second by u and subtract, we obtain:

$$v^2 \frac{d}{dt}\left(\frac{u}{v}\right) = f_0 V^2.$$

But $u/v = \cot\alpha$, where α is the angle formed by $\mathbf{u}$ with respect to the x-direction, so $v^2 = V^2 \sin^2\alpha$. Consequently

$$\frac{d \cot\alpha}{dt} = \frac{f_0}{\sin^2\alpha},$$

from which it follows

$$\frac{d\alpha}{dt} = -f_0. \tag{16.17}$$

This implies that every fluid element, wherever it is in the (x, y) plane, is subject to a uniform circular motion with anticyclonic rotation (thus with clockwise—anti-clockwise—motion in the northern—southern—hemisphere), linear speed V and period $T_i = 2\pi/f_0$: this is precisely the inertial period introduced in Sect. 12.3 and is equal to half the rotation period $T = 2\pi/\Omega$ of the reference frame (or $T = 2\pi/\Omega_{en}(\varphi)$ in the case of the plane tangent to the Earth at latitude φ). The radius of the circle is called *radius of inertia* r_i and can be determined by equating the centrifugal force V^2/r_i to the Coriolis force $f_0 V$, which act perpendicularly—and in the opposite direction—to the velocity of the fluid element, balancing each other. Therefore,

$$r_i = \frac{V}{f_0}. \tag{16.18}$$

The signal of inertia currents is omnipresent in time series of marine currents obtained from current measurements and manifests itself with oscillations of the velocity components with amplitudes of $10 - 20\,\text{cm s}^{-1}$, or more, and with a peak at $\sigma = f_0$ in the corresponding spectrum; naturally, the amplitude of the peak depends on

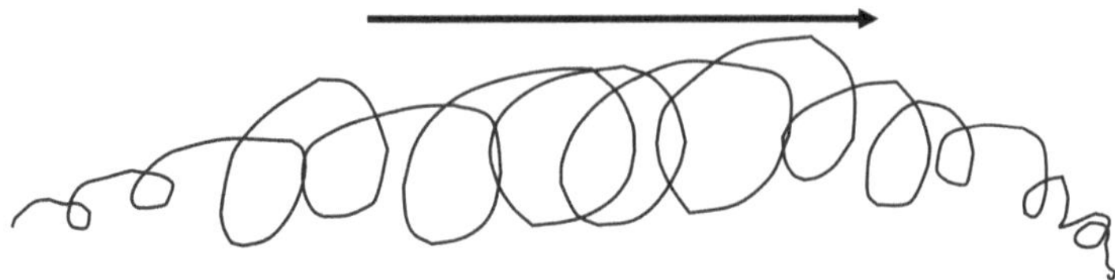

Fig. 16.4 Schematic representation of the trajectory of a fluid element resulting from the superposition of a mean flow (from left to right) and inertial oscillations (northern hemisphere)

the variations of the surface wind. Given that, as already seen, at $\varphi \cong 30°$, $T_i \cong 24h$, so, at that latitude the inertial peak can be confused with the diurnal frequencies of the tidal component if the time series is not long enough.

The presence of inertia currents is also revealed by Lagrangian measurements carried out with drifting buoys. In Fig. 16.4 a schematic representation of a drifter's trajectory is shown, in which it is possible to distinguish a mean component that carries the drifter (from left to right) and inertia circles, whose radius, initially increasing, is an indication of an intensification of the wind. The subsequent decrease in r_i indicates the dissipation of the previously generated inertia currents. Both Eq. (16.18) that links r_i to V, and the inertial period are accurately verified by experimental data. As an order of magnitude, for a typical speed of $V = 10\,\mathrm{cm\,s^{-1}}$, at mid-latitudes $r_i \cong 1$ km.

Finally, it is useful to mention the relationship that exists between inertia currents and long gravity waves. The latter change substantially for very large wavelengths, and therefore for very large periods (very small frequencies). For these waves, called *Poincaré waves* (or *inertia-gravity waves*) the following dispersion relation can be demonstrated that generalises Eq. (9.24):

$$\sigma^2 = gDk^2 + f_0^2. \tag{16.19}$$

For wavelengths $\lambda \gg R_e$, where

$$R_e = \frac{\sqrt{gD}}{f_0} \tag{16.20}$$

is the so-called *external Rossby deformation radius*, σ tends to identify with f_0, which therefore represents the lower limit for this wave field (this result can also be extended to internal waves). In the limit $\sigma \to f_0$, the plane of polarisation on which the fluid motion takes place goes from being vertical for $\lambda \ll R_e$ (gravity waves in shallow water without the effect of the Coriolis force) to the horizontal plane. At that point, the effect of the restoring force due to gravity ceases and we are reduced precisely to inertia currents; in short, these can be seen as the degeneration of Poincaré waves in the aforementioned limit.

16.3 Oceanic Bottom Ekman Layer

The inclination of the free surface is associated with an absolute pressure field and therefore with a geostrophic current which, in a state of barotropic motion, is independent of depth. It has also been seen (Sect. 14.3) how stratification, if associated with a state of baroclinic motion, involves a relative pressure gradient force that induces a vertical variation of geostrophic currents (the phenomenon of thermal wind). In some cases, especially in the deep ocean, this effect results in the substantial cancellation of currents well above the seabed. However, this is not always the case.

In the deep oceans, there can be currents of dense water (such as, for example, the North Atlantic Deep Water in the North Atlantic) that flow over the seabed. Currents of this type are also present in the Mediterranean Sea. On the continental shelf, non-zero bottom currents are more the norm than the exception. Therefore, in these cases, the geostrophic current near the seabed will have to rapidly decrease within the boundary layer until it cancels out to satisfy the no-slip condition, but in doing so it will deflect towards low pressures. This is the bottom Ekman layer analysed in the different context of an SPBL and described by Eq. (15.12); the same formulas are also valid for the ocean (for which they were, in fact, originally derived) provided that its depth is $d \gg D^*$. In particular, the ageostrophic transport Q in the direction perpendicular to the geostrophic current (to the left in the northern hemisphere, to the right in the southern) is given by Eq. (15.14):

$$Q = \frac{\rho_w u_g D^*}{2}. \tag{16.21}$$

Ekman also considered cases in which this condition is not met. In the limit where the depth is less than the theoretical thickness of the layer, the effect of the Coriolis force tends to be negligible compared to friction forces, resulting in the disappearance of geostrophic balance. In this case, the flow above the boundary layer will tend to be oriented along the pressure gradient rather than along the isobars.

16.4 Generation of Geostrophic Currents from Ageostrophic Flows

We are now able to briefly discuss the two main mechanisms responsible for the generation of geostrophic currents by the wind. The first mechanism requires the existence of a coast, the second instead does not depend on the presence of coasts but requires that the wind has a spatial variation. Obviously, both mechanisms act in combination in semi-enclosed or coastal seas (e.g., this is the norm in Mediterranean sub-basins).

In the first generation mechanism (Fig. 16.5) one can for simplicity assume that a wind (blue arrow), constant in time and space, flows parallel to a straight coast to its

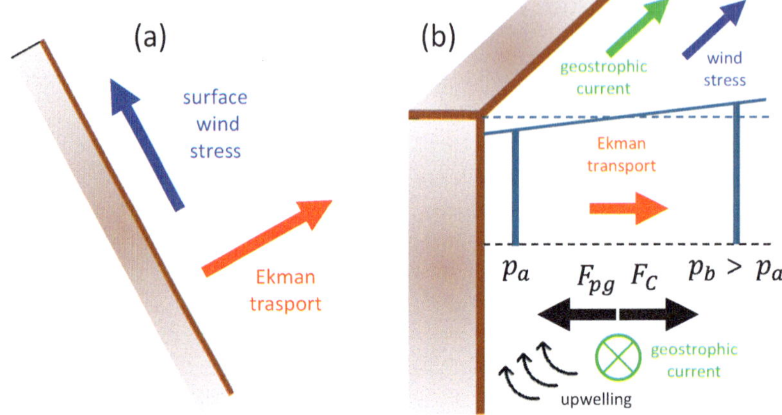

Fig. 16.5 Scheme of generation of geostrophic currents by the wind along a straight coast in a homogeneous ocean (northern hemisphere; the vertical displacement of the free surface in (b) is not to scale)

left in the northern hemisphere. The ageostrophic currents thus generated produce a transport **S** in the surface Ekman layer perpendicular to the coast (red arrow): this has two effects. The first is that the continuous mass flow that moves water from the coast in the Ekman layer must be balanced by an influx of deep waters to the surface, called *upwelling* (Fig. 16.5b). This phenomenon has a significant ecological relevance, as nutrient-rich waters are introduced into the photic zone increasing primary production, with important consequences for the local fish fauna. The second effect involves a tilt of the free surface, which will adjust to an asymptotically constant value as shown in the figure: this tilt will be associated with an absolute pressure gradient (independent of depth) that will produce a barotropic geostrophic current (if the ocean is homogeneous) directed along the coast and with the same direction as the wind (green arrow).

In summary, the presence of the coast induces (i) a divergence of Ekman transport, (ii) a consequent tilt of the sea surface, (iii) a horizontal pressure gradient and, therefore, (iv) a geostrophic current. It is clear that if the wind blows with the coast to its right the opposite scenario emerges: downwelling, sea level higher near the coast, etc.. It should be emphasised that if the wind is not parallel to the coast the phenomenon still occurs, but only the component of the wind parallel to the coast itself will count.

The second generation mechanism is schematically described in Fig. 16.6. Consider a stratified sea and a surface wind stress field as represented by the blue arrows: this will produce a surface transport in the Ekman layer indicated by the red arrows and a local mass accumulation, with consequent displacement of the free surface upwards and the production of a downwelling associated with a downward vertical velocity at the base of the Ekman layer (purple arrow), called *Ekman pumping*. The effect is to move the isopycnals downwards producing an inclination

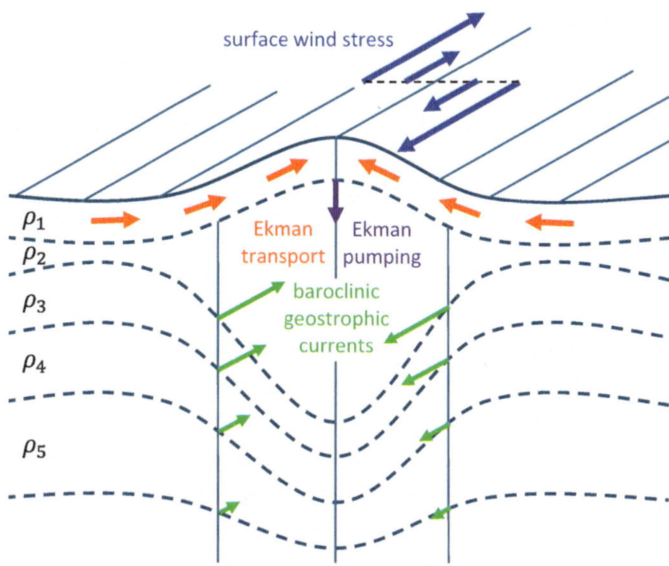

Fig. 16.6 Scheme of generation of geostrophic currents by the wind in the open sea (northern hemisphere; vertical displacements are not to scale)

opposite to that of the free surface (but with angles even three orders of magnitude larger). Geostrophic currents will be associated with the surface topography, represented in the figure by the green arrows, with decreasing amplitude with depth due to the baroclinic compensation.

With a system of winds in the opposite direction, there would be a lowering of the free surface, an upward vertical speed (called *Ekman suction*), a raising of the isopycnals and therefore opposite inclinations and currents to those shown in the figure, but the baroclinic compensation would remain. It should be noted that, both in this example and in the previous one (Fig. 16.5), the parallelism between winds and geostrophic currents is due to the particular symmetry of the cases considered, this is not at all a general characteristic of the geostrophic circulation indirectly generated by the wind. Moreover, the direction of the wind is irrelevant. However, the example shown in Fig. 16.6 can well represent schematically the predominantly zonal circulation in the oceanic interior at medium and low latitudes: in this case the direction of the winds, as well as the currents, is zonal, with a latitude of convergence around 30°N, in which case the winds to the south represent the Trade Winds and those to the North the Westerlies (naturally, in an ocean limited by continental margins there will also be a meridional component of the geostrophic current, as discussed in Sect. 20.3).

In summary, the convergence (divergence) of Ekman transport in the open sea generates a raising (lowering) of the free surface, an Ekman pumping (suction) at the base of the Ekman layer and a deformation of the isopycnals, whose corresponding field of relative pressure gradient force, superimposed on the absolute one associated with the surface topography, produces baroclinic geostrophic currents that decrease with depth. It is important to note that, for this scenario to occur it is not necessary for the wind to change direction, as happens in this example: it is enough that the wind stress presents a spatial variation. A simple analytical treatment will clarify this aspect.

It is recalled that $\overline{\mathbf{u}}_E^{(s)}$ (Eq. 16.10) represents the vertical average, within the surface Ekman layer of thickness D_E, of the current given by Eq. (16.5) (now also dependent on the horizontal coordinates). Integrating over the Ekman layer the equation of continuity for an incompressible fluid, we get

$$\int_{-D_E}^{0} \frac{\partial w(x, y, z)}{\partial z} dz = w(x, y, -D_E) = D_E \nabla_H \cdot \overline{\mathbf{u}}_E^{(s)} \qquad (16.22)$$

as w must be zero at the surface. Equation (16.12) provides

$$\frac{1}{\rho_w f_0}\left(\tau_{23}^{(s)}, -\tau_{13}^{(s)}\right) = D_E \overline{\mathbf{u}}_E^{(s)}.$$

Taking the horizontal divergence of this expression and considering Eq. (16.22) we get

$$\frac{1}{\rho_w f_0}\left(\nabla \times \boldsymbol{\tau}^{(s)}\right)_3 = \nabla_H \cdot \left(D_E \overline{\mathbf{u}}_E^{(s)}\right) = w_E \qquad (16.23)$$

where $w_E = w(x, y, -D_E)$.

This expression is full of meaning. First of all, the divergence of the Ekman volume transport is equal to the vertical velocity w_E at the base of the Ekman layer. Moreover, such divergence is proportional to the curl (more correctly, to the vertical component of the curl, the only non-zero one) of the surface wind stress. Therefore, if the curl is negative (anticyclonic atmospheric circulation, clockwise in the northern hemisphere) there is an Ekman pumping ($w_E < 0$), to which corresponds a downwelling (the example of Fig. 16.6). On the contrary, if the curl is positive there is an Ekman suction ($w_E > 0$), to which corresponds an upwelling. In Sect. 20.3 it will be seen that the isoline $(\nabla \times \boldsymbol{\tau})_3 = 0$ enjoys an important property in the context of the wind-driven ocean circulation. Finally, as for the order of magnitude of w_E, consider that the average surface wind stress records a variation of $\sim$ 0.1 Pa over a latitude variation of $\Delta\varphi \sim 10°$, corresponding to $\sim$ 1000 km. Substituting these values into Eq. (16.23) we obtain, at mid-latitudes, $|w_E| \sim 10^{-6}\,\mathrm{m\,s^{-1}} \sim 30\,\mathrm{m\,y^{-1}}$. Although the vertical volume flux corresponding to this speed appears very small, it is still much greater than that, for example, associated with average evaporation and precipitation, which instead is of the order of $\sim 1\,\mathrm{m\,y^{-1}}$.

16.5 Elementary Current System in a Coastal Ocean

This chapter concludes by considering the combination of the surface and bottom Ekman ageostrophic currents and the geostrophic one in the coastal situation described in the previous paragraph; this combination is known as the *elementary current system*. Consider a straight coast, a flat bottom, $d \gg D_E, D^*$ and a homogeneous sea where, therefore, only the absolute pressure gradient and barotropic geostrophic currents are present. This situation is very particular; for example, at mid-latitudes it can occur in autumn or winter in a coastal sea in the presence of strong wind-induced mixing, which makes the density almost constant on the continental shelf. However, the value of this example lies in its usefulness in summarising in a simple way the various components of the coastal circulation, rather than in its applicability to real situations.

In Fig. 16.7 the geostrophic (green arrows) and ageostrophic (red and purple arrows) transports are schematically shown in the three layers in which it is possible to divide the dynamics vertically. In layer 1 (surface Ekman layer) both the Ekman transport (red arrow) and the transport of the geostrophic current (green arrow) associated with the inclination of the free surface are present. Layer 2 represents the *deep layer*, in which only the barotropic geostrophic current is present, equal to that in layer 1 (the length of the green arrow is greater than that in 1 due to the greater thickness of the deep layer). Finally, in layer 3 (bottom Ekman layer) the geostrophic current tends to zero, generating an ageostrophic transport (purple arrow) equal and opposite to the surface one.

Figure 16.8 shows the current velocity as a function of depth. On the surface, the vector sum of the geostrophic and Ekman current gives a resultant (black arrow) whose angle with respect to the wind direction is less than 45°. The green arrow indicates the purely geostrophic current within the deep layer. Finally, the current cancels out at the bottom following the profile shown in Fig. 15.3.

From Eq. (16.8) we have $S = \tau^{(s)}/f_0$; moreover, from Eq. (16.21) we have $Q = \rho_w u_g D^*/2$. The stationarity of the system requires that these transports are equal and opposite, therefore:

$$u_g = \frac{2\tau^{(s)}}{f_0 \rho_w D^*}. \tag{16.24}$$

This formula establishes a linear relationship between wind stress and the geostrophic current it generates through the mechanism described in Sect. 16.4. This is the simplest quantitative relationship between wind stress and geostrophic current that can be imagined but, as already pointed out, its applicative relevance is limited due to the many assumptions required for its derivation. This relationship is nevertheless very significant for its conceptual value, because it includes some essential dynamic aspects of the coastal circulation induced by the wind.

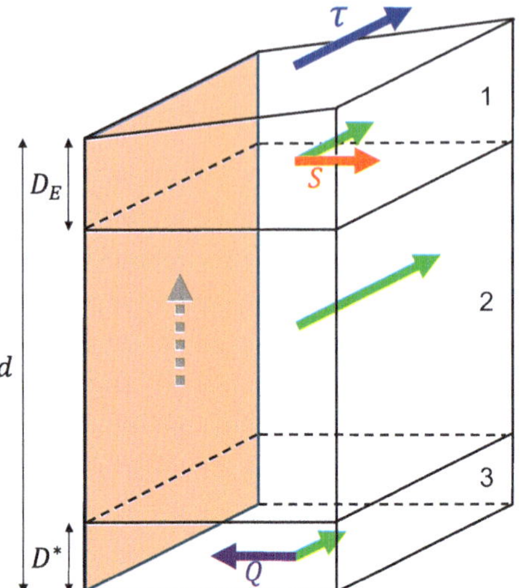

Fig. 16.7 Scheme of the elementary current system in a constant density coastal ocean. Layers 1, 2 and 3 represent, respectively, the surface Ekman layer, the deep layer and the bottom Ekman layer. The blue arrow represents the surface wind stress (parallel to the coast), the green arrows the transport of the geostrophic current in each layer and the red and purple arrows the Ekman transport in the surface and bottom layers. Finally, the dashed grey arrow indicates the upwelling necessary to balance Ekman transports. The thicknesses of the layers and the lengths of the arrows are only indicative, but the $\mathbf{S} = -\mathbf{Q}$ condition for the system to be stationary is also graphically respected

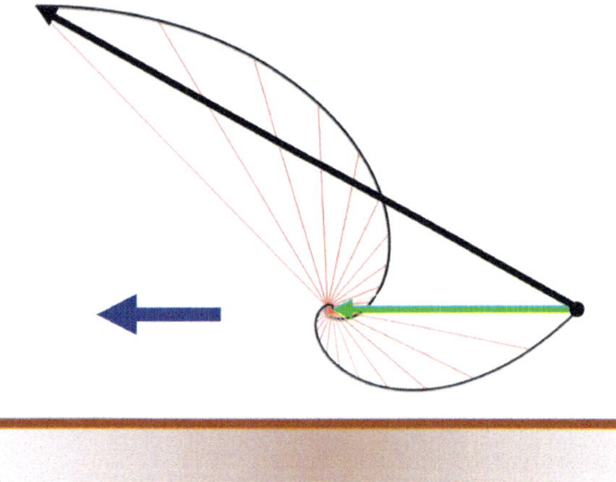

Fig. 16.8 Diagram of the elementary current system in a coastal ocean with constant density. Blue arrow: surface wind stress. Black arrow: surface current. Green arrow: barotropic geostrophic current in the deep layer

Bibliography

Ekman, V.W.: On the influence of the Earth's rotation on ocean-currents. Arkiv Foer Matematik, Astronomi Och Fysik. **11**, 1–53 (1905)

Further Recommended Reading

Abarbanel, H.D.I., Young, W.R. (eds.): General Circulation of the Ocean. Springer-Verlag, New York (1987)
Cushman-Roisin, B.: Introduction to Geophysical Fluid Dynamics. Prentice-Hall, Englewood Cliffs, New Jersey (1994)
Defant, A.: Physical Oceanography, vol. I. Pergamon Press, New York (1961)
Gill, A.E.: Atmosphere-Ocean Dynamics. Academic Press, New York (1982)
Marshall, J., Plumb, R.A.: Atmosphere, Ocean, and Climate Dynamics. Elsevier, Amsterdam (2008)
Neumann, G., Pierson Jr, W.J.: Principles of Physical Oceanography. Prentice-Hall, Englewood Cliffs, New Jersey (1966)
Pedlosky, J.: Geophysical Fluid Dynamics. Springer-Verlag, New York (1987)
Pond, S., Pickard, G.L.: Introductory Dynamical Oceanography. Pergamon Press, Oxford (1983)
Vallis, G.K.: Atmospheric and Oceanic Fluid Dynamics. Cambridge University Press, Cambridge (2006)
Warren, B.A., Wunsch, C. (ed.): Evolution of Physical Oceanography; Scientific Surveys in Honor of Henry Stommel. The Massachusetts Institute of Technology, Cambridge, Massachusetts (1981)

Chapter 17
The Shallow-Water Approximation

This chapter introduces the so-called shallow-water approximation, which allows for a simplified treatment of motions with horizontal spatial scales much larger than the vertical one. After dealing with a single layer of homogeneous fluid, which limits the dynamics to the barotropic case, we move on to a multilayer fluid, capable of describing baroclinic motions. In this context, the discussion of a simulation of a significant oceanographic phenomenon allows the introduction of the concepts of intrinsic oceanic variability and chaotic dynamical system.

17.1 The Shallow-Water Equations

This paragraph deals with the so-called shallow-water approximation, which is very useful as it allows for the analysis and modelling of a series of complex phenomena in a simplified way, separating the dynamics from the thermodynamics of the fluid system. This approximation implies motions on large spatial scales—in a sense that will be specified—therefore it is relevant for oceanic and meteorological dynamics from the mesoscale to planetary scale motions. Both the noun "water" and the adjective "shallow" are to be understood in a relative sense: the fluid can also be air and can also be "deep" in an absolute sense. However, even if applicable to the atmosphere with appropriate generalisations, the shallow-water approximation will be discussed here in the oceanic context because in this case the theory is simpler and, therefore, more suitable for an introductory educational treatment. The equations derived in this paragraph and the next refer to a layer of fluid with constant density, so baroclinic effects are excluded; however, in Sects. 17.3, 17.4 and 17.5 generalisations will be presented that allow for the consideration of such effects.

In general, shallow-water dynamics allows for the obtaining of a series of fundamental information about the functioning of the oceanic and atmospheric systems in a very simple way, especially if the problem is reformulated in terms of vorticity

S. Pierini, *Oceanic and Atmospheric Fluid Dynamics*, UNITEXT for Physics,
https://doi.org/10.1007/978-3-031-77991-6_17

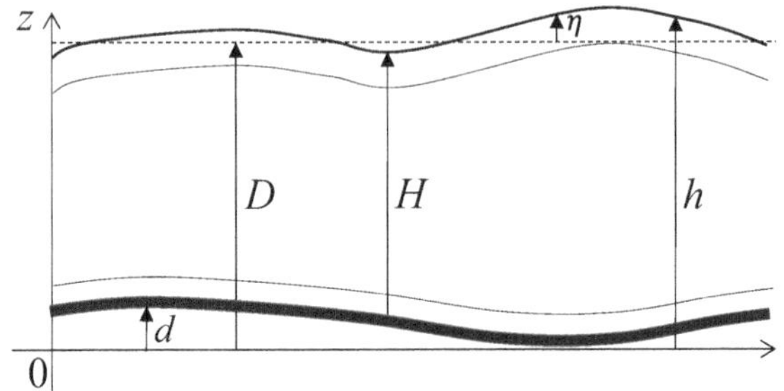

Fig. 17.1 Definition of the variables that appear in the shallow-water theory (η and d are not to scale and η is enormously amplified). The thin line below the free surface and the one above the bottom schematise the boundary of the two Ekman layers, whose thicknesses are assumed to be less than D

(Chap. 18) and if one limits oneself to considering subinertial motions (Chap. 19). From a modelling point of view, the shallow-water equations have played—and still play today—an important role in the numerical simulation of oceanographic processes, such as tides (in which context the equations are known as *Laplace's tidal equations*), storm surges, circulation in lagoons and fjords, tsunamis, etc.. Even in the presence of the most realistic three-dimensional models, the large-scale wind-driven ocean circulation can—in some significant cases—make use of the shallow-water equations, especially in their multilayer version.

Consider a layer of homogeneous and incompressible fluid of density $\rho = const$ on a plane tangent to the Earth's surface at latitude φ in the beta-plane approximation ($f = f_0 + \beta y$). The layer is bounded below by a bottom bathymetry $d(x, y)$ defined with respect to a given horizontal reference level (corresponding to $z = 0$) and, above, by the sea surface represented by the deviation $\eta(x, y, t)$ from the undisturbed level (the geoid) located at $z = D$, as shown in Fig. 17.1. The variables H and h are defined as $h(x, y, t) = D + \eta(x, y, t)$ and $H(x, y, t) = h(x, y, t) - d(x, y)$; therefore, H represents the thickness of the layer at point $\mathbf{x}$ at time t.

The shallow-water approximation consists in assuming that the *aspect ratio* δ between the typical length scale of vertical motions (essentially given by D) and that of horizontal motions (here indicated by L), is much less than unity:

$$\delta = \frac{D}{L} \ll 1 \tag{17.1}$$

(the same condition must also be satisfied by the bottom bathymetry). Essentially, the shallow-water approximation limits the dynamics to large-scale spatial motions but does not impose any limitation on the temporal scale. For example, as far as surface gravity waves are concerned, this approximation filters out deep water waves but preserves those in shallow water (in the limit of wavelengths so large non-hydrostatic effects are negligible, therefore only non-dispersive shallow water waves are allowed): these waves, while verifying Eq. (17.1), represent superinertial motions (see Sect. 12.3).

The first consequence of the approximation given by Eq. (17.1) concerns the vertical component of the motion. Denoting with U the typical scale of horizontal motions and with W that of vertical motion, from the continuity equation $\nabla \cdot \mathbf{u} = 0$ we get $W \leq U\delta$ (the symbol $\leq$ indicates the possibility that the terms u_x and v_y are in partial balance). This implies that, for $\delta \ll 1$, the acceleration in the third equation of motion is negligible compared to the acceleration of gravity. By scaling arguments, it can indeed be shown (Pedlosky 1987) that the hydrostatic equation is accurate to the second order in the parameter δ. Ultimately, in the shallow-water approximation, the hydrostatic balance is accurately verified; the reverse is also true, that is, for hydrostatic motions the scales satisfy the shallow-water condition. Consequently, the equations of motion will contain only the two horizontal components which, therefore, will describe a two-dimensional motion.

In relation to this, for Eq. (3.7) the horizontal pressure gradient is determined solely by the inclination of the free surface η, and is therefore independent of z. Consequently, this must hold also for the accelerations as well as for the velocities if this occurs at an initial time. This implies that the motion is two-dimensional in the same sense implied by the Taylor-Proudman theorem, as defined by Eq. (13.5). In that context, the two-dimensionality was based on the assumption $\varepsilon \ll 1$, which is not required here, but the two-dimensionality in the shallow-water approximation is nonetheless valid; even in this case, it is therefore convenient to imagine the fluid motion as consisting of vertical columns in horizontal motion. These will, however, have to modify their height and the area of their horizontal section if they flow over a variable bottom ($\nabla d \neq 0$), in order to conserve their mass.

The substantial two-dimensionality of the motion entails the following simplifications. First of all, among the nonlinear terms in the acceleration, we will have $wu_z = wv_z = 0$, therefore:

$$\frac{du}{dt} = u_t + uu_x + vu_y; \quad \frac{dv}{dt} = v_t + uv_x + vv_y \tag{17.2}$$

(in this chapter and in the next one, we will often resort to the compact notation for partial derivatives in order to facilitate the reading of the equations). Equation (17.2) can be rewritten in compact form as $d\mathbf{u}/dt = \mathbf{u}_t + (\mathbf{u} \cdot \nabla_H)\mathbf{u}$ and the forces of turbulent viscosity are reduced to

$$\mathbf{F}_V = K_H \nabla_H^2 \mathbf{u}, \tag{17.3}$$

meaning, from now on, $\mathbf{u} = (u, v)$. Finally, it is possible to transform the pressure gradient force using the variable η. The hydrostatic balance (Eq. 3.7) provides the pressure p at the level z:

$$p(x, y, z, t) = p_a + \rho g\big[h(x, y, t) - z\big],$$

where p_a is the atmospheric pressure at sea level, which is assumed to be constant both spatially and temporally. Taking the horizontal gradient of both members we get:

$$\nabla_H p = \rho g \nabla_H h \equiv \rho g \nabla_H \eta, \tag{17.4}$$

Using Eqs. (17.2–17.4) and the expression of the Coriolis force on the beta-plane (Eq. 12.15), from Eq. (12.8) we obtain the following equations of motion in the shallow-water approximation:

$$\begin{cases} u_t + uu_x + vu_y - fv = -g\eta_x + K_H \nabla_H^2 u \\ v_t + uv_x + vv_y + fu = -g\eta_y + K_H \nabla_H^2 v \end{cases} \tag{17.5}$$

where in general $f = f_0 + \beta y$. The same equations can be written in vector form as follows,

$$\frac{d\mathbf{u}}{dt} + f\mathbf{k} \times \mathbf{u} = -g\nabla_H \eta + K_H \nabla_H^2 \mathbf{u}, \tag{17.6}$$

provided that it is remembered that $\mathbf{u}$ indicates the horizontal velocity and that the acceleration is given by Eq. (17.2). Given the two-dimensionality of the motion the absolute vorticity will be reduced to $\omega_{a3} = \zeta + f$ (cf. Eq. 13.7) with $\zeta = v_x - u_y$.

Equation (17.16) will now be generalised to take into account the ageostrophic flows present in the two Ekman layers, whose thicknesses are assumed to be less than D as schematised in Fig. 17.1. The velocity in Eq. (17.6) (independent of z as $\rho = const$) is associated with the pressure gradient force and can therefore be denoted $\mathbf{u}_p(x, y, t)$ (in the quasigeostrophic approximation treated in Chap. 19, $\mathbf{u}_p$ coincides with the geostrophic velocity, but in this context it can also represent non-geostrophically balanced flows such as, for example, non-dispersive shallow water waves).

On the other hand, there is an ageostrophic motion field $\mathbf{u}_E(x, y, z, t)$, null at every depth except in the two thin surface and bottom Ekman layers. Equation (17.6) can be generalised to take into account these flows (i) considering the velocity as dependent on z, (ii) adding $\partial/\partial z(K_v \partial \mathbf{u}/\partial z)$ to the second member, (iii) vertically averaging over the layer of thickness H and (iv) imposing the continuity conditions of the stress (see the procedure followed to obtain Eq. (16.12) through Eqs. (16.13 and 16.14)). In this way we obtain

$$\frac{d\mathbf{u}}{dt} + f\mathbf{k} \times \mathbf{u} = -g\nabla_H \eta + \frac{\boldsymbol{\tau}^{(s)} - \boldsymbol{\tau}^{(b)}}{\rho H} + K_H \nabla_H^2 \mathbf{u} \tag{17.7}$$

where now $\mathbf{u}(x, y, t)$ should be interpreted as the vertically averaged horizontal velocity,

$$\mathbf{u} = \mathbf{u}_p + \overline{\mathbf{u}}_E, \tag{17.8}$$

$\boldsymbol{\tau}^{(s)}$ is the surface wind stress given by Eq. (15.21) and $\boldsymbol{\tau}^{(b)}$ is the bottom stress (b stands for *bottom*) parameterisable with a similar formula. As for the nonlinear terms, the vertical average generally modifies their structure. However, $O(|\overline{\mathbf{u}}_E|)$ can be up to an order of magnitude smaller than $O(|\mathbf{u}_p|)$ (see the discussion in Sect. 20.3); moreover $\boldsymbol{\tau}^{(s)}$, and therefore $\overline{\mathbf{u}}_E$, can vary on spatial scales (of the order of several hundred km) much larger than those of $\mathbf{u}_p$ (which, in particular in the baroclinic mesoscale described by the multilayer versions of Eq. (17.7), cf. Sects. 17.3, 17.4 and 17.5, are of the order of the internal Rossby deformation radius). Consequently, the average of the nonlinear terms includes practically only $\mathbf{u}_p$ and this justifies its presence in Eq. (17.7).

17.2 Integrated Continuity Equation in Shallow-Water

In the two evolution equations in Eq. (17.7) the three dependent variables u, v and η appear; to make the mathematical problem well posed, a third evolution equation must therefore be derived in which η must appear as a prognostic variable (i.e., containing the term η_t). This equation will be based on the continuity equation $\nabla \cdot \mathbf{u} = 0$ but will also require the imposition of boundary conditions on the sea bottom and on the free surface. This will allow us to consider in a simple and elegant way the vertical structure of the fluid layer.

Since we intend to obtain an equation in which the derivative of η with respect to time appears, we will have to relax the assumption adopted in the previous paragraph, and here excessively drastic, $w = 0$. The continuity equation can be integrated with respect to w taking into account the independence of u and v with respect to z:

$$w(x, y, z, t) = -z\big(u_x + v_y\big) + B(x, y, t), \tag{17.9}$$

where B is an integration constant (with respect to z) that can however depend on x, y and t. Its expression can be obtained by imposing the free-slip conditions on the seabed (Fig. 17.2).

On the bottom, for a variation of d along x, we will have $(w/u)|_{z=d} = \tan\alpha = d_x$ for which, also considering a variation along y, we get:

$$w\big[x, y, d(x, y), t\big] = ud_x + vd_y. \tag{17.10}$$

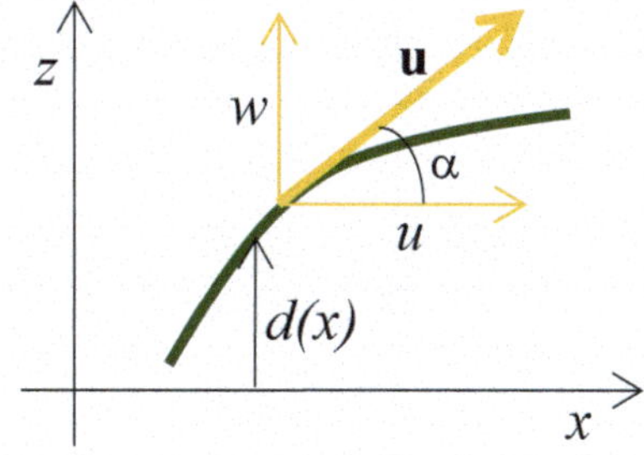

Fig. 17.2 Imposition of the free-slip boundary condition on the seabed (the gradient of d is amplified)

Evaluating Eq. (17.9) in $z = d$ and eliminating w with Eq. (17.10) we get B,

$$B = ud_x + vd_y + d\big(u_x + v_y\big),$$

which, substituted into Eq. (17.9), provides:

$$w(x, y, z, t) = (d - z)\big(u_x + v_y\big) + ud_x + vd_y. \tag{17.11}$$

Now we impose the free-slip condition on the free surface:

$$w\big[x, y, h(x, y), t\big] = uh_x + vh_y + h_t,$$

where the additional term h_t (absent on the bottom) is the contribution to the vertical velocity from the free surface. Substituting this equation into Eq. (17.11) evaluated at $z = h$ and taking into account that $h_t = \eta_t$ we finally get:

$$\eta_t + (Hu)_x + (Hv)_y = 0. \tag{17.12}$$

This prognostic equation in η is called the *integrated continuity equation in shallow water*.

In conclusion, the three evolution equations Eqs. (17.7 and 17.12) for the variables u, v and η describing the dynamics in the shallow-water approximation ($\delta \ll 1$, Fig. 17.1) are summarised here:

$$\begin{cases} \frac{d\mathbf{u}}{dt} + f\mathbf{k} \times \mathbf{u} = -g\nabla_H \eta + \frac{\boldsymbol{\tau}^{(s)} - \boldsymbol{\tau}^{(b)}}{\rho H} + K_H \nabla_H^2 \mathbf{u} \\ \frac{\partial H}{\partial t} + \nabla_H \cdot (H\mathbf{u}) = 0 \end{cases} \tag{17.13}$$

where $H = D + \eta - d$.

17.3 The Two-Layer Model

The shallow-water equations are extremely useful but suffer from a strong limitation: baroclinic flows are absent. This is certainly a significant limitation as regards the description of the dynamics in stratified oceans (Sect. 14.3). The shallow-water theory can however be generalised to include also baroclinic effects, and therefore relative currents, considering the so-called *multilayer model*.

In this model, different immiscible layers of homogeneous and incompressible fluid, each with its own density, are superimposed on each other (obviously the density of each layer must be less than that of the layer below in order for the stratification to be stable). In each layer the shallow-water dynamics applies, in particular, the speed is independent of z; however, the inclination of the surfaces that separate the layers, called *interfaces*, produce relative pressure gradients which in turn induce

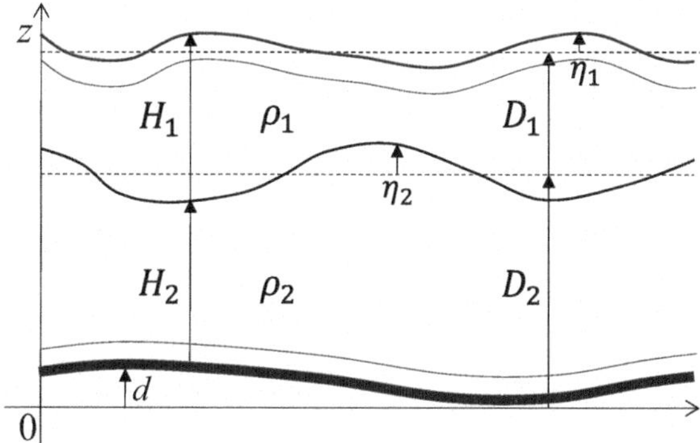

Fig. 17.3 Definition of the variables that appear in the two-layer model. The vertical deviations are not to scale, in particular $O(|\eta_1|) = O(|\eta_2|(\rho_2 - \rho_1)/\overline{\rho}) \ll O(|\eta_2|)$. The thin line under the free surface and the one above the bottom schematise the boundary of the two Ekman layers, whose thicknesses are assumed to be less than D_1 and D_2 respectively

different velocities in the different layers. In this paragraph, we will consider the simplest version of the multilayer model, the *two-layer model*, which includes all the essential features of the more general version and whose structure is shown in Fig. 17.3. The extension to a model with an arbitrary number of layers is immediate.

The dynamics is described by the following equations:

$$\begin{cases} (a)\frac{d\mathbf{u}_1}{dt} + f\mathbf{k} \times \mathbf{u}_1 = -g\nabla_H \eta_1 + \frac{\boldsymbol{\tau}^{(s)}}{\rho_1 H_1} + K_H \nabla_H^2 \mathbf{u}_1 \\ (b)\frac{\partial H_1}{\partial t} + \nabla_H \cdot (H_1 \mathbf{u}_1) = 0 \end{cases} \tag{17.14}$$

$$\begin{cases} (a)\dfrac{d\mathbf{u}_2}{dt} + f\mathbf{k} \times \mathbf{u}_2 = -\alpha\nabla_H \eta_1 - g'\nabla_H \eta_2 + \dfrac{\boldsymbol{\tau}^{(b)}}{\rho_2 H_2} + K_H \nabla_H^2 \mathbf{u}_2 \\ (b)\dfrac{\partial H_2}{\partial t} + \nabla_H \cdot (H_2 \mathbf{u}_2) = 0 \end{cases} \tag{17.15}$$

Equations (17.14–17.15) are the transposition of Eq. (17.13) for each of the two layers (the indices 1 and 2 refer respectively to the upper and lower layer), but with some differences.

First of all, in the equations of the upper layer, the wind stress $\boldsymbol{\tau}^{(s)}$ is present, while in those of the lower layer, the bottom stress $\boldsymbol{\tau}^{(b)}$ is present. Moreover, retracing the derivation of the integrated continuity equation (Eq. 17.12), as far as the upper layer is concerned, in Eq. (17.10) instead of d there will be η_2 but, being this time-dependent, the term η_{2t} will also be present on the right-hand side: from this it will result $H_1 = D_1 + \eta_1 - \eta_2$. Instead, in the second layer H_2 is similar to the H of the barotropic case: $H_2 = D_2 + \eta_2 - d$. The presence of η_2 in the equations of both layers constitutes a first coupling between them.

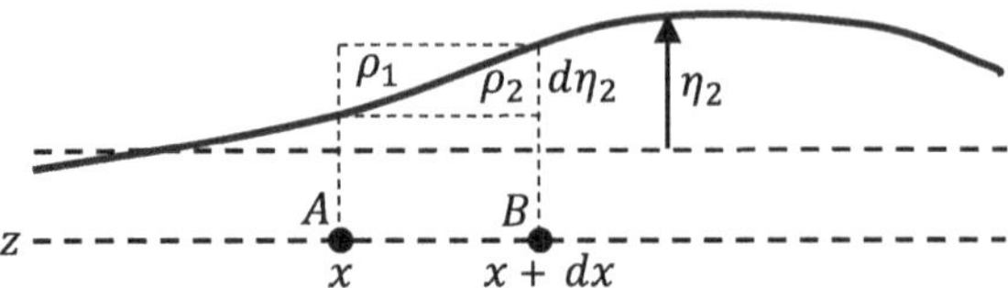

Fig. 17.4 Origin of the relative pressure gradient force acting on the lower layer. The solid line is the interface while the thick dashed line is its undisturbed level

A further coupling concerns the relative pressure gradient force $-g'\nabla_H\eta_2$ present in Eq. (17.15a), with the reduced gravity $g' = g(\rho_2 - \rho_1)/\rho_2$: at the generic level z the hydrostatic pressure in B (Fig. 17.4) differs from that in A by the term $(\rho_2 - \rho_1)gd\eta_2$, hence:

$$-\frac{1}{\rho_2}\frac{\partial p}{\partial x} = -\frac{1}{\rho_2}\frac{\rho_2 g d\eta_2 - \rho_1 g d\eta_2}{dx} = -g'\frac{\partial \eta_2}{\partial x}.$$

In Eq. (17.15a) there is also the absolute pressure gradient force $-\alpha\nabla_H\eta_1$ determined by the inclination of the free surface η_1, where $\alpha = g\rho_1/\rho_2 \cong g$.

The dissipative effects due to the discontinuous velocity transition between one layer and the other through the interface are generally parameterised in both layers by an additional term called *interfacial friction* (here omitted for simplicity), with quadratic dependence on the difference between the two velocities.

It is worth noting that this model, net of the Coriolis force and the terms containing the surface and bottom stresses, coincides with the one introduced in Sect. 11.2 in the context of internal gravity waves. In particular, the linearised model supports the two barotropic and baroclinic oscillation modes shown in Fig. 11.4.

17.4 The Reduced-Gravity Model

In Sect. 16.4 it was seen that the Ekman pumping/suction velocity (proportional to the curl of wind stress, cf. Eq. 16.23) applied at the base of the surface Ekman layer, produces an inclination of the isopycnals opposite to that of the free surface (but with much larger amplitudes), resulting in a reduction of the geostrophic speed in depth due to the thermal wind relationship (see, for example, Fig. 16.6).

In some important cases related to large-scale oceanic circulation, the deformation of a thin thermocline can produce a nearly perfect baroclinic compensation that makes the underlying ocean practically quiescent. This dynamic situation is denoted as *reduced gravity*. An interesting example of this phenomenon is provided by the *Kuroshio Extension* (KE, see for example, Pierini et al. 2014, for a review on the subject). The *Kuroshio* is a western boundary current (Sect. 20.4) that flows along the western edge of the North Pacific Ocean; at latitude $\varphi \cong 35°N$ this joins an analogous, but weaker current coming from the North (the Oyashio) and separates from the coasts of Japan forming an intense meandering jet, the KE, which advances into the ocean in a practically zonal direction (Fig. 17.5a).

In Fig. 17.5b, the vertical section of the potential density along a transect that orthogonally cuts the KE shows a strong lowering of the thermocline for a meridional extension of about $1°N$, reaching about 600 m in depth; this results in the cancellation of the current below the same depth (Fig. 17.5c). Therefore, the KE constitutes a significant example of reduced-gravity dynamics.

In these cases the two-layer model is subject to an interesting simplification. Consider a model with $D_1 \ll D_2$ in which the pressure gradient force in the lower layer (present in the second term of Eq. (17.15a)) vanishes in order to simulate the reduced-gravity dynamics (we set $\alpha \cong g$):

$$-g\nabla_H\eta_1 - g'\nabla_H\eta_2 = 0. \tag{17.16}$$

In this case the lower layer is quiescent ($\mathbf{u}_2 = 0$) if it is at an initial time, so we are reduced to a single active layer, the upper one. If we define $\tilde{\eta} = -\eta_2$, the Eq. (17.14) with Eq. (17.16) become:

$$\begin{cases} \frac{d\mathbf{u}_1}{dt} + f\mathbf{k}\times\mathbf{u}_1 = -g'\nabla_H\tilde{\eta} + \frac{\boldsymbol{\tau}^{(s)}}{\rho_1 H} + K_H\nabla_H^2\mathbf{u}_1 \\ \frac{\partial H}{\partial t} + \nabla_H\cdot(H\mathbf{u}_1) = 0 \end{cases} \tag{17.17}$$

where $H \cong D_1 + \tilde{\eta}$ as, given that the relative variation of the density between the two layers is very small ($g(\rho_2 - \rho_1)/\rho_2 = O(10^{-3})$), $|\eta_1| \ll |\tilde{\eta}|$: this is called the *reduced-gravity model* (also referred to as the *1.5-layer model*; see Fig. 17.6). Note that in this version, the intersection of the thermocline with the surface (the so-called *outcropping*), possible in real situations, is not allowed; a more advanced version of the model can however take this phenomenon into account.

Therefore, the reduced-gravity dynamics is described by a single active layer in the shallow-water approximation. It is interesting to note that Eqs. (17.17) are formally identical to the equations for a homogeneous fluid (Eq. 17.13), which return the first with the simple transformation

$$g \to g';\ \eta \to \tilde{\eta}. \tag{17.18}$$

On the other hand, once $\tilde{\eta}$ is obtained, η_1 can be determined by the following diagnostic relationship derived from Eq. (17.16):

$$\eta_1 = \frac{\rho_2 - \rho_1}{\rho_2}\tilde{\eta}. \tag{17.19}$$

In the next paragraph, an application of the reduced-gravity dynamics to the problem of KE variability will be presented.

We conclude by noting that it is possible to generalise this model to include an arbitrary number of active layers instead of the single one considered here; in

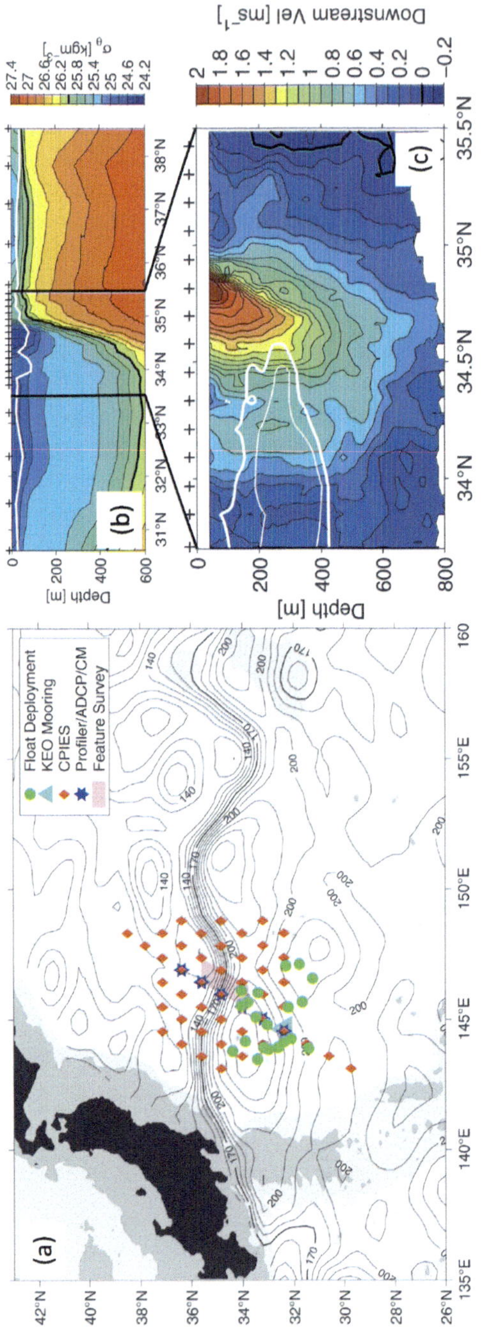

Fig. 17.5 **(a)** The observing system of the "Kuroshio Extension System Study" (KESS) implemented in 2004 (see Qiu et al. 2006); the isolines represent the sea level obtained from altimetric data (in *cm*). **(b)** Vertical section of the potential density σ_θ along a transect that cuts the KE. **(c)** Vertical section of the current velocity along the jet measured by hull-mounted ADCP; the black lines connect in latitude the sections **(b)** and **(c)**. Figures adapted from Qiu et al. (2006, J. Phys. Oceanogr. 36, 457–473, © American Meteorological Society. Used with permission)

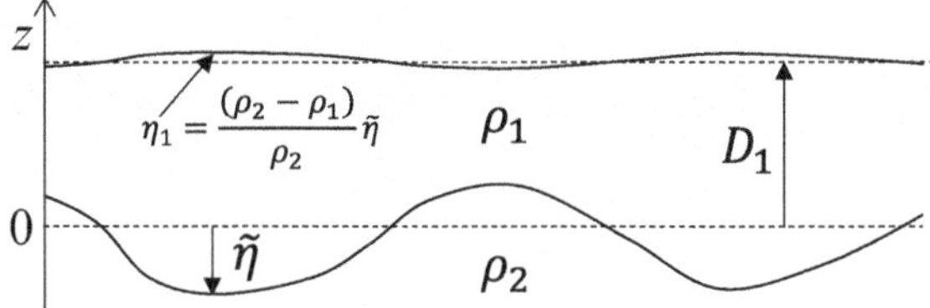

Fig. 17.6 Definition of the variables that appear in the reduced-gravity model (vertical deviations are not to scale). It is emphasised that, in the model, only $\tilde{\eta}$ (positive downwards) represents a prognostic variable, from which η_1 (positive upwards) can be derived using the formula indicated in the figure

this case, it will be referred to as the *n-layer reduced-gravity model.* In general, the disappearance of the dynamics of surface gravity waves makes numerical integration using explicit finite difference methods much faster than that of a multilayer model (Sect. 17.3) due to the Courant-Friedrichs-Lewy condition required for its stability.

17.5 An Example of Intrinsic and Chaotic Variability of a Reduced-Gravity Flow

In this paragraph, a modelling study on the intrinsic and chaotic variability of the KE, an ocean current that, as seen in the previous paragraph, constitutes an interesting example of reduced-gravity dynamics, will be briefly discussed. This will provide an opportunity to touch on the important concepts of *intrinsic variability* and *chaos* in *nonlinear dynamical systems*, the mathematical treatment of which, although simplified, is beyond the scope of this text. The simple qualitative treatment presented in this paragraph may stimulate further study.

In a process modelling study (Pierini 2006), the dynamics of the KE was studied using Eq. (17.17) (with the further addition of interfacial friction). Figure 17.7 shows the domain within which the equations were numerically solved. Note the schematic nature of the coasts, suitable for a process study aimed at simulating the main characteristics of the flow net of secondary effects that could obscure the interpretation of the most salient ones. It has been shown (Pierini 2008) that the presence of the western coast, albeit schematic, is essential for the generation of a realistic flow; the wide zonal extension of the domain is also necessary to allow an input of momentum, vorticity and energy of the necessary magnitude for the formation of a sufficiently realistic jet.

A significant aspect of the study involved forcing the oceanic system with a time-independent surface wind stress field $\boldsymbol{\tau}(x, y)$ (its curl is shown in Fig. 17.7), as the aim of the study was to determine whether the decadal variability of the KE observed with remote sensing altimetric data was attributable to internal dynamical mechanisms rather than atmospheric forcing. To understand this point, it is first necessary to specify what is meant by a *nonlinear and dissipative dynamical system* (NDDS; for

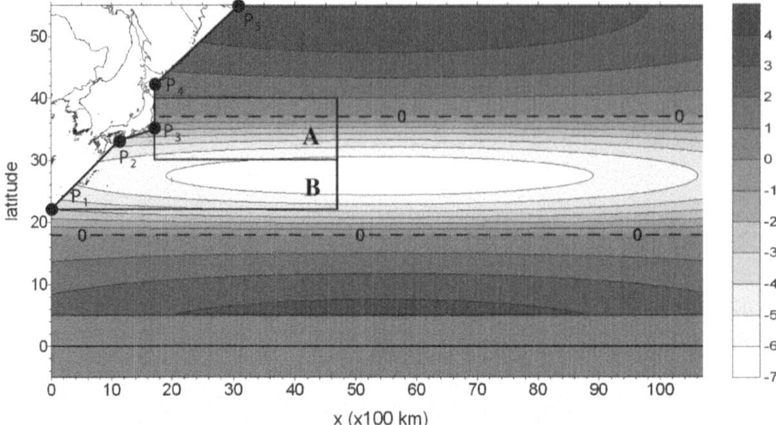

Fig. 17.7 Integration domain of the reduced-gravity model (Eq. 17.17) used for the numerical simulation of the intrinsic variability of the KE. The isolines represent the curl of the climatological surface wind stress (units: $10^{-8}\ N\ m^{-3}$). Figure adapted from Pierini (2006, J. Phys. Oceanogr. 36, 1605–1625, © American Meteorological Society. Used with permission)

further insights within a vast scientific literature on the subject, one can refer, for example, to Kautz 2011, for an essentially qualitative introductory treatment, to Baker and Gollub 1990, Strogatz 1994 and Hilborn 2000, for introductory mathematical treatments and to Ghil 2019, for an updated review in the field of Earth sciences, also including historical sections).

To put it very briefly and qualitatively, the ocean, in its mathematical schematisation, can be imagined as a *dynamical system*, composed of a space Σ (also known as *phase space* or *state space*), each point of which completely represents the oceanic state (in this specific case the fields u, v and $\tilde{\eta}$ at a certain time t) and in which a well-defined evolution law (induced by the equations of motion) is defined: in this abstraction, the evolution of the system is described by a trajectory in Σ. This dynamical system is also *nonlinear* (as the equations describing its evolution are) and *dissipative* (as viscosity dissipates its mechanical energy). Because of this latter characteristic, in order for the dynamics in an NDDS not to asymptotically reduce to the absence of motion (in the case in question $u = v = \tilde{\eta} = 0$ for every t), an external forcing must continuously replace the dissipated energy. If the forcing is time-independent, the dynamical system is said to be *autonomous*, otherwise *nonautonomous*.

An autonomous NDDS can admit a plurality of equilibrium states, also known as *fixed points*: these correspond to stationary solutions of the equations of motion (in the case in question u, v and $\tilde{\eta}$ would not depend on time). In this case, if $P \in \Sigma$ is a fixed point of the system, the evolution (thus the trajectory) with P as the initial condition reduces to P itself. Another possible solution is the *limit cycle*, represented by a closed curve $C \in \Sigma$; in this case the evolution is periodic, i.e., every point on C is revisited after a certain period T. A fixed point or a limit cycle are called *attractors* of the autonomous NDDS if they are stable; in this case, every point lying in a certain subset of Σ, called the *basin of attraction* of that attractor, is asymptotically attracted to that type of evolution (there can also be quasi-periodic attractors but, above all, aperiodic attractors with a very special character, as will be seen shortly). Reflect on the fact that a limit cycle corresponds to a periodic variation (called *self-sustained*) despite the external forcing being time-independent; this means that it is the system itself, through nonlinear interactions, that produces that variability, which is therefore called *intrinsic* (or also internal or natural).

With reference to the ocean system, this exhibits significant variability that is partly induced by similar variations in the atmospheric forcing, but partly is of an intrinsic nature. The *intrinsic oceanic variability* has a high-frequency component associated with mesoscale dynamics generated by barotropic and baroclinic instability processes (see Sects. 20.1 and 20.2 for notes on these important dynamical mechanisms) and a low-frequency component that is induced by the first through an inverse kinetic energy cascade due to nonlinear interactions between mesoscale vortices (see, for example, Arbic et al. 2012; Sérazin et al. 2018) and also by other less general mechanisms that depend on the specific area of interest (see, for example, Penduff et al. 2014, 2018; Rubino et al. 2023).

How to highlight the intrinsic variability of an ocean system? One can resort to ensemble simulations (Penduff et al. 2014), but the simplest way is to consider the ocean as an autonomous NDDS (Dijkstra 2005; Dijkstra and Ghil 2005), thus subjecting it to a constant atmospheric forcing, as done in the study in question. Indeed, any variability that should emerge in this case would necessarily be attributable to intrinsic mechanisms. The modelling study of the KE mentioned here indeed shows a relevant intrinsic variability of the jet on a decadal time scale, as shown in Fig. 17.8. The significant qualitative agreement with altimeter data (Pierini et al. 2009) supports the hypothesis that the decadal variability of the KE is substantially of an intrinsic nature.

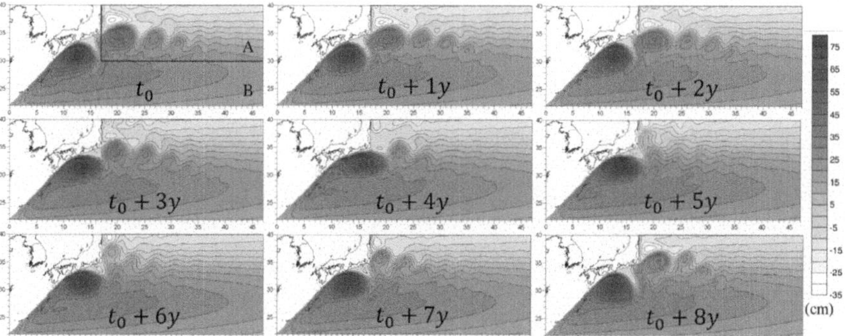

Fig. 17.8 Nine successive snapshots of the sea level η separated by one year ($t_0 = 143$ y, cf. Fig. 17.9), obtained from the numerical simulation of the intrinsic variability of the KE and shown in the subdomain A⋃B of Fig. 17.7. Figures adapted from Pierini (2006, J. Phys. Oceanogr. 36, 1605–1625, © American Meteorological Society. Used with permission)

The fact that the real variability of the KE is synchronised (Qiu and Chen 2010) with that of the North Pacific Oscillation (one of the main modes of oscillation of the North Pacific Ocean) is not in contradiction with the intrinsic nature of the jet's variability. Indeed, it has been shown (Pierini 2014) that this complex phenomenon can be described as a case of *intrinsic variability paced by an external deterministic forcing*; in this process, the westward propagation of baroclinic Rossby waves (Sects. 19.2 and 19.5) provides the teleconnection between the central Pacific, where the waves are generated, and the KE area (see Sect. 19.5 and Fig. 19.4). It is interesting to note that, within the climate sciences this same dynamic process also manifests itself in phenomena very different from the one considered here, such as, for example, in the Atlantic Multidecadal Variability (Otterå et al. 2010) and in the late Pleistocene glacial terminations (Pierini 2023).

It is important to note that the variability obtained from the simulation in question is not periodic but aperiodic, as shown in Fig. 17.9 by the time series of the kinetic energy integrated in subdomains A and B. Each complete oscillation of the KE (one of which is shown in Fig. 17.8) manifests itself with its specific duration and spatial structure, two oscillations can be similar but also quite different, however, some distinctive features that characterise them all are recognizable. Note that a low-order spectral ocean model, based on a severe truncation of the Galerkin projection of the original nonlinear partial differential equations, helped to formulate the same problem in the framework of dynamical systems theory (Pierini 2011).

Furthermore, a phenomenon of utmost importance occurs: if during its evolution the system is subject to any perturbation, however small, the subsequent evolution completely loses coherence with the unperturbed one, manifesting itself in a completely different way, while still retaining its distinctive characteristics: such

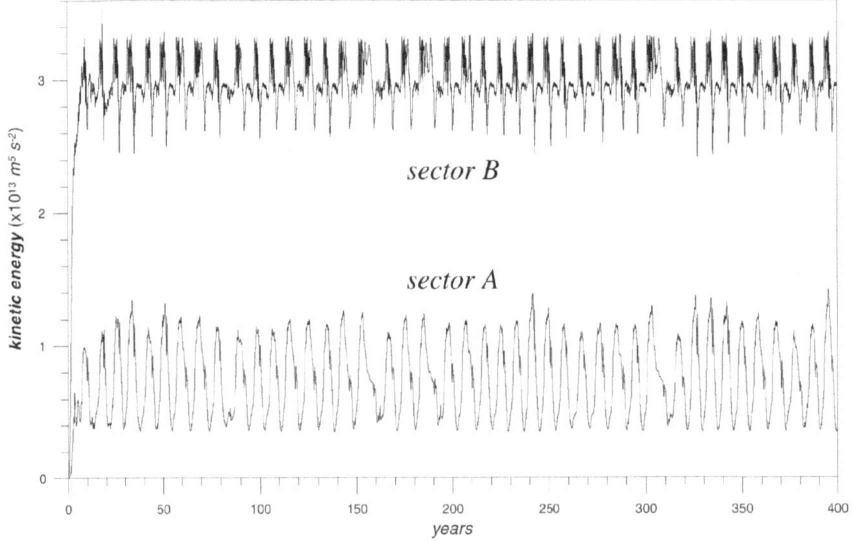

Fig. 17.9 Evolution of the kinetic energy integrated in sectors A and B shown in Figs. 17.7 and 17.8. Figure adapted from Pierini (2006, J. Phys. Oceanogr. 36, 1605–1625, © American Meteorological Society. Used with permission)

behaviour is called *chaotic*. In general, chaos occurs when in a deterministic dynamical system the evolution is *aperiodic* and exhibits *extreme dependence on initial conditions*, thus limiting the predictability of the evolution to a short time interval, called the *predictability time* of the system. In other words, in a dynamical system (chaotic or not chaotic) governed by deterministic evolution equations, an initial condition uniquely determines the future with absolute accuracy; however, if the system is chaotic, the approximate knowledge of the initial condition (a situation entirely normal in nature) does not allow to predict the future approximately beyond the predictability time of the system (Lorenz 1963, 1993; in addition to the fundamental contribution of this author to the discovery of chaos, it is also worth mentioning the equally fundamental contribution of the great French mathematician and physicist Henri Poincaré, 1854–1912, whose study of the three-body problem dating back to the end of the nineteenth century implied an extreme dependence on the initial conditions: Poincaré 1892). Having said this, the aperiodic trajectory of a chaotic NDDS nonetheless invades a limited sector, however large, of Σ, known as a *strange attractor*. This attractor, and the non-chaotic ones mentioned above, can be different manifestations of the same NDDS corresponding to different values of one or more *control parameters*. The reader interested in studying the fascinating phenomenon of chaos at the same mathematical level as the present text can refer, for example, to the texts on dynamical systems cited above. For a popular reading, one can refer to the book by Gleick (1987).

Ultimately, beyond the example discussed here, the intrinsic oceanic variability in general is chaotic in its manifestation of both high and low frequency, and this significantly influences the predictability of the ocean system. The same applies to the atmospheric system and, more generally, to the climate system. It is for this reason that the analysis of the intrinsic variability of the climate system and its subsystems is one of the main research topics currently underway in the field of climate physics.

Bibliography

Arbic, B.K., Scott, R.B., Flierl, G.R., Morten, A.J., Richman, J.G., Shriver, J.F.: Nonlinear cascades of surface oceanic geostrophic kinetic energy in the frequency domain. J. Phys. Oceanogr. **42**, 1577–1600 (2012)

Baker, G.L., Gollub, J.P.: Chaotic Dynamics, an Introduction. Cambridge University Press, Cambridge (1990)

Dijkstra, H.A.: Nonlinear physical oceanography. Springer, Dordrecht (2005)

Dijkstra, H.A., Ghil, M.: Low-frequency variability of the large-scale ocean circulation: a dynamical systems approach. Rev. Geophys. **43**, RG3002 (2005)

Ghil, M.: A century of nonlinearity in the geosciences. Earth Space Sci. **6**, 1007–1042 (2019)

Gleick, J.: Chaos, Making a New Science. Viking Books (1987)

Hilborn, R.C.: Chaos and Nonlinear Dynamics. Oxford University Press, Oxford (2000)

Kautz, R.: Chaos, the Science of Predictable Random Motion. Oxford University Press, Oxford (2011)

Lorenz, E.N.: Deterministic nonperiodic flow. J. Atmos. Sci. **20**, 130–141 (1963)

Lorenz, E.N.: The Essence of Chaos. University of Washington Press (1993)

Otterå, O.H., Bentsen, M., Drange, H., Suo, L.: External forcing as a metronome for Atlantic multidecadal variability. Nat. Geosci. **3**, 688–694 (2010)

Pedlosky, J.: Geophysical Fluid Dynamics. Springer-Verlag, New York (1987)
Penduff, T., Barnier, B., Terray, L., Bessieres, L., Serazin, G., Gregorio, S., Brankart, J.M., Moine, M.P., Molines, J.M., Brasseur, P.: Ensembles of eddying ocean simulations for climate. In: CLIVAR Exchanges, Special Issue on High Resolution Ocean Climate Modelling 19 (2014)
Penduff, T., Serazin, G., Leroux, S., Close, S., Molines, J.M., Barnier, B., Bessieres, L., Terray, L., Maze, G.: Chaotic variability of ocean heat content: climate-relevant features and observational implications. Oceanogr. **31**(2), 63–71 (2018)
Pierini, S.: A Kuroshio Extension System model study: decadal chaotic self-sustained oscillations. J. Phys. Oceanogr. **36**, 1605–1625 (2006)
Pierini, S.: On the crucial role of basin geometry in double-gyre models of the Kuroshio Extension. J. Phys. Oceanogr. **38**, 1327–1333 (2008)
Pierini S.: Low-frequency variability, coherence resonance and phase selection in a low-order model of the wind-driven ocean circulation. J. Phys. Oceanogr. **41**, 1585–1604 (2011)
Pierini, S.: Kuroshio Extension bimodality and the North Pacific Oscillation: a case of intrinsic variability paced by external forcing. J. Climate **27**, 448–454 (2014)
Pierini, S.: The deterministic excitation paradigm and the late Pleistocene glacial terminations. Chaos **33**, 033108 (2023)
Pierini, S., Dijkstra, H.A., Riccio, A.: A nonlinear theory of the Kuroshio Extension bimodality. J. Phys. Oceanogr. **39**, 2212–2229 (2009)
Pierini, S., Dijkstra, H.A., Mu, M.: Intrinsic low-frequency variability and predictability of the Kuroshio Current and of its extension. Adv. Oceanogr. Limnol. **5**, 79–122 (2014)
Poincaré, H.: Les méthodes nouvelles de la mécanique céleste. Gauthier-Villars et Fils (1892)
Qiu, B., Chen, S.: Eddy–mean flow interaction in the decadally modulating Kuroshio Extension system. Deep-Sea Res. **II**(57), 1098–1110 (2010)
Qiu, B., Hacker, P., Chen, S., Donohue, K.A., Randolph Watts, D., Mitsudera, H., Hogg, N.G., Jayne, S.R.: Observations of the subtropical mode water evolution from the Kuroshio Extension system study. J. Phys. Oceanogr. **36**, 457–473 (2006)
Rubino, A., Pierini, S., Rubinetti, S., Gnesotto, M., Zanchettin, D.: The skeleton of the Mediterranean Sea. J. Mar. Sci. Eng. **11**, 2098 (2023)
Sérazin, G., Penduff, T., Barnier, B., Molines, J.-M., Arbic, B.K., Müller, M., Terray, L.: Inverse cascades of kinetic energy as a source of intrinsic variability: a Global OGCM study. J. Phys. Oceanogr. **48**, 1385–1408 (2018)
Strogatz, S.H.: Nonlinear Dynamics and Chaos. Westview Press, Cambridge, Massachusetts (1994)

Further Recommended Readings

Cushman-Roisin, B.: Introduction to Geophysical Fluid Dynamics. Prentice-Hall, Englewood Cliffs, New Jersey (1994)
Gill, A.E.: Atmosphere-Ocean Dynamics. Academic Press, New York (1982)
Holton, J.R.: An Introduction to Dynamic Meteorology. Elsevier, Amsterdam (2004)
Kantha, L.H., Clayson, C.A.: Numerical Models of Oceans and Oceanic Processes. Academic Press, San Diego (2000)
Kowalik, Z., Murty, T.S.: Numerical Modelling of Ocean Dynamics. World Scientific, Singapore (1993)
Marshall, J., Plumb, R.A.: Atmosphere, Ocean, and Climate Dynamics. Elsevier, Amsterdam (2008)
Monin, A.S.: Theoretical Geophysical Fluid Dynamics. Kluwer Academic Publishers, Dordrecht (1990)
Vallis, G.K.: Atmospheric and Oceanic Fluid Dynamics. Cambridge University Press, Cambridge (2006)
Visconti, G., Ruggieri, P.: Fluid Dynamics: Fundamentals and Applications. Springer Nature Switzerland AG (2020)

Chapter 18
Potential Vorticity and Its Applications

In this chapter, the shallow-water evolution equations derived in the previous chapter are reformulated in terms of vorticity, thus obtaining an evolution equation for the so-called potential vorticity. The Lagrangian conservation of this quantity, valid in particular contexts, allows in a simple and intuitive way to obtain important dynamical information.

18.1 Evolution Equation of Potential Vorticity in Shallow Water

In this chapter, the equations of motion in shallow water (Eq. 17.13) will be reformulated in terms of vorticity without requiring further assumptions beyond the $\delta \ll 1$ condition adopted in the aforementioned approximation. The concept of potential vorticity, of great relevance in the context of oceanic and atmospheric fluid dynamics, will also be introduced.

We start from the complete set of equations of motion in the shallow-water approximation (Eq. 17.13), which are reported here by specifying all the terms:

$$\begin{cases} \text{(a)}\, u_t + uu_x + vu_y - fv = -g\eta_x + \frac{\tau_1}{\rho H} + K_H \nabla_H^2 u \\ \text{(b)}\, v_t + uv_x + vv_y + fu = -g\eta_y + \frac{\tau_2}{\rho H} + K_H \nabla_H^2 v \\ \text{(c)}\, \frac{dH}{dt} + H\left(u_x + v_y\right) = 0 \end{cases} \tag{18.1}$$

where for simplicity $\boldsymbol{\tau} = \boldsymbol{\tau}^{(s)} - \boldsymbol{\tau}^{(b)}$; note that the integrated continuity equation (c) is rewritten introducing the Lagrangian derivative of H. By deriving Eq. (18.1b) with respect to x, Eq. (18.1a) with respect to y and subtracting, we obtain:

S. Pierini, *Oceanic and Atmospheric Fluid Dynamics*, UNITEXT for Physics,
https://doi.org/10.1007/978-3-031-77991-6_18

$$
\begin{aligned}
&(v_x - u_y)_t + u(v_x - u_y)_x + v(v_x - u_y)_y \\
&= -vf_y + (u_x + v_y)\left[-f - (v_x - u_y)\right] + \frac{(\nabla \times \boldsymbol{\tau})_3}{\rho H} + K_H \nabla_H^2 (v_x - u_y). \quad (18.2)
\end{aligned}
$$

Now, recalling the expression of relative vorticity (Sect. 13.3) $v_x - u_y = \zeta = \omega_{R3}$ and noting that $vf_y = \mathbf{u} \cdot \nabla f = df/dt$, Eq. (18.2) gives:

$$
\zeta_t + u\zeta_x + v\zeta_y \equiv \frac{d\zeta}{dt} = -\frac{df}{dt} - (\zeta + f)(u_x + v_y) + \frac{(\nabla \times \boldsymbol{\tau})_3}{\rho H} + K_H \nabla_H^2 \zeta.
$$

Now using Eq. (18.1c) we get:

$$
\frac{d\zeta}{dt} + \frac{df}{dt} - \frac{\zeta + f}{H}\frac{dH}{dt} = \frac{(\nabla \times \boldsymbol{\tau})_3}{\rho H} + K_H \nabla_H^2 \zeta.
$$

Dividing by H we get:

$$
\frac{1}{H}\frac{d(\zeta + f)}{dt} - \frac{\zeta + f}{H^2}\frac{dH}{dt} = \frac{(\nabla \times \boldsymbol{\tau})_3}{\rho H^2} + \frac{K_H}{H}\nabla_H^2 \zeta,
$$

so, finally, we obtain:

$$
\frac{d}{dt}\left(\frac{\zeta + f}{H}\right) = \frac{(\nabla \times \boldsymbol{\tau})_3}{\rho H^2} + \frac{K_H}{H}\nabla_H^2 \zeta. \quad (18.3)
$$

The ratio between the absolute vorticity and the thickness of the fluid,

$$
\Pi = \frac{\zeta + f}{H} \quad (18.4)
$$

is called *potential vorticity* and the equation just obtained describes the *evolution of potential vorticity in the shallow-water approximation*. As already emphasised, no further approximation was imposed in deriving this equation; therefore, the dynamics is perfectly equivalent to that of the original equations in shallow water. The important reformulation of these equations for subinertial motions will be developed in Sect. 19.1.

In conclusion, Eq. (18.3) expresses the Lagrangian variation of the absolute vorticity—normalised with respect to H—due to (i) the input of vorticity from the surface wind stress curl and (ii) the dissipative effects produced by turbulent viscosity. In Sect. 16.4 (Eq. 16.23) we saw that the vertical Ekman pumping velocity w_E (equal to the divergence of Ekman transport which, in turn, modulating the free surface generates geostrophic currents) is proportional to the curl of the wind stress. The latter now naturally emerges as a forcing of potential vorticity in Eq. (18.3).

18.2 Conservation of Potential Vorticity

In some cases of great interest, it is possible to assume with good approximation that the turbulent viscosity and the input of vorticity from the wind are negligible, in which case Eq. (18.3) is reduced to:

$$\frac{d\Pi}{dt} = 0. \tag{18.5}$$

In this case, we speak of *conservation of potential vorticity* (or, the angular momentum of the fluid column) in the Lagrangian sense. The next two paragraphs will consider cases in which the imposition of this condition is plausible or even substantially verified (as in the case of Rossby waves); this will allow us to deduce simple but fundamental information on relevant aspects of meteorological and oceanographic dynamics. It is also emphasised that there are cases of absolute interest, such as, for example, the Sverdrup balance (Sect. 20.3) and western boundary currents (Sect. 20.4), for which Π is not conserved.

It should be noted that, while the evolution equation Eq. (18.3) refers specifically to the ocean, as it includes the wind forcing (which has no direct analogy for the atmosphere), the conservation law expressed by Eq. (18.5) can also be applied to the atmosphere. See Sect. 19.4 for notes on the generalisation of the conservation of potential vorticity in the quasigeostrophic approximation in a stratified fluid, also with reference to the atmosphere.

18.3 Zonality of Prevailing Winds, Topographic Effects

It has already been mentioned in Sect. 14.1 that the average zonal component of the winds prevails over the meridional one: in Fig. 14.1 that component is shown as a function of latitude and altitude. This property can be deduced from the conservation of Π, which can be approximately invoked for winds in the free atmosphere. Given that such dynamics is typically subinertial ($\varepsilon \ll 1$), Eq. (18.4) reduces to $\Pi \cong f/H$. Consider a flat-bottom beta-plane atmosphere ($d = 0$, orographic effects negligible on large scales) in the northern hemisphere: since $|\eta| \ll D$ we have $H \cong D = const$, hence in this case the conservation of Π is equivalent to the Lagrangian conservation of the planetary vorticity of the generic fluid column. But the constancy of f implies that the motion occurs along the parallels ($y = const$), from which derives the zonal character of the prevailing winds.

If we relax the assumption of ζ being completely negligible compared to f, we obtain an interesting result that highlights the different behaviour of westerly and easterly winds. Consider for simplicity a fluid column initially with $\zeta = 0$ in purely zonal motion in the northern hemisphere along $y = 0$, therefore with initial absolute vorticity equal to f_0. Any deviation to the North of a *westerly* wind will lead to the birth of a positive curvature vorticity and to an increase of planetary vorticity. Consequently, in such a process both ζ and f undergo a positive variation; Π cannot therefore be conserved (the same happens, with opposite sign, for a deviation to the South). On the contrary, for *easterly* winds, a deviation to the North leads to the birth of a negative curvature vorticity, therefore, the positive variation of f can be compensated by the negative one of ζ in such a way that the absolute vorticity remains constant and equal to f_0 (again, the same happens with the opposite sign for a shift towards the South).

Among the various topographic effects that can be treated with the conservation of potential vorticity, we consider here the case of an oceanic flow of reduced spatial scale for which the f-plane approximation applies ($f = f_0 = const$); also consider the northern hemisphere ($f_0 > 0$), a variation of the topography as that illustrated in Fig. 18.1a and an upwind uniform flow ($\zeta_0 = 0$) perpendicular to the isobaths (Fig. 18.1b). Before the interaction with the topographic variation, each fluid column possesses the potential vorticity $\Pi_0 = f_0/H_0$. On the ramp, with H decreasing (it is assumed that the variation of H due to the free surface is negligible compared to the topographic variation) we will have $\Pi = (\zeta + f_0)/H$; putting $\Pi = \Pi_0$ we get

$$\zeta = f_0\left(\frac{H}{H_0} - 1\right) < 0.$$

In essence, the decrease in depth produces an induction of relative vorticity by the planetary vorticity which, in the case in question, is negative (anticyclonic); Fig. 18.1b shows the deviation of the flow due to the negative curvature vorticity. In a complete treatment of this process (which is beyond the scope of this text) it is possible to establish an analogy between this phenomenon and the electromagnetic

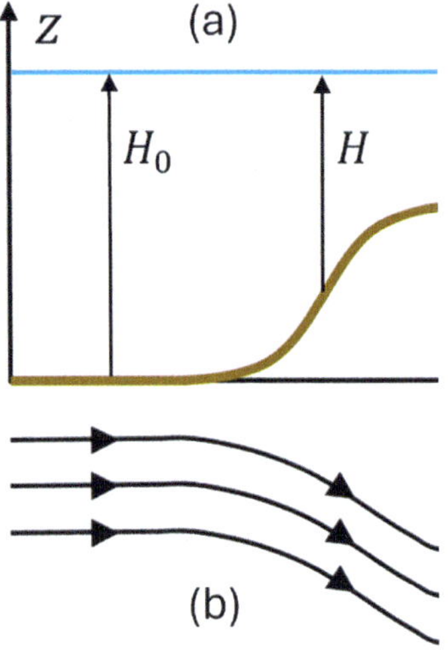

Fig. 18.1 Topographic effect on the f-plane. **a** Topographic variation (not to scale) with rectilinear isobaths perpendicular to the vertical plane shown in the figure. **b** Deflection on the horizontal plane of a homogeneous upwind flow due to the conservation of potential vorticity

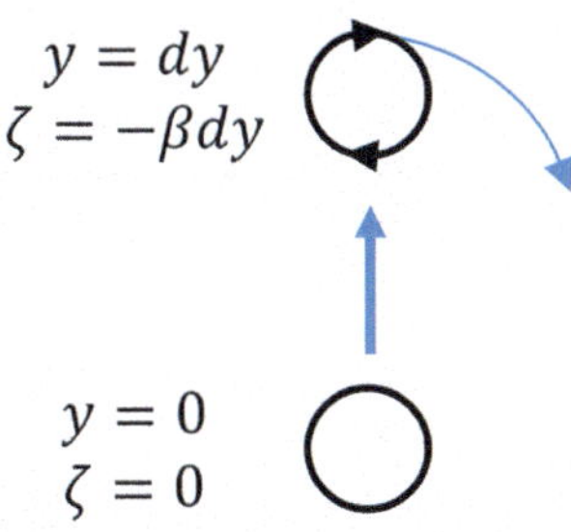

Fig. 18.2 Diagram of the Rossby wave restoring mechanism

induction described by Faraday-Neumann's law (see Appendix A, Application 2); for this in-depth study see Pedlosky (1987).

18.4 The Restoring Mechanism of Rossby Waves

In Sect. 19.2 Rossby waves will be introduced; it is anticipated that these consist of meridional oscillations of large-scale (quasi) geostrophic flows whose phase propagates westward (see Fig. 19.2). Every wave field in a mechanical system owes its existence to a certain *restoring mechanism*; for example, it has been seen how gravity and reduced gravity constitute the restoring forces that allow the existence, respectively, of surface and internal gravity waves. What is the restoring mechanism at the base of the existence of Rossby waves? To identify it one can exploit the conservation of potential vorticity which, in the specific case, is verified with good approximation.

Consider a layer of fluid on the beta-plane with a flat bottom in the northern hemisphere and imagine a column of fluid with zero relative vorticity initially located at latitude $y = 0$ (Fig. 18.2): its absolute vorticity is, therefore, initially equal to f_0. A shift towards the North of the fluid column of a certain $dy > 0$ will entail, for the conservation of Π, an induction of relative vorticity by the planetary vorticity equal to $\zeta = -\beta dy < 0$. To this must be associated a local negative curvature vorticity that will tend to move the fluid column towards the South (Fig. 18.2), bringing it back to its original position. However, its inertia will make it progress further South of $y = 0$, where the opposite mechanism will occur: this will produce a meridional oscillation.

In conclusion, the restoring mechanism that gives rise to the Rossby waves derives from the induction of relative vorticity by the planetary vorticity as a consequence of a meridional displacement of the fluid column, and necessarily requires that f depends on the latitude, effect taken into account in the beta-plane approximation. Ultimately, the sphericity of the earth is essential in this mechanism.

Bibliography

Pedlosky, J.: Geophysical Fluid Dynamics. Springer-Verlag, New York (1987)

Further Recommended Reading

Cushman-Roisin, B.: Introduction to Geophysical Fluid Dynamics. Prentice-Hall, Englewood Cliffs, New Jersey (1994)
Gill, A.E.: Atmosphere-Ocean Dynamics. Academic Press, New York (1982)
Holton, J.R.: An introduction to Dynamic Meteorology. Elsevier, Amsterdam (2004)
Marshall, J., Plumb, R.A.: Atmosphere, Ocean, and Climate Dynamics. Elsevier, Amsterdam (2008)
Monin, A.S.: Theoretical Geophysical Fluid Dynamics. Kluwer Academic Publishers, Dordrecht (1990)
Vallis, G.K.: Atmospheric and Oceanic Fluid Dynamics. Cambridge University Press, Cambridge (2006)

Chapter 19
Quasigeostrophic Approximation, Rossby Waves

In this chapter, the potential vorticity evolution equation derived in the previous chapter is reformulated for subinertial motions within the so-called quasigeostrophic approximation. In this dynamical context, it will be easy to study Rossby waves, whose main atmospheric and oceanic characteristics will be analysed.

19.1 Quasigeostrophic Approximation

The shallow-water approximation allows the study of large-scale meteorological and oceanographic motions through simpler equations than those of Navier–Stokes (with Coriolis force). This advantage, however, comes at a cost, which in that context is nevertheless acceptable: the new system only admits hydrostatic motions. On the other hand, if—always for large-scale spatial motions—one wants to focus on large-scale temporal motions (subinertial motions, $T \gg T_i$), one can adopt a further approximation, called *quasigeostrophic*, which filters out superinertial motions. If this limitation is acceptable, this approximation offers the great advantage of leading to an even simpler mathematical problem as, as will be seen, it includes only one prognostic variable.

In the *quasigeostrophic approximation*, therefore, in addition to the assumption $\delta \ll 1$ that characterises the shallow-water approximation, it is also required that the flows are subinertial, $\varepsilon \ll 1$, but such as not to exclude the presence of inertial acceleration in the dynamical balance. In this way, a prognostic equation will be obtained for a nonlinear flow (an exactly geostrophic flow is stationary and linear), while essentially maintaining the geostrophic balance (Eq. 12.22).

Figure 19.1 summarises the steps leading from the Navier–Stokes equations (with Coriolis force) to the quasigeostrophic ones. The heuristic derivation that will now be presented has the advantage of being simple and intuitive; for a rigorous mathematical derivation the reader is referred to the text by Pedlosky (1987).

S. Pierini, *Oceanic and Atmospheric Fluid Dynamics*, UNITEXT for Physics,
https://doi.org/10.1007/978-3-031-77991-6_19

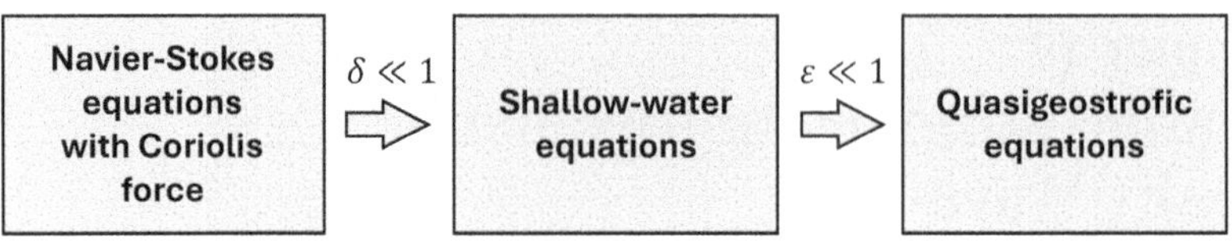

Fig. 19.1 Diagram summarising the transition from the primitive Navier–Stokes equations (with Coriolis force) to the quasigeostrophic equations

The condition $\varepsilon \ll 1$ implies, with good approximation, the geostrophic balance, expressed by the diagnostic relations

$$\begin{cases} u \cong -\frac{\partial \psi}{\partial y} \\ v \cong \frac{\partial \psi}{\partial x} \end{cases} \tag{19.1}$$

where (Sect. 12.4) the streamfunction is defined by $\psi = p/(\rho f) = \eta g/f$ (here it is convenient to consider the more general beta-plane); the continuity equation is therefore automatically verified. The prognostic equation that ψ must satisfy is provided by the potential vorticity evolution equation in shallow water (Eq. 18.3), which will now be approximated for subinertial motions.

First of all, the potential vorticity is approximated taking into account that $\varepsilon \ll 1$. Defining $\theta = (d - \eta)/D$ ($|\theta| \ll 1$) we have:

$$\Pi = \frac{\zeta + f}{H} = \frac{\zeta + f}{D(1-\theta)} \cong \frac{\zeta + f}{D}(1+\theta) \cong$$
$$\cong \frac{1}{D}\left(\zeta - \frac{f_0 \eta}{D} + f + \frac{f_0 d}{D}\right) = \frac{1}{D}\left(\nabla^2 \psi - \frac{\psi}{R_e^2} + f + \frac{f_0 d}{D}\right),$$

where $R_e = \sqrt{gD}/f_0$ is the external Rossby deformation radius already defined by Eq. (16.20) and $\zeta = v_x - u_y = \psi_{xx} + \psi_{yy}$; moreover, the terms $\zeta\theta$ have been neglected because $|\zeta\eta/D| \ll |\zeta|$ and, for subinertial motions, $|\zeta| \ll |f|$ (see Sect. 13.3). The term $-\psi/R_e^2$ represents the effect of the horizontal divergence of the flow, to which is associated an Ekman pumping/suction (see Sect. 16.4) that deforms the free surface (the effect associated with this term is called *vertical stretching*). Note that this term has the same weight as relative vorticity for horizontal length scales L for which $L \approx R_e$; for smaller scales relative vorticity prevails, and vice versa. Finally, the term $f_0 d/D$ represents topographic effects; see Sect. 20.5 for the discussion of a particularly significant case.

Now inserting this approximate expression of Π into the evolution equation of potential vorticity in the shallow-water approximation (Eq. 18.3) we get:

$$\frac{\partial q}{\partial t} + J(\psi, q) = \frac{(\nabla \times \boldsymbol{\tau})_3}{\rho D} + K_H \nabla_H^4 \psi, \tag{19.2}$$

where $\nabla_H^4 \psi = \psi_{xxxx} + 2\psi_{xxyy} + \psi_{yyyy}$, the scaled potential vorticity $q = D\Pi$ is given by

$$q = \nabla^2 \psi - \frac{\psi}{R_e^2} + f_0 + \beta y + \frac{f_0 d}{D} \tag{19.3}$$

and where the bilinear Jacobian differential operator J is defined as $J(a, b) = a_x b_y - a_y b_x$ (note the approximation $H^2/D \sim D$ at the denominator of the wind forcing and $H/D \sim 1$ at the denominator of the viscous term). The Jacobian expresses the advective terms, among which the nonlinear $J\left(\psi, \nabla^2\psi\right)$ (obviously $J\left(\psi, -\psi/R_e^2\right) = 0$). Taking into account that the isolines of ψ coincide with the streamlines, in a closed domain A bounded by the solid lateral boundary ∂A, the free-slip condition is given by

$$\psi|_{\partial A} = C(t), \tag{19.4}$$

where $C(t)$ is an unknown function depending *only* on time that will be determined by imposing mass conservation:

$$\frac{d}{dt} \iint\limits_{A} \psi \, ds = 0. \tag{19.5}$$

Note that for non-divergent flows for which the term ψ/R_e^2 is negligible in q, the constant C in Eq. (19.4) is arbitrary (the geostrophic balance Eq. (19.1) is not affected by this arbitrariness), so the condition Eq. (19.5) is not required.

In conclusion, the prognostic equation Eq. (19.2), called the *evolution equation of potential vorticity in the quasigeostrophic approximation* for a homogeneous and incompressible fluid, supplemented by the conditions Eqs. (19.4–19.5), allows to determine the evolution of the single variable ψ from the initial condition $\psi(x, y, 0) = \psi_0(x, y)$. In turn, ψ provides the horizontal velocity field through the geostrophic balance (Eq. 19.1).

It is interesting to re-obtain the Taylor-Proudman theorem (Sect. 13.1) in the quasigeostrophic context. Imposing the conditions of stationarity and linearity ($\varepsilon \cong 0$), absence of turbulent viscosity effects ($E_H \cong 0$), considering the wind stress curl null and being on the f-plane, Eq. (19.2) reduces to:

$$J(\psi, d) = 0. \tag{19.6}$$

But this equation is verified if

$$\psi = F(d) \tag{19.7}$$

for any choice of the real function F. Therefore, the streamlines must coincide with the isobaths, that is, each fluid column must always remain at the same depth: these are nothing but the Taylor columns of the Taylor-Proudman theorem. On the other hand, the quasigeostrophic Eq. (19.2) allows to describe the more realistic dynamics associated with the deviation of the fluid columns from the isobaths.

In conclusion, it is worth noting that if the hypothesis $\delta \ll 1$ filters out deep water (non-hydrostatic) waves while preserving shallow water waves, the further hypothesis $\varepsilon \ll 1$ also filters out the latter, as they are superinertial, so that the surface gravity waves are completely absent in the quasigeostrophic dynamics. In this approximation, the acceleration of gravity g appears only in the Rossby deformation radius. If the gravity waves are absent, the dynamics of the Rossby waves are present, whose mathematical treatment is greatly facilitated by the quasigeostrophic equations, as will be seen in the next paragraph.

19.2 Rossby Waves

Rossby waves (named after the Swedish-American meteorologist Carl-Gustaf Rossby, 1898–1957, who first explained their nature), also known as *planetary waves*, play a fundamental role in the subinertial adjustment in both the atmosphere and the oceans. There are Rossby waves at mid-high latitudes and Rossby waves at low latitudes included in the equatorial belt: in this case we speak of equatorial Rossby waves. Although this second category of waves is of great importance in climate (think, for example, of the dynamics of El Niño in the context of the coupled ocean–atmosphere phenomenon El Niño-Southern Oscillation—ENSO—, whose mechanism is based on the activation of Kelvin and equatorial Rossby waves), here we will limit ourselves to dealing with the first category of waves which, in the case of the atmosphere, helps to explain the nature of synoptic disturbances at mid-latitudes. There are also topographic Rossby waves, whose existence derives from the topographic beta effect discussed in Sect. 20.5; this topic will also be omitted here.

As an example of the phenomenology of atmospheric Rossby waves at mid-latitudes, one can consider the *polar jet stream* in the northern hemisphere: this intense westerly wind system is located near the tropopause (around $250-300\ mbar$) between the Ferrel cell and the polar cell (averaging around $60°N$) and is much more irregular and variable than the subtropical one shown in Fig. 14.1, actively participating in the weather at mid-latitudes (which is why it does not appear in a climatological dataset like the one shown in that figure). This jet tends to develop large-scale meridional undulations like those shown in Fig. 19.2, which propagate westward: these are the Rossby waves, whose restoring mechanism was revealed in Sect. 18.4. If their phase velocity is westward, it is also true that the signal propagates in a fluid that moves eastward (the westerly winds) with a comparable speed in modulus, so that, relative to the Earth, the wave moves more slowly—and predominantly—eastward, but eventually also westward or presenting persistent blocking phenomena.

A detailed analysis of Rossby waves in the atmospheric and oceanic context is very complex and beyond the scope of this text. In the remaining paragraphs, we will limit ourselves to deriving the dispersion relation for barotropic (Sect. 19.3) and

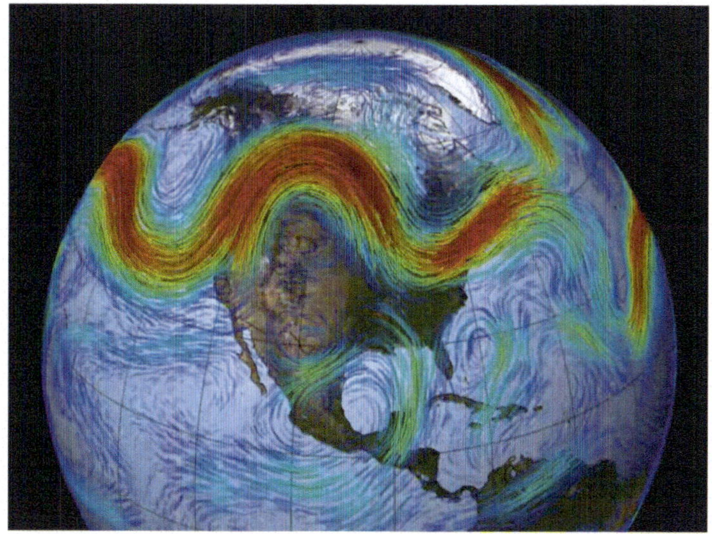

Fig. 19.2 Dynamics of Rossby waves on the polar jet stream. Image taken from an animation of NASA's Goddard Space Flight Center (NASA/GSFC/JPL/LaRC, MISR Science Team; https://misr.jpl.nasa.gov/) available on the website of the National Ocean Service of the National Oceanic and Atmospheric Administration (NOAA, https://oceanservice.noaa.gov/facts/rossby-wave.html)

baroclinic (Sect. 19.4) Rossby waves from the equation of conservation of potential vorticity in the quasigeostrophic approximation. Finally, in Sect. 19.5 we will derive the phase and group velocities and apply them to the different atmospheric and oceanic contexts.

19.3 Dispersion Relation for Barotropic Waves

We begin by considering barotropic Rossby waves, so we can directly refer to the results of Sect. 19.1. Rossby waves are quasigeostrophic and satisfy with good approximation the conservation of potential vorticity; therefore, Eq. (19.2) gives:

$$\frac{\partial q}{\partial t} + J(\psi, q) = 0. \tag{19.8}$$

In the linear approximation (negligible nonlinear terms $J\left(\psi, \nabla^2\psi\right)$) and for a flat bottom ($d = 0$) Eq. (19.8) reduces to:

$$\frac{\partial}{\partial t}\left(\nabla^2\psi - \frac{\psi}{R_e^2}\right) + \beta\frac{\partial \psi}{\partial x} = 0. \tag{19.9}$$

Searching for a solution of the type

$$\psi = ae^{i(k_1x+k_2y-\sigma t)}$$

as already done for gravity waves, we obtain the following dispersion relation:

$$\sigma = -\frac{\beta k_1}{k_1^2 + k_2^2 + 1/R_e^2}. \tag{19.10}$$

Remember that R_e is the external Rossby deformation radius (Eq. 16.20). At mid-latitudes we have $R_e \approx 2000$ km in the ocean and $R_e \approx 3000$ km in the troposphere.

19.4 Dispersion Relation for Baroclinic Waves

To refer to Rossby waves in the atmosphere and in the ocean under realistic dynamic and parametric conditions, it is necessary to consider baroclinic waves in a stratified fluid. The generalisation of Eq. (19.8) to the case of a continuously stratified fluid goes beyond the scope of this discussion; here we will limit ourselves to presenting the generalisation of the dispersion relation Eq. (19.10) (for the mathematical derivation, see, for example, the texts by Le Blond and Mysak 1978; Gill 1982; Cushman-Roisin 1994). In the application to the atmosphere, the fluid layer corresponds to the troposphere, the free surface to the tropopause and the lower boundary to the Earth's surface.

The analysis of the behaviour of linear Rossby waves in a stratified fluid requires the introduction of a streamfunction with the following form:

$$\psi(x, y, z, t) = a(z)\cos(k_1x + k_2y - \sigma t). \tag{19.11}$$

This expression must be inserted in the generalisation of Eq. (19.8) for a stratified fluid and the vertical boundary conditions treated in Sect. 17.2 must be imposed. In this way, we reduce to an eigenvalue problem similar to that obtained for internal gravity waves (cf. Eqs. 11.8 and 11.9). Considering a constant Brunt-Väisälä frequency (Eq. 11.5), $N = const$, the dispersion relation turns out to be

$$\sigma = -\frac{\beta k_1}{k_1^2 + k_2^2 + k_3^2 f_0^2/N^2} \tag{19.12}$$

where the vertical wave number k_3 appears in the expression of $a(z)$:

$$a(z) = A\cos(k_3 z). \tag{19.13}$$

The eigenvalue problem admits the barotropic mode and an infinite and discrete set of baroclinic modes.

In the barotropic mode ψ is independent of z, therefore $k_3 D \ll 1$. In this case we obtain

$$k_3 = \frac{N}{\sqrt{gD}}$$

Consequently $k_3^2 f_0^2 / N^2 = 1/R_e^2$, so that Eq. (19.12) returns Eq. (19.10). Therefore, the barotropic Rossby mode (also called *external*) in a stratified fluid coincides with the only (barotropic) mode present in a fluid of constant density, as indeed happens for long internal waves (see Sect. 11.2).

In *baroclinic modes* we have

$$k_3^{(n)} = \frac{n\pi}{D}$$

where $n = 1, 2, 3, \ldots$; in this case ψ presents strong variations along the vertical and multiple sign changes given by Eq. (19.13). Consequently, $k_3^2 f_0^2 / N^2 = (n\pi)^2 / R_i^2$, where

$$R_i = \frac{ND}{f_0} = \frac{\sqrt{g'D}}{f_0} \tag{19.14}$$

is the *internal Rossby deformation radius* (the expression in terms of the reduced gravity g' is obtained from Eq. (11.5) considering $N = const$). Similarly to what was observed in relation to potential vorticity for a fluid of constant density (Eq. 19.3), for horizontal length scales L such that $L \approx R_i$, the vertical stretching of the fluid column—here due to buoyancy—has the same weight as the relative vorticity. At mid-latitudes, for a typical value $N \approx 0.01\,\text{rad}\,\text{s}^{-1}$ for the standard atmosphere in the troposphere, we get $R_i \approx 1000\,\text{km}$. On the other hand, in the oceans, for a typical value $N \approx 0.002\,\text{rad}\,\text{s}^{-1}$ we get $R_i \approx 50 - 200\,\text{km}$. In general, $R_i \ll R_e$ since $g' \ll g$.

In conclusion, the dispersion relation of Rossby waves assumes the same analytical form for barotropic and baroclinic waves:

$$\sigma = -\frac{\beta k_1}{k_1^2 + k_2^2 + 1/R^2} \tag{19.15}$$

where R is defined as follows:

$$\begin{cases} \textit{barotropic mode} \rightarrow R = R_e \\ \textit{baroclinic modes} \rightarrow R = R_i/(n\pi) \end{cases}$$

19.5 Rossby Waves in the Atmosphere and in the Ocean

Before analysing Rossby waves in the atmospheric and oceanic context, the phase and group velocities derived from Eq. (19.15) must be analysed. For this purpose, it should be noted that this dispersion relation is profoundly different from those for surface gravity waves (Eq. 9.21) or internal waves for a two-layer fluid (Eq. 11.12). Indeed, in those cases σ depends on $k = \left(k_1^2 + k_2^2\right)^{1/2}$, therefore, symmetrically on its two components: this implies that those wave fields are isotropic and that the phase velocity is always parallel to the group velocity.

On the contrary, in Eq. (19.15) σ presents a dependence on the zonal component of the wave number k_1 different from that of the meridional component k_2 due to the anisotropy introduced by the beta-effect. The first consequence is that the modulus of $\mathbf{c_p}$ e $\mathbf{c_g}$ will depend on the dirèction of $\mathbf{k}$ for the same k; the second is that $\mathbf{c_p}$ and $\mathbf{c_g}$ will generally not be parallel vectors. For their calculation it is convenient to reformulate the definitions introduced in Chap. 9 ($c_p = \sigma/k$, $c_g = \partial\sigma/\partial k$) as follows:

$$\begin{cases} c_{pi} = \sigma \frac{k_i}{k^2} \\ c_{gi} = \frac{\partial \sigma}{\partial k_i} \end{cases} \tag{19.16}$$

(in the first, $\mathbf{k}/k$ is the unit vector in the direction of wave propagation). Substituting the expression of σ (Eq. 19.15) into Eq. (19.16) we get:

$$\begin{cases} \mathbf{c_p} = \frac{1}{k^2(k^2+1/R^2)}\left(-\beta k_1^2, -\beta k_1 k_2\right) \\ \mathbf{c_g} = \frac{\beta}{\left(k^2+1/R^2\right)^2}\left(k_1^2 - k_2^2 - 1/R^2, 2k_1 k_2\right) \end{cases} \tag{19.17}$$

For a detailed analysis of these expressions, refer to specialist texts. Here we will limit ourselves to considering the particularly simple case of waves with purely zonal propagation, for which $\mathbf{c_p}$ and $\mathbf{c_g}$ are parallel, so that it is possible to compare them in a graph. For $\mathbf{k} = (k_1, 0)$, Eq. (19.17) are reduced to:

$$\begin{cases} c_{p1} = -\frac{\beta}{k_1^2+1/R^2} \\ c_{g1} = \beta \frac{k_1^2-1/R^2}{\left(k_1^2+1/R^2\right)^2} \end{cases} \tag{19.18}$$

Figure 19.3 shows c_{p1} (thick blue line) and c_{g1} (red line).

Since c_{p1} is negative definite, the phase of the Rossby waves always propagates westward. For very short waves ($k_1 \gg 1/R$) the phase velocity tends to

$$c_{p1} \cong -\frac{\beta}{k_1^2} \tag{19.19}$$

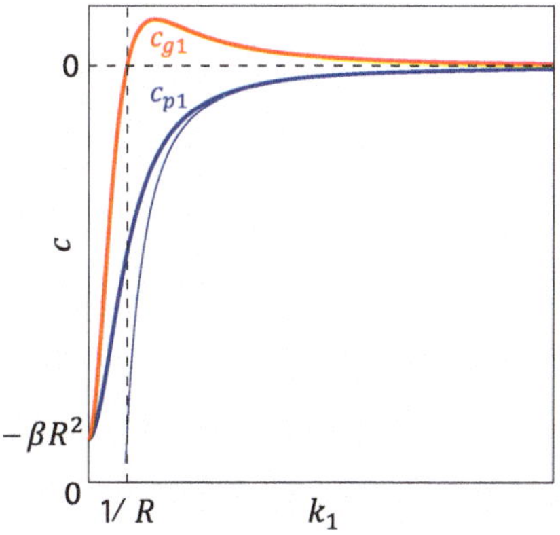

Fig. 19.3 Blue line (red line): phase velocity c_{p1} (group velocity c_{g1}) for Rossby waves with purely zonal propagation: $\mathbf{k} = (k_1, 0)$ (Eq. 19.18). The thin blue line shows the phase velocity for short waves (Eq. 19.19)

(thin blue line). As for the group velocity, c_{g1} is negative for $k < 1/R$ but is positive for shorter wavelengths; in this range Rossby wave packets, along with their energy, travel eastward while within them the wave phase shifts westward. Conversely, for very long waves ($k \to 0$) the waves are non-dispersive, with $c_{p1} = c_{g1} = -\beta R^2$.

In atmospheric applications, for barotropic waves the condition $k_1 > 1/R$ is verified, so Eq. (19.19) can be used, which implies $c_{p1} \propto \lambda^2$. In order to take into account the westerly winds to which the Rossby waves are superimposed (see Sect. 19.2), it is simple to generalise Eq. (19.9) by considering the advection associated with a constant zonal velocity U. From the new dispersion relation it follows that the wave propagation speed relative to the Earth, $\overline{c}_{p1}$, is simply the sum of the phase velocity given by Eq. (19.19) (negative) and U (positive):

$$\overline{c}_{p1} = c_{p1} + U. \tag{19.20}$$

Depending on the wavelength λ, the sign of $\overline{c}_{p1}$ will be positive if $U > |c_{p1}|$, and vice versa. In reality $\overline{c}_{p1}$ is generally positive (eastward propagation) but there can be persistent blocking situations during which $\overline{c}_{p1} \approx 0$. For baroclinic waves Eq. (19.19) may not be a good approximation, but Eq. (19.20) will still apply, where of course c_{p1} will be given by Eq. (19.18) (or by Eq. (19.17) in the more general case).

Moving to the ocean, Fig. 19.4 shows an example of baroclinic Rossby waves propagating zonally in the North Pacific Ocean; the signal represents the sea surface anomaly measured by the Ssalto/Duacs multimission radar altimeter. In this regard, it should be noted that, although the most distinctive feature of Rossby waves is the horizontal oscillation of geostrophic currents, this is nevertheless associated with an oscillation of the free surface $\eta(= \psi f/g)$, the observation of which can therefore reveal the waves. The westward propagation is highlighted by the inclination of the positive and negative bands in the longitude-time diagram. These waves provide the teleconnection between the central Pacific, where they are generated, and the KE area, determining the synchronisation between the intrinsic low-frequency variability

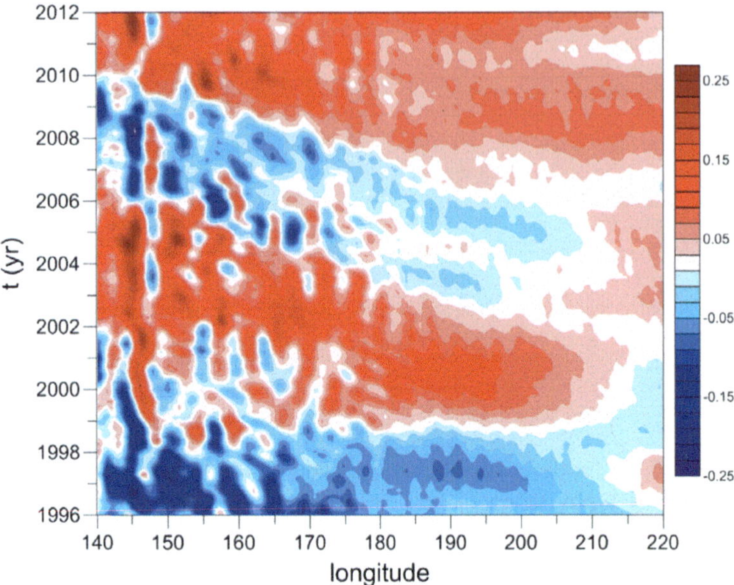

Fig. 19.4 Example of propagation of oceanic Rossby waves in the North Pacific Ocean. Longitude-time diagram of the anomaly of Absolute Dynamic Topography (ADT) (filtered over 1 year and averaged in the $31°N - 36°N$ band) obtained from altimetric data of the Ssalto/Duacs multimission and distributed by Archiving, Validation and Interpretation of Satellite Oceanographic data (AVISO). Figure taken from Gentile et al. (2018, Ocean Model. 131, 24–39, © Elsevier. Used with permission)

of that current and the oscillation modes of the North Pacific (see the discussion in Sect. 17.5).

From Fig. 19.4 it appears that the wavelength λ is much greater than the internal Rossby deformation radius ($R_i \approx 50-100$ km, cf. Sect. 19.4): $k^2 \ll R_i/(n\pi)$. Therefore, the waves fall within the non-dispersive limit for which $c_{p1} = c_{g1} = -\beta R^2$. For the first baroclinic mode ($n = 1$, by far the most energetic), from Eqs. (12.14 and 19.14) we get:

$$c_{p1} = c_{g1} = -\frac{2\Omega_e}{r_e} \cos\varphi_0 \frac{N^2D^2}{\pi^2 f_0^2} = -\frac{N^2D^2}{2\pi^2\Omega_e r_e} \frac{1}{\sin\varphi_0 \tan\varphi_0} \tag{19.21}$$

from which an interesting dependence on mid-low latitudes emerges: $c_{p1} \propto \varphi_0^{-2}$. In fact, the propagation speed of oceanic Rossby waves revealed by altimetric data is in good agreement with that predicted by this linear theory (Chelton and Schlax 1996). For a discussion on nonlinear effects, refer to Sect. 20.2 .

Bibliography

Chelton, D.B., Schlax, M.G.: Global observations of oceanic Rossby waves. Science **272**, 234–238 (1996)
Cushman-Roisin, B.: Introduction to Geophysical Fluid Dynamics. Prentice-Hall, Englewood Cliffs, New Jersey (1994)
Gentile, V., Pierini, S., de Ruggiero, P., Pietranera, L.: Ocean modelling and altimeter data reveal the possible occurrence of intrinsic low-frequency variability of the Kuroshio Extension. Ocean Model **131**, 24–39 (2018)
Gill, A.E.: Atmosphere-Ocean Dynamics. Academic Press, New York (1982)
Le Blond, P.H., Mysak, L.A.: Waves in the Oceans. Elsevier, New York (1978)

Further Recommended Reading

Holton, J.R.: An Introduction to Dynamic Meteorology. Elsevier, Amsterdam (2004)
Marshall, J., Plumb, R.A.: Atmosphere, Ocean, and Climate Dynamics. Elsevier, Amsterdam (2008)
Monin, A.S.: Theoretical Geophysical Fluid Dynamics. Kluwer Academic Publishers, Dordrecht (1990)
Pedlosky, J.: Geophysical Fluid Dynamics. Springer, New York (1987)
Vallis, G.K.: Atmospheric and Oceanic Fluid Dynamics. Cambridge University Press, Cambridge (2006)

Chapter 20
Atmospheric and Oceanic Vortices, Wind-Driven Ocean Circulation

In this concluding chapter, we discuss atmospheric motions on the synoptic scale and mesoscale oceanic motions, both of which originate from baroclinic and barotropic instability mechanisms. We then focus on the main aspects of large-scale wind-driven ocean circulation, such as the Sverdrup balance and western boundary currents. We conclude by discussing some laboratory simulations of these currents, also in connection with potential climate changes.

20.1 Atmospheric Dynamics on the Synoptic Scale

The climatological atmospheric circulation, that is, averaged over large temporal scales and all longitudes (zonal average) shows regular structures, such as, for example, the subtropical jet stream shown in Fig. 14.1. However, the actual unaveraged circulation is quite different, as highlighted, for example, by the map of winds and geopotential heights at 500 mbar over the European continent in Fig. 20.1. The circulation presents significant undulations of the subpolar jet stream (which, as seen in Sect. 19.5, supports Rossby waves), highlighting a variety of geostrophic motions in the form of meanders and coherent cyclonic and anticyclonic structures (respectively on the British Islands and the Baltic Sea in the example in the figure), which structure the weather. The evolution of these circulations occurs on temporal scales from hours to days and its predictability is limited (see Sect. 17.5). It is important to note that the spatial scale in play of O(1000 km), *called synoptic scale*, coincides with that of the Rossby deformation radius (Eq. 19.14).

This coincidence is not accidental. In fact, the process, called *cyclogenesis*, that leads to these dynamics is mainly due to the *baroclinic instability* of the jet stream, which manifests itself on spatial scales $L \approx R_i$ (Charney 1947; Eady 1949). The horizontal flow with vertical shear in geostrophic and hydrostatic balance that involves

S. Pierini, *Oceanic and Atmospheric Fluid Dynamics*, UNITEXT for Physics,
https://doi.org/10.1007/978-3-031-77991-6_20

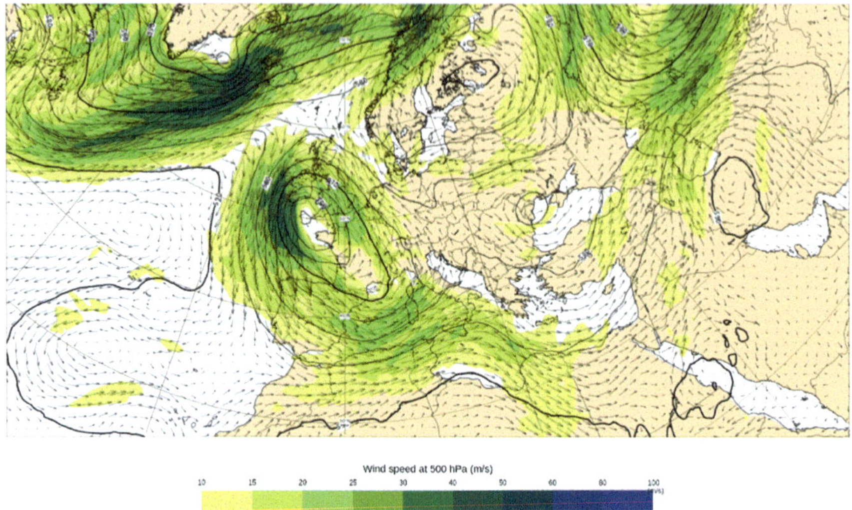

Fig. 20.1 Winds (arrows) and geopotential heights (isolines) at 500 mbar over the European continent from high-resolution forecast for the day 5 September 2024, 00 UTC (the colour indicates the wind amplitude in $\mathrm{m\,s^{-1}}$. Image obtained from © European Centre for Medium-Range Weather Forecasts (ECMWF, https://www.ecmwf.int/, published under a Creative Commons Attribution 4.0 International, CC BY 4.0)

the thermal wind (Chap. 14) is intrinsically unstable, as the potential energy associated with the inclination of the isopycnals tends to reconvert into the kinetic energy of motions like those described above. The linear theory shows that the instability is preferentially active for perturbations with horizontal length scales L such that their kinetic energy (KEn) equals the potential energy available for reconversion (*available potential energy*, APE). It can be shown that

$$\frac{KEn}{APE} \approx \left(\frac{L}{R_i}\right)^2. \tag{20.1}$$

Therefore, the equivalence between KEn and APE implies $L \approx R_i$. It is worth remembering that, as observed in Sect. 19.1, this condition also involves a balance, in the quasigeostrophic potential vorticity, between the relative vorticity and the vertical stretching due to buoyancy: this highlights the essential role played by the Rossby deformation radius in atmospheric dynamics (and, as will be seen, also in ocean dynamics).

In addition to baroclinic instability, there is also the mechanism of *barotropic instability*, in which there is an extraction of kinetic energy from a horizontally sheared jet stream to the advantage of meanders of the current itself, which, as for baroclinic instability, can then give rise to vortices. An important consequence of the motions produced by these two mechanisms, denoted *geostrophic turbulence*, is the horizontal mixing of momentum, which contributes to the horizontal component of eddy viscosity (Sect. 6.4). For a detailed discussion of the mechanisms of baroclinic and barotropic instability and geostrophic turbulence, see, for example, the texts by Gill (1982), Pedlosky (1987), Cushman-Roisin (1994) and Vallis (2006).

20.2 Mesoscale and Sub-basin Oceanic Vortices

The situation in the oceans presents differences but also similarities compared to the atmospheric one. The average sub-surface current velocity in the oceans based on Lagrangian data is shown in Fig. 20.2.

It should first be noted that the only current system similar to the atmospheric jet streams is the *Antarctic Circumpolar Current* (ACC, shown in the figure between $40°S$ and $60°S$), as it is the only one that embraces all longitudes, flowing eastward without interruption. In contrast, the other current systems are zonally limited by the lateral boundaries of the great oceans. *The western boundary currents* stand out; for a brief discussion of them, see Sect. 20.4. Here it is appropriate to focus on the *Gulf Stream* (Stommel 1965; Fofonoff 1981), described in Fig. 20.2 by a rather regular current that originates in the Gulf of Mexico, flows near the coast up to Cape Hatteras (in North Carolina) where it detaches from the continent and, in part recirculates in the North Atlantic and in part extends towards the North-East at high latitudes in the form of the *North Atlantic Drift*, which contributes significantly to the Earth's climate (Sect. 20.5).

Despite the significant phenomenological difference, just like in the atmosphere, eliminating the average makes the regularity of the current disappear and a rich

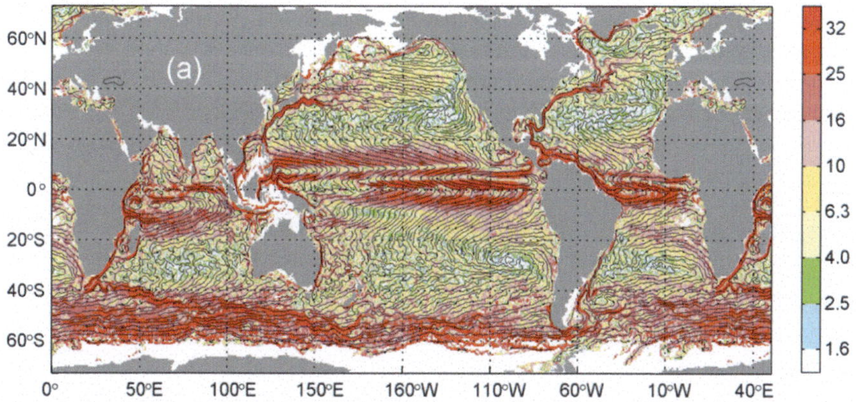

Fig. 20.2 Mean streamlines calculated from drifters at $15-m$ depth (Maximenko et al. 2009; the colour scale indicates the average speed of the drifters in cm s^{-1}). Figure adapted from Maximenko et al. (2009, J. Atmos. Oceanic Technol. 26, 1910–1919, © American Meteorological Society. Used with permission)

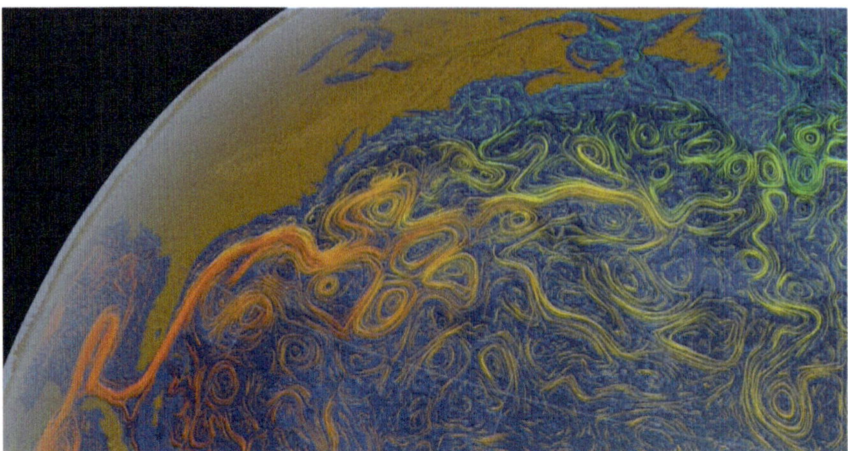

Fig. 20.3 Image adapted from an animation of NASA's Goddard Space Flight Center (NASA/GSFC/JPL/LaRC, MISR Science Team, https://misr.jpl.nasa.gov/. Animators: G. Shirah and H. Mitchell—NASA/GSFC. Scientists: H. Zhang—UCLA, D. Menemenlis—NASA/JPL CalTech) available at https://svs.gsfc.nasa.gov/3913/. The animation, produced by the MIT/JPL "ECCO2" project (https://ecco-group.org/home.cgi), uses the outputs of the MIT general circulation model (MITgcm) to synthesize satellite and in situ data

field of vortices appears. For example, Fig. 20.3 shows an instantaneous snapshot of the surface current obtained from satellite and in-situ data synthesized by an ocean circulation model. As can be seen, the regularity of the current is limited to its first section; after the separation from the continent a variety of vortices appear, both cyclonic and anticyclonic, whose size is given by the internal Rossby deformation radius which, as seen (Sect. 19.4) is of the order of 50–200 km.

These baroclinic vortices, called *mesoscale eddies*, correspond to the atmospheric ones (whose length scale is preferably called *synoptic*) because, although being an order of magnitude smaller, they both verify the condition $L \approx R_i$. In fact, the mechanisms of baroclinic and barotropic instability active in the atmosphere also operate in the oceans in similar ways, extracting energy mainly from the western boundary currents and from the ACC. These vortices, detected mainly by satellite altimetry, have a lifespan that can reach even a few weeks and explain more than half of the oceanic variability (Chelton and Schlax 2007), which is typically of an intrinsic nature (Sect. 17.5).

A salient feature of mesoscale oceanic vortices is their high degree of nonlinearity, which causes these structures to transport mass, momentum and heat in their movement; this underlines their importance on the global heat balance in the ocean. A measure of nonlinearity is given by the ratio r between the fluid velocity inside the vortex and its propagation speed: nonlinearity is significant if $r > 1$. Altimetric data show that r varies from 1 to 4 at extra-tropical latitudes. In the oceanic geostrophic turbulence, a typically nonlinear phenomenon occurs, that of the *inverse energy cascade* from small scales to larger ones (Sect. 17.5); this contributes to giving life to oceanic coherent structures.

Another important feature of mesoscale eddies concerns their propagation: these vortices travel zonally westward with a speed close to that of non-dispersive baroclinic Rossby waves (Sect. 19.5, Eq. 19.21). It is interesting to note that the only mesoscale field that propagates eastward (but at a slower speed) is the one associated with the ACC. The reason is clear: just as happens for the Rossby waves that modulate the polar jet stream, the average eastward flow of the ACC prevails over the westward phase speed of the eddies.

The oceanic mesoscale is not only present in the large oceans but also in the more limited Mediterranean seas. Figure 20.4 shows a snapshot of the surface current in the Mediterranean Sea of the same type as that of Fig. 20.3: mesoscale eddies are omnipresent also in this semi-enclosed sea but with smaller scales than the oceanic ones ($R_i \approx 10 - 40\,\text{km}$). However, it should be emphasised that here, in addition to the mesoscale vortices, there are also a variety of other coherent structures specific to the sea area in question, whose existence depends on the local basin geometry as well as the complex global thermohaline and wind-driven circulation in the Mediterranean Sea (Malanotte-Rizzoli et al. 2014; Pinardi et al. 2015).

In this context, we talk about *sub-basin scale circulation* to refer to those circulations, in many cases wind-driven (wind-driven gyres), whose horizontal structure is strongly conditioned by the shape of the coasts. For example, the Tyrrhenian Sea

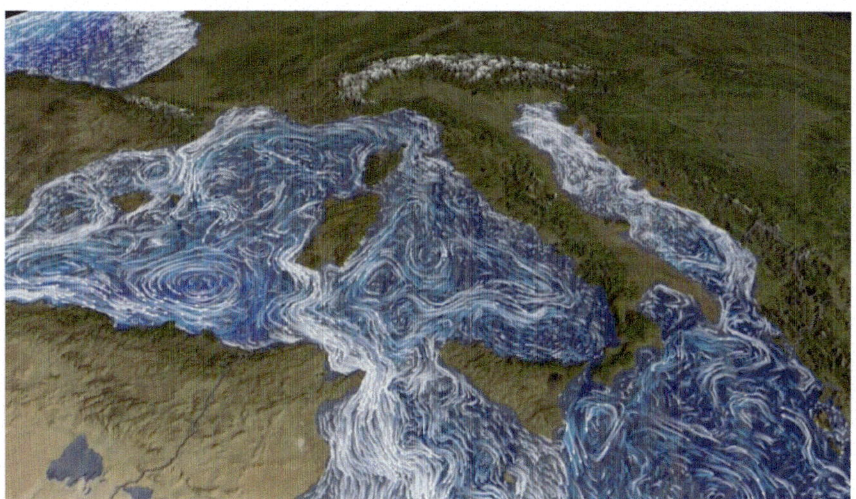

Fig. 20.4 Image taken from an animation of NASA's Goddard Space Flight Center (NASA/GSFC/JPL/LaRC, MISR Science Team, https://misr.jpl.nasa.gov/. Animators: G. Shirah and H. Mitchell—NASA/GSFC. Scientists: H. Zhang—UCLA, D. Menemenlis—NASA/JPL CalTech) available at https://svs.gsfc.nasa.gov/3820/. The animation, produced by the MIT/JPL "ECCO2" project (https://ecco-group.org/home.cgi), uses the outputs of the MIT general circulation model (MITgcm) to synthesise satellite and in situ data

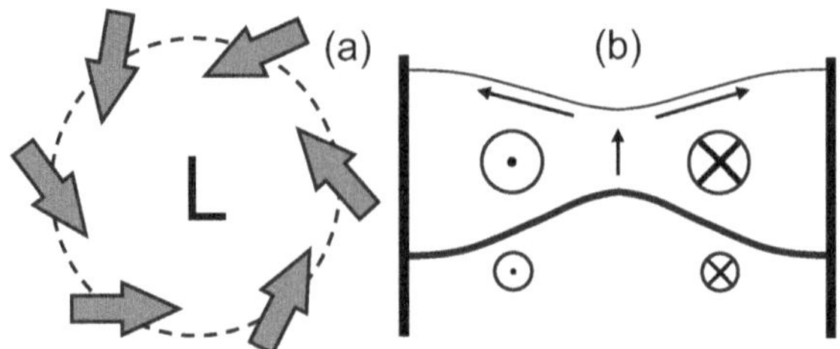

Fig. 20.5 Schematic representation of the effect of a surface wind with cyclonic vorticity on a closed sea with a two-layer structure (northern hemisphere). (**a**) Dashed line: isobar; arrows: surface wind stress. (**b**) Vertical section in the sea; thin line: free surface; thick line: thermocline; side arrows: Ekman transport; vertical arrow: upwelling (arrows and inclinations of surfaces not to scale)

is generally home to a cyclonic circulation, more intense in winter then in summer, which is essentially driven by a typically cyclonic surface wind stress field (Pierini and Simioli 1998).

The mechanism that determines the circulation is described schematically for a two-layer ocean in Fig. 20.5. In Fig. 20.5a the dashed line indicates an isobar at sea level around a low pressure centre; the arrows indicate the surface wind stress and have an ageostrophic component directed towards the centre (Chap. 15). The wind-driven Ekman transport is divergent: this results in a displacement of surface waters (side arrows in Fig. 20.5b) and a lowering of the free surface in the centre of the basin, with corresponding pressure gradient force associated with a surface cyclonic vortex. This displacement also produces an upwelling (Eq. 16.23) and a lifting of the thermocline at the centre of the vortex. The inclination of the thermocline opposite to that of the free surface produces a baroclinic compensation which, in the proposed example, reduces the intensity of the current in the lower layer without altering its direction of rotation. If the wind forcing is constant over time, the circulation of the sea will reach a steady state when the dissipative effects (such as horizontal eddy viscosity and bottom friction) exactly balance the continuous input of momentum, energy and vorticity from the wind.

In relation to this circulation scheme, it is worth highlighting three aspects. Firstly, the symmetry of the surface wind field considered in the example is only a convenient—but not necessary—assumption (in fact, in the specific case of the Tyrrhenian Sea this assumption is not verified); it is the limitation and shape of the basin that provide the horizontal structuring of the circulation. Moreover, baroclinic compensation can produce different scenarios; for example, if the relative pressure gradient force is sufficiently intense, the direction of circulation in the lower layer may be opposite to that of the upper layer. On the other hand, almost perfect baroclinic compensation can occur, in which case the speed in the lower layer is zero (this is the reduced gravity dynamics discussed in Sects. 17.4 and 17.5). Finally, an anticyclonic atmospheric circulation produces a similar circulation in the ocean: in this case, in Fig. 20.5, arrows and inclinations would have the opposite direction.

20.3 Wind-Driven Ocean Circulation, the Sverdrup Balance

The schematic example of Fig. 20.5 relating to a circulation induced in a closed basin by a surface wind stress field with cyclonic vorticity (or anticyclonic in the specular case) can be applied to the case of large-scale wind-driven ocean circulation. Figure 20.6a schematically shows the anticyclonic circulation induced in a rectangular basin in the northern hemisphere by the surface stress field associated with the westerly winds at mid-latitudes (dark grey arrow) and the trade winds at low latitudes (light grey arrow; the fact that the surface winds deviate markedly from the zonal direction is irrelevant in this context). A spatially and temporally constant negative wind stress curl would induce, at regime, a perfectly symmetric stationary anticyclonic oceanic circulation on the f-plane, as shown in the figure.

However, the experimental evidence is in stark contrast with this result. For example, consider the subtropical circulation in the North Atlantic Ocean: as shown in Fig. 20.2, the anticyclonic circulation is strongly asymmetric, despite the wind forcing presenting no zonal asymmetry that could explain the oceanic one. Three important features are noted: (i) along the western boundary of the ocean flows a very intense and narrow current, the Gulf Stream (Fig. 20.3; a schematic of this current is shown in area *A* in Fig. 20.6b), (ii) there is no analogous current on the opposite margin and (iii) the substantial mass transport to the north provided by the

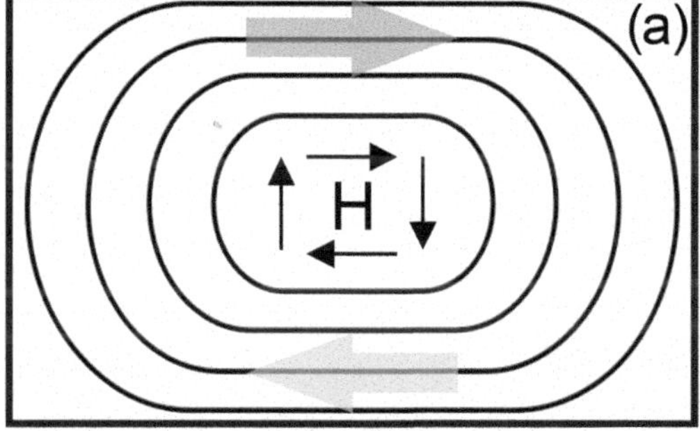

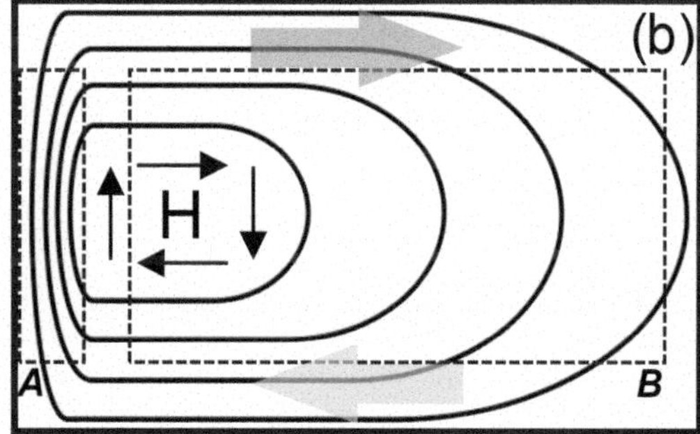

Fig. 20.6 Scheme of circulation in a large-scale rectangular oceanic basin subjected to an anticyclonic surface wind stress field. The large dark grey arrow represents the westerly winds at mid-latitudes while the light grey one represents the trade winds at low latitudes; the thin arrows indicate the anticyclonic direction of the oceanic circulation. **(a)** Circulation on the f-plane. **(b)** Circulation on the beta-plane; area *A* includes the western boundary current while area *B* denotes the oceanic interior in which the Sverdrup balance applies (the extension of the two areas is not to scale)

Gulf Stream recirculates in the *oceanic interior* (net of the North Atlantic Drift) through very weak currents that do not present any significant zonal asymmetry (a schematic of this circulation is shown in area B in Fig. 20.6b). More generally, similar patterns of anticyclonic circulation at mid-latitudes (the *subtropical gyres*) are found in all the major oceans in both the northern and southern hemispheres. In addition, in the northern hemisphere there are also cyclonic circulations at high latitudes (the *subpolar gyres*). In all cases, an intensified current emerges to the west (*western boundary current*), the specific characteristics of which obviously depend on the ocean in question.

One of the main problems of physical oceanography in the first half of the twentieth century was to theoretically interpret the dynamics in the oceanic interior and to identify the dynamic mechanism that produces the western intensification of the currents. The fundamental works of Sverdrup (1947), Stommel (1948) and Munk (1950) answered these questions, highlighting the fundamental role played by the beta-effect (and therefore by the Earth's sphericity) in the wind-driven oceanic circulation on a large scale. The dynamics in the oceanic interior will be considered below; the mechanism of western intensification will be analysed in the following Sect. 20.4.

In relation to this, it should be noted that, although the wind-driven oceanic circulation is limited to the upper ocean, in this paragraph and the next, the evolution equations of potential vorticity for a homogeneous fluid will be used. These lead to a very simple and intuitive treatment of the complex problem and also highlight the essential dynamic mechanisms that regulate real phenomena.

Consider the evolution equation of potential vorticity (Eq. 18.3) for an ocean of constant depth ($H \cong D$) on the beta-plane:

$$\frac{d}{dt}(\zeta + f) = \frac{(\nabla \times \boldsymbol{\tau})_3}{\rho D} + K_H \nabla_H^2 \zeta. \tag{20.2}$$

In the oceanic interior (i.e., sufficiently east of the western boundary current: area B in Fig. 20.6b) the weak horizontal shear of the currents allows us to neglect both the dissipative effects and the nonlinear terms (this last assumption may not be appropriate in correspondence with the western boundary current extension, as in the case of the Kuroshio, cf. Sect. 17.5). Moreover, it is advantageous and appropriate to consider the problem stationary. Under these assumptions, relative vorticity is conserved ($d\zeta/dt = 0$) and, considering that $df/dt = \beta v$, Eq. (20.2) is reduced to:

$$M_{Sv} = \frac{(\nabla \times \boldsymbol{\tau})_3}{\beta} \tag{20.3}$$

where $M_{Sv} = \rho D v$ is the vertically integrated mass transport along y, known as *Sverdrup transport*.

Equation (20.3), known as the *Sverdrup balance* (Sverdrup 1947) is considered one of the milestones of dynamical oceanography. Referring to the example in Fig. 20.6b, a column of fluid moving south in the oceanic interior of a subtropical gyre in the northern hemisphere sees its planetary vorticity decrease due to the beta-effect. Under the assumption of conservation of potential vorticity, there would be an induction of positive relative vorticity, in contrast with the supposed Lagrangian conservation of ζ. This does not happen as the variation in planetary vorticity is balanced by the input of anticyclonic vorticity provided by the surface wind stress curl.

It is interesting to note that along the lines $(\nabla \times \boldsymbol{\tau})_3 = 0$ the meridional transport is null; within the validity limits of Eq. (20.3) these lines will therefore constitute the limits of the subtropical and subpolar gyres. For example, the field of $(\nabla \times \boldsymbol{\tau})_3$ in the North Pacific shown in Fig. 17.7 is negative in the band $\varphi = 18°N - 37°N$ and positive outside. The anticyclonic oceanic circulation produced by the surface stress field is indeed substantially limited within those latitudes, but the front that separates the anticyclonic circulation from the cyclonic one at the northern side of the Kuroshio Extension does not follow the line $\varphi = 37°N$ (see Fig. 17.8): this is because there the nonlinear effects are not negligible.

In Sect. 16.4 we saw how geostrophic currents are *indirectly* generated by the wind through the divergence of Ekman transport; consequently, there cannot be a precise relationship between wind stress and geostrophic current, except in particular cases (for example, see Eq. (16.24) for a coastal sea). This fact makes the Sverdrup balance a particularly fascinating relationship because, although limited to the meridional component of the flow and valid only far from the western boundary, it establishes a local connection between the wind forcing—in the form of the surface wind stress curl—and the meridional velocity v. Remember that, as discussed in Sect. 17.1 (Eq. 17.8), v is the sum of the geostrophic current (independent from z in the schematisation of a constant density fluid) and of the vertically averaged Ekman current. Therefore, in terms of meridional mass transport we will have

$$M_{Sv} = M_p + M_E, \tag{20.4}$$

where $M_p = \rho D v_p$ and $M_E = \rho D \overline{v}_E$.

We now move on to estimate these transports in a realistic case. With reference to the climatological surface wind stress field in the North Pacific Ocean, whose curl is reported in Fig. 17.7, we can consider the following values of the zonal stress (Pierini 2006; figure 1a therein):

$$\tau_1(\varphi = 34°N) \cong 0.036\ Pa;\ \tau_1(\varphi = 28°N) \cong 0,$$

to which corresponds a wind stress curl at the intermediate latitude $\varphi = 31°N$ equal to

$$(\nabla \times \boldsymbol{\tau})_3 = -\frac{\partial \tau_1}{\partial y} \cong -\frac{0.036}{6 \times 111 \times 10^3} \cong -5.4 \times 10^{-8}\,\mathrm{N\,m^{-3}}.$$

The corresponding meridional Sverdrup transport (towards the south), with $\beta \cong 2 \times 10^{-11} rad\ m^{-1} s^{-1}$, is given by:

$$M_{Sv} = \frac{(\nabla \times \boldsymbol{\tau})_3}{\beta} \cong -2.7 \times 10^3\ \mathrm{kg\,m^{-1}\,s^{-1}}. \tag{20.5}$$

On the other hand, given that at the intermediate latitude, $\tau_1(\varphi = 31°N) \cong 0.022$ Pa, the local meridional Ekman transport (towards the south), with $f_0 \cong 10^{-4}$ rad s^{-1}, is equal to (Eq. 16.11):

$$M_E = -\frac{\tau_1}{f_0} \cong -0.22 \times 10^3\ \mathrm{kg\,m^{-1}\,s^{-1}}. \tag{20.6}$$

From Eqs. (20.4–20.6) it follows that the geostrophic meridional transport due to the convergence of the surface Ekman transport is $M_p \cong -2.5 \times 10^3\ \mathrm{kg\,m^{-1}\,s^{-1}}$: therefore, M_p is an order of magnitude greater than the ageostrophic Ekman transport M_E. This is, in fact, a general characteristic of the large-scale wind-driven oceanic circulation.

Finally, it is useful to evaluate the mass transport $T_{Sv} = LM_{Sv}$ and volume transport $T_{Sv}^{(vol)} = T_{Sv}/\rho$ integrated over the entire zonal extent of the ocean, where L is the width of the basin. At $\varphi = 31°N$, for the Pacific Ocean we have $L \cong 11,000$ km, from which we obtain $T_{Sv} \cong -30 \times 10^9$ kg s^{-1} and

$$T_{Sv}^{(vol)} \cong -30 Sv$$

where Sv indicates the *Sverdrup* unit of measurement: 1 $Sv = 10^6\ \mathrm{m^3\,s^{-1}}$. This estimate is compatible with the transport obtained from observational data; however, it should be emphasised that the Sverdrup balance in its barotropic, flat bottom, stationary and linear version (Eq. 20.3) suffers from a series of limitations that do not generally allow a satisfactory direct comparison with observations.

Even admitting the validity of the barotropic dynamics, the effects associated with topography and bottom stress may not be negligible. The presence of stratification can cause significant baroclinic effects, especially at low latitudes. As already pointed out in this paragraph, nonlinear effects are not negligible in correspondence with western boundary current extensions. Finally, the variability of circulation (on various time scales) may not be a simple adiabatic sequence of stationary states, in which case it will be necessary to consider the time-dependent version of the Sverdrup balance (*time-dependent Sverdrup balance*). That said, the Sverdrup balance has marked an important step in the evolution of dynamical oceanography and its conceptual significance remains very relevant.

20.4 Western Boundary Currents

A detailed treatment of western boundary currents requires the solution of the dynamic equations with appropriate parameterisations of the dissipative effects, the imposition of the no-slip conditions along the lateral boundary and the consideration of nonlinear effects, certainly not negligible in currents with such intense velocity shear. For a detailed treatment of this problem reference is made to texts like those already mentioned. Here we will limit ourselves to analysing the dynamic mechanism that produces the western intensification of the currents. Finally, in the next paragraph, some results obtained from the physical simulation of the phenomenon using a rotating platform will be illustrated.

As already discussed, Fig. 20.2 clearly shows the average western boundary currents. On the other hand, the oceanic mesoscale produced by the mechanisms of baroclinic and barotropic instability (Sect. 20.2) can only concentrate, in addition to the ACC, precisely in correspondence with these intense current systems: this is shown in Fig. 20.7 by the difference in the variance of the sea level anomaly between two altimetric products for the period 1993–2012. Figures 20.2 and 20.7 are therefore complementary, in the sense that the first shows the average flows and the second the intense mesoscale variability associated with them.

The Gulf Stream and the Brazil Current can be distinguished respectively in the North and South Atlantic Ocean, the Kuroshio Current and the East Australian

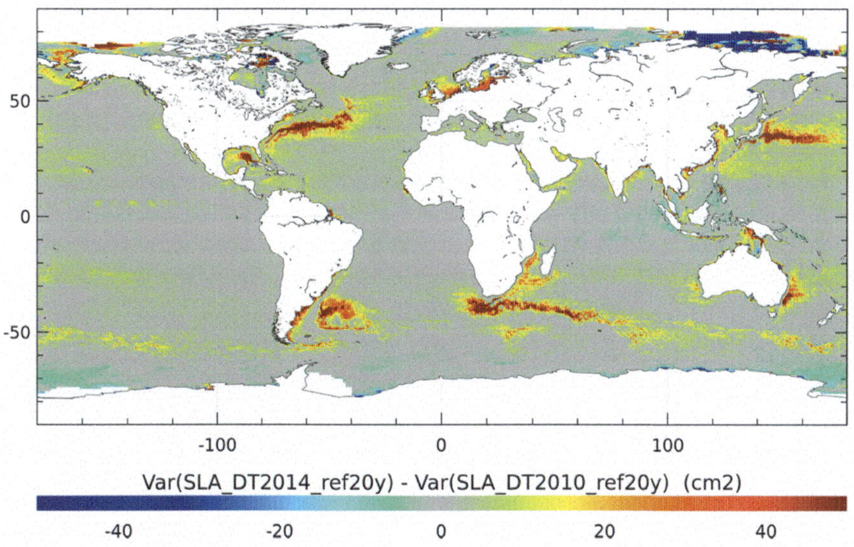

Fig. 20.7 Difference between the variance of the sea level anomaly obtained from the DT2014 altimetric product and that obtained from the DT2010 altimetric product, for the twenty-year period 1993–2012. Figure taken from Pujol et al. (2016, Ocean Sci. 12, 1067–1090; published under a Creative Commons Attribution 4.0 Licence)

Current respectively in the North and South Pacific Ocean and the Agulhas Current along the south-eastern coasts of the African continent. Less visible in these maps is the cold Oyashio Current (already mentioned in Sect. 17.4), part of the cyclonic subpolar circulation in the North Pacific.

The western intensification of the currents can be explained qualitatively by referring to the same Eq. (20.2) used to derive the Sverdrup balance: we will focus again on the potential vorticity (here proportional to the absolute vorticity) that, once again, will prove to be a valuable diagnostic tool.

Imagine simulating the oceanic circulation on the beta-plane in the northern hemisphere induced by a field of surface stress with a spatially and temporally constant negative curl, starting from vanishing initial conditions. At the beginning of the transient phase (called *spinup*), the weak circulation is well described by the streamlines of Fig. 20.6a; the negative relative vorticity induced by the wind will predominantly be in the form of curvature vorticity. This symmetrical pattern will soon turn into an asymmetric circulation like the one outlined in Fig. 20.6b due to the different vorticity balance that occurs for the fluid columns moving south and north, as we will now discuss. Initially, it is possible to neglect the dissipative effects, so Eq. (20.2) gives

$$\frac{d\zeta}{dt} = -\beta v + \frac{(\nabla \times \boldsymbol{\tau})_3}{\rho D}, \tag{20.7}$$

with $\zeta < 0$.

For a fluid column moving south in the oceanic interior, during the spinup the wind forcing (with negative curl) prevails over the variation of planetary vorticity (v is still too weak), progressively intensifying the flow: this implies that the term $-\beta v$ (positive) will grow over time. Asymptotically the two terms will balance (Eq. 20.3), reaching stationarity in the oceanic interior.

For a fluid column moving north on the western side of the basin, the intensification of the anticyclonic vorticity—and therefore of the flow—will progress due to the continuous input of vorticity, which can never be balanced as the terms on the right-hand side of Eq. (20.7) are both negative. From a certain point onwards the first term will even prevail over the second:

$$\frac{d\zeta}{dt} \cong -\beta v. \tag{20.8}$$

The increase in $|\zeta|$ will manifest itself in the form of intense negative shear vorticity (Eq. 4.21) due to the rectifying effect of the western margin of the ocean, producing a zonal narrowing of the area in which $v > 0$ (area A in Fig. 20.6b) and a v-profile like the one shown schematically in Fig. 20.8. Note the positive relative vorticity along the lateral boundary generated by the no-slip condition. Ultimately, the western intensification of the currents is due to the induction of negative relative vorticity in the western part of the basin due to the beta-effect.

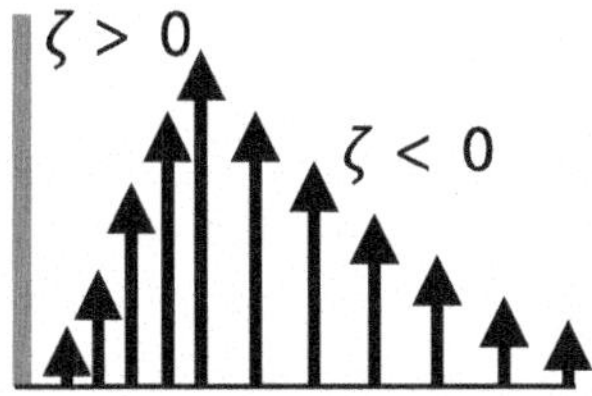

Fig. 20.8 Schematic representation of the profile of the meridional velocity (v) along the western margin of the ocean; the grey line represents the coastline. The zonal extensions where $\zeta > 0$ and $\zeta < 0$ are not to scale

The western intensification will stop growing when the dissipative effects, which in the meantime will have become important due to the strong horizontal velocity shear, will contribute to the balance that, in the western margin will take the form:

$$\mathbf{u} \cdot \nabla \zeta \cong -\beta v + K_H \nabla_H^2 \zeta. \tag{20.9}$$

Note that, in reaching the steady state ($\partial \zeta / \partial t = 0$), the nonlinear advective terms will have become important, so $d\zeta/dt = \mathbf{u} \cdot \nabla \zeta$ in Eq. (20.9). Asymptotically the Sverdrup balance (Eq. 20.3) will be reached in area B of Fig. 20.6b and that given by Eq. (20.9) in area A.

20.5 Laboratory Simulations

Fluid dynamic phenomena can be simulated through laboratory experiments by reproducing them on a smaller scale: think, for example, of aerodynamic simulations in wind tunnels or those in ship model basins. To this end, it is necessary to scale both the geometry and the dynamics of the phenomenon; this leads to the concept of *dynamic similarity* (already mentioned in Sect. 6.2, which cannot be further explored here). This methodology is defined as *physical modelling*; naturally, in recent decades, mathematical/numerical simulation of the same phenomena has also become possible. What is the relationship between these two modelling approaches?

The interaction between physical and mathematical modelling in fluid dynamics is a rather complex issue. Until the '70s–'80s, physical modelling played an irreplaceable role. Subsequently, with the rapid development of digital computers, and therefore of mathematical modelling, laboratory simulations were seen more as support for the improvement of mathematical models, but these could in turn assist experimental analysis. In a more balanced and constructive approach, the integrated use of physical and mathematical/numerical models, theoretical analysis and experimental investigations of real phenomena can lead to a deeper understanding of fluid dynamic processes. The real challenge is therefore to achieve cooperation between physical and mathematical modelling in order to move from competition to synergy. In this perspective, even today laboratory simulations of fluid dynamic processes play an important role, especially in process studies. Moreover, a laboratory experiment has a unique impact from an educational point of view.

In this context, this concluding paragraph will briefly consider the laboratory simulation of western boundary currents. This requires, of course, the use of rotating tanks in order to include the Coriolis force in the dynamics, but how to simulate the sphericity of the earth (the beta-effect) which, as we have seen, is essential for western intensification? One must resort to an equivalent topographic beta-effect.

In the quasigeostrophic approximation given by Eq. (19.3), if we define $d = \alpha y$, the term $f_0 d/D = \beta^* y$ (with $\beta^* = f_0 \alpha/D$) becomes similar to the βy variation of f with latitude: for this reason $\beta^* y$ is called the *topographic beta-effect*. Therefore, in a rotating tank with counterclockwise revolution (northern hemisphere) there must be a linear variation of the bottom topography in such a way as to produce a decrease in fluid depth in the positive y-direction which, in this case, corresponds to the coordinate aligned with the meridians of the real case.

The classic approach in this context is based on the so-called *sliced cylinder model* (e.g., Pedlosky and Greenspan 1967; Beardsley 1969; Beardsley and Robbins 1975; Griffiths and Kiss 1999), in which a homogeneous fluid contained in a cylinder with an inclined flat bottom is forced on the surface by a rigid rotating slab that mimics the effect of an anticyclonic wind field, in order to bypass the technical problems associated with the use of real wind in the laboratory. In addition, a generalisation of this model that takes into account the effect of the inclination of the side walls is known as the *sliced cone model* (for example, Griffiths and Veronis 1997). For a video format example of this type of simulations, one can refer to the laboratory experiment GFD XIII by Marshall and Plumb (2008).

It is also possible to adopt an alternative modelling setup if one wishes to simulate a very wide western boundary current (in large diameter rotating tanks) without however reproducing the circulation in the entire basin (Pierini et al. 2008, 2011, 2022). The basic idea is the following. Given that in the dynamic balance in the area of western intensification the wind forcing is negligible (Eq. 20.9), the intensified current can be generated by resorting to an appropriate lateral forcing rather than a surface one, as shown by the setup (Fig. 20.9a) adopted in experiments carried out with the 5 m-diameter rotating platform of SINTEF (Trondheim, Norway).

The motion of a piston (P) in a side channel produces a flow that will be found, practically without horizontal shear, at the mouth of the wide channel in which the topographic beta-effect is present. The induction of anticyclonic relative vorticity and the rectifying effect of the "western" coast will produce an intensified current, like the one shown in Fig. 20.10, obtained with a rotation period of the tank $T = 30$ s and a piston speed $u_p = 4$ mm s^{-1} (the currents were measured using the photogrammetric method). In a subsequent series of experiments (Pierini et al. 2022) carried out with the 13 m-diameter rotating platform of LEGI-CNRS (Grenoble) to study the interaction of a western boundary current with a lateral gap, instead of the piston, a pump system was used that allowed for simulations of virtually unlimited duration;

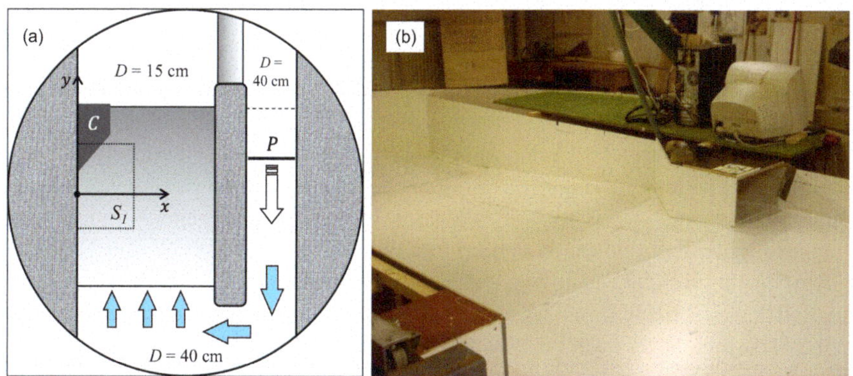

Fig. 20.9 **(a)** Diagram of the experimental apparatus used by Pierini et al. (2011) to simulate the western boundary currents using the 5 m-diameter rotating platform of SINTEF (Trondheim, Norway). **(b)** Photo of the topography adopted in some experiments. Figures adapted from Pierini et al. (2011, J. Phys. Oceanogr. 41, 2063–2079, © American Meteorological Society. Used with permission)

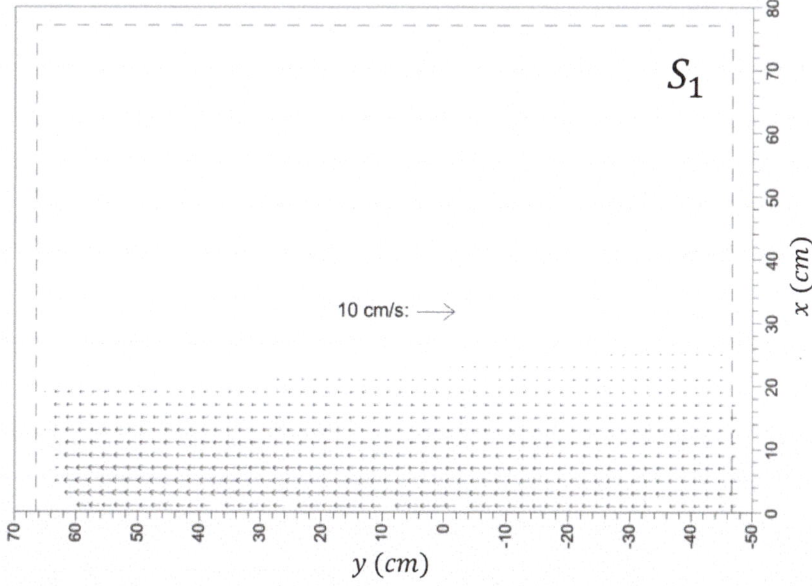

Fig. 20.10 Currents measured using the photogrammetric method in the S_1 area indicated in Fig. 20.9a in the absence of continental profile C for $T = 30\,\text{s}$ and $u_p = 4\,\text{mm}\,\text{s}^{-1}$. Figure adapted from Pierini et al. (2011, J. Phys. Oceanogr. 41, 2063–2079, © American Meteorological Society. Used with permission)

moreover, the speeds were measured using the more sophisticated *Particle Imaging Velocimetry* (PIV) method.

We will now focus on experiments carried out in the presence of a continental profile (C in Fig. 20.9a; also see the photo in Fig. 20.9b) that mimics the shape of the American continent. Under what conditions does the current detach itself or continue to flow along the coast even after the end of the inclined stretch? These simple experiments are, as we will see, connected to a phenomenon—much more complex—potentially linked to a significant climate change.

Among the various sensitivity experiments carried out, in Fig. 20.11 two are reported that effectively summarise the main result: detachment occurs if the current's inertia is sufficient to overcome its tendency to flow along the coast due to the mechanism described in the previous paragraph. In Fig. 20.11a a flow is shown that almost exactly corresponds to that—in the absence of a continent—of Fig. 20.10 ($u_p = 4 - 5\,\mathrm{mm\,s^{-1}}$, respectively in Figs. 20.10 and 20.11a); the path of the western boundary current (evidenced by a dye stream) does not detach from the continent, continuing to flow along the coast even after the end of the inclined stretch of the continental profile. On the other hand, if the intensity of the current is increased by increasing the piston speed, detachment occurs: this is what happens, for example, for $u_p = 20\,\mathrm{mm\,s^{-1}}$ (Fig. 20.11b).

In Fig. 20.12 the angle (determined based on the velocity field measured with the photogrammetric method) that the current forms with the y-axis as a function of y itself for different values of u_p is reported. As can be seen, for $u_p \leq 5\,\mathrm{mm\,s^{-1}}$ there is no detachment which, on the other hand, occurs for $u_p \geq 10\,\mathrm{mm\,s^{-1}}$. It is therefore conceivable that a sudden transition occurs at a precise intermediate value $5\,\mathrm{mm\,s^{-1}} < \overline{u}_p < 10\,\mathrm{mm\,s^{-1}}$ which, using a terminology now widely used in climate sciences (and not only), can be defined as a *tipping point*. This term refers to a threshold value of a certain control parameter that, if exceeded, leads to a sudden and dramatic change in the dynamical system (for examples of climate tipping points see, for example, Lenton et al. 2008; for further insights into this concept in the context

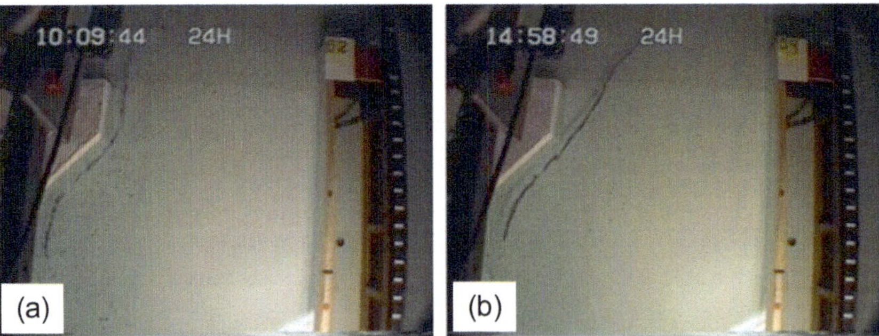

Fig. 20.11 (**a**) Path of the current evidenced by a dye stream in the presence of continental profile C for $T = 30\,\mathrm{s}$ and $u_p = 5\,\mathrm{mm\,s^{-1}}$. (**b**) As in (**a**) but for $u_p = 20\,\mathrm{mm\,s^{-1}}$; note that the current detaches from the continent, unlike what happens in (**a**). Figures adapted from Pierini et al. (2011, J. Phys. Oceanogr. 41, 2063–2079, © American Meteorological Society. Used with permission)

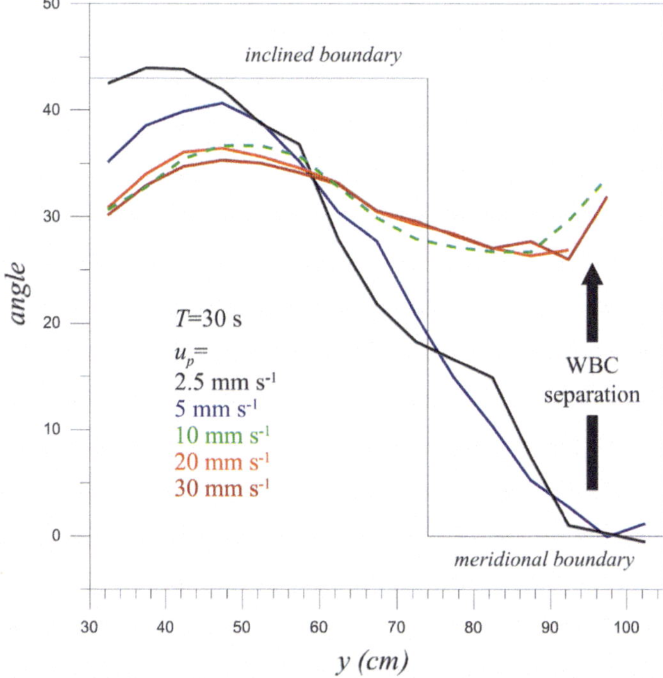

Fig. 20.12 Angle formed by the western boundary currents with the y-axis for different values of u_p. Figure adapted from Pierini et al. (2011, J. Phys. Oceanogr. 41, 2063–2079, © American Meteorological Society. Used with permission)

of climate physics see, for example, Ghil and Lucarini 2020, for multistable systems and Wieczorek et al. 2011, and Pierini and Ghil 2021, for excitable systems).

The idealised result shown in Figs. 20.11 and 20.12 provides the opportunity to mention the *thermohaline oceanic circulation*, the *Atlantic Meridional Overturning Circulation* (AMOC) and its possible weakening/collapse. Similarly to what was pointed out in Sect. 17.5, the simple qualitative treatment that follows can stimulate further insights.

In addition to the surface wind, oceanic circulation is also generated by temperature and salinity gradients—and therefore density—determined by heat and freshwater fluxes (evaporation and precipitation) at the air-sea interface: this component of the circulation is called *thermohaline*. One of its most prominent features is the AMOC, or the Atlantic branch of the global thermohaline circulation (Fig. 20.13), composed of warm surface currents (including the Gulf Stream) and cold deep currents formed from the transformation of the former at high latitudes. This global current system has the structure of a *conveyor belt* that contributes significantly to the global thermal balance.

Using a simple model of thermohaline convection, Stommel (1961) showed that surface heat and freshwater fluxes have an opposite effect on the formation of dense

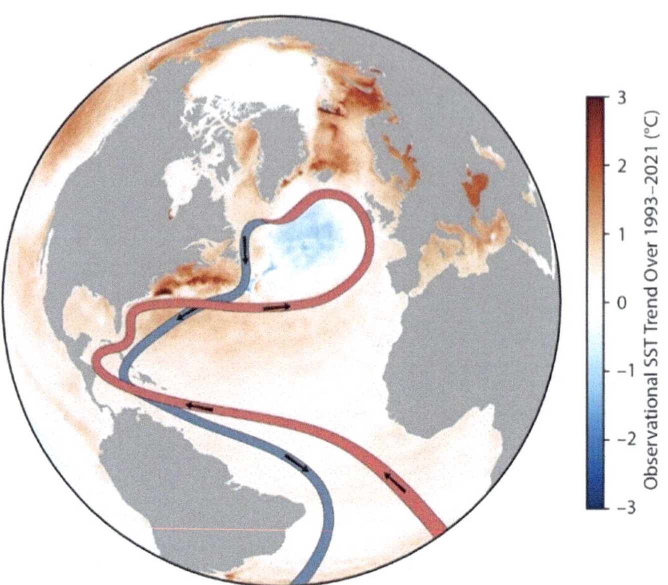

Fig. 20.13 Diagram of the Atlantic Meridional Overturning Circulation (AMOC). Figure taken from Rahmstorf (2024, Oceanography, https://doi.org/https://doi.org/10.5670/oceanog.2024.501) and produced by Copernicus Climate Change Service (https://climate.copernicus.eu/, image credit: Ruijian Gou)

water at high latitudes, where the transformation of warm surface waters into waters with greater density occurs, which sink and flow southward in depth (Fig. 20.13): while temperature has a stabilising effect, salinity tends to destabilise the system. These competitive effects produce a feedback mechanism, called the *salt transport feedback*, which in turn implies the existence of two stable states corresponding, respectively, to the presence and absence of the AMOC. This idealised result, which implies the existence of a tipping point (the so-called *Stommel bifurcation*), has been confirmed by various studies based on much more sophisticated climate models.

In this context, some recent studies suggest that there may be an imminent risk of exceeding this tipping point: this would result in a collapse (*shutdown*) of the AMOC with very significant consequences for the Earth's climate (see, for example, Caesar et al. 2018; Ditlevsen and Ditlevsen 2023; Van Westen et al. 2024; see Rahmstorf 2024, for a summary of the problem). It is therefore important to identify indicators that can reveal a possible trend towards this climate change.

A composite indicator of the weakening of the AMOC consists of the shift of the Gulf Stream towards the coast north of the separation point (Cape Hatteras), a positive temperature anomaly in that region and a negative temperature anomaly (*cold blob*) in the northern Atlantic (Zhang 2008; Caesar et al. 2018). The latter is associated with a net heat loss through the ocean surface that has occurred since the mid-twentieth century. The analysis of these phenomena suggests that a weakening of the AMOC may be underway (Caesar et al. 2018; Todd and Ren 2023; Rahmstorf 2024) and may also be of anthropogenic origin. Concurrently, a weakening of the Gulf Stream in its initial stretch, the Florida Current, has been observed (Piecuch and Beal 2023).

In light of all this, the simple experimental result of the laboratory highlighted by Figs. 20.11 and 20.12 can be seen as a schematic—but, of course, not realistic—representation of an aspect of this complex phenomenon. Let's start from the state in which the western boundary current separates from the coast at the end of the inclined stretch (Fig. 20.11b, $u_p > \overline{u}_p$ in Fig. 20.12); a sufficient weakening of the current ($u_p < \overline{u}_p$) produced by a decrease in piston speed (corresponding to the weakening of the Florida Current) leads to a decisive shift of the current towards the coast north of the inclined stretch of the continental profile (corresponding to the region north of Cape Hatteras). This would necessarily correspond to a positive temperature anomaly in the same coastal region (the cold blob, on the other hand, would not be represented in this simple scheme).

It is superfluous to list the many effects absent in the laboratory experiment that contribute to determining that particular aspect of the climatic phenomenon. Nonetheless, a question naturally arises: could the separation mechanism analyzed in the experiments,called *inertial overshooting* (e.g., Dengg 1993; Dengg et al. 1996; Matsumoto and Lynch-Stieglitz 2003), have a counterpart in the real ocean? If so, it could have a major effect on the functioning of the AMOC and could even be associated with a tipping point. This is therefore a mechanism that deserves to be studied carefully.

It is believed that this hint at a cutting-edge research topic concerning climate physics, arising—in line with the spirit of this text—from a simple model, can mark the right conclusion of this introductory treatment to oceanic and atmospheric fluid dynamics, paving the way for more complex studies.

Bibliography

Beardsley, R.C.: A laboratory model of the wind-driven ocean circulation. J. Fluid Mech. **38**, 255–271 (1969)

Beardsley, R.C., Robbins, K.: The 'sliced-cylinder' laboratory model of the wind-driven ocean circulation. Part 1. Steady forcing and topographic Rossby wave instability. J. Fluid Mech. **69**, 27–40 (1975)

Caesar, L., Rahmstorf, S., Robinson, A., Feulner, G., Saba, V.: Observed fingerprint of a weakening Atlantic Ocean overturning circulation. Nature **556**, 191–196 (2018)

Charney, J.G.: The dynamics of long waves in a baroclinic westerly current. J. Meteorol. **4**, 136–162 (1947)

Chelton, D.B., Schlax, M.G.: Global observations of large oceanic eddies. Geophys. Res. Lett. **34**, L15606 (2007)

Cushman-Roisin, B.: Introduction to Geophysical Fluid Dynamics. Prentice-Hall, Englewood Cliffs, New Jersey (1994)

Dengg, J.: The problem of Gulf Stream separation: A barotropic approach. J. Phys. Oceanogr. **23**, 2182–2200 (1993)

Dengg, J., Beckmann, A., Gerdes, R.: The Gulf Stream separation problem. In The Warmwatersphere of the North Atlantic Ocean, edited by W. Krauss, Gebrueder Borntraeger, Berlin, 253–290 (1996)
Ditlevsen, P., Ditlevsen, S.: Warning of a forthcoming collapse of the Atlantic meridional overturning circulation. Nat. Commun. **14**, 4254 (2023)
Eady, E.T.: Long waves and cyclone waves. Tellus **1**, 33–52 (1949)
Fofonoff, N.P.: The Gulf Stream system. In: Warren, B.A., Wunsch, C. (ed.) Evolution of Physical Oceanography; Scientific Surveys in Honour of Henry Stommel. The Massachusetts Institute of Technology, Cambridge, Massachusetts (1981)
Ghil, M., Lucarini, V.: The physics of climate variability and climate change. Rev. Mod. Phys. **92**, 035002 (2020)
Gill, A.E.: Atmosphere-Ocean Dynamics. Academic Press, New York (1982)
Griffiths, R.W., Kiss, A.E.: Flow regimes in a wide 'sliced-cylinder' model of homogeneous beta-plane circulation. J. Fluid Mech. **399**, 205–236 (1999)
Griffiths, R.W., Veronis, G.: A laboratory study of the effects of a sloping side boundary on wind-driven circulation in a homogeneous ocean model. J. Mar. Res. **55**, 1103–1126 (1997)
Lenton, T.M., et al.: Tipping elements in the earth's climate system. Proc. Natl. Acad. Sci. USA. **105**, 1786–1793 (2008)
Malanotte-Rizzoli, P., Artale, V., Borzelli-Eusebi, G.L., Brenner, S., et al.: Physical forcing and physical/biochemical variability of the Mediterranean Sea: a review of unresolved issues and directions for future research. Ocean Sci. **10**, 281–322 (2014)
Matsumoto, K., Lynch-Stieglitz, J.: Persistence of Gulf Stream separation during the Last Glacial Period: Implications for current separation theories. J. Geophys. Res. **108**, 3174 (2003)
Munk, W.H.: On the wind-driven ocean circulation. J. Meteorol. **7**, 80–93 (1950)
Pedlosky, J.: Geophysical Fluid Dynamics. Springer, New York (1987)
Pedlosky, J., Greenspan, H.P.: A simple laboratory model for the oceanic circulation. J. Fluid Mech. **27**, 291–304 (1967)
Piecuch, C.G., Beal, L.M.: Robust weakening of the Gulf Stream during the past four decades observed in the Florida Straits. Geophys. Res. Lett. **50**, e2023GL105170 (2023)
Pierini, S.: A Kuroshio Extension System model study: decadal chaotic self-sustained oscillations. J. Phys. Oceanogr. **36**, 1605–1625 (2006)
Pierini, S., Simioli, A.: A wind-driven circulation model of the Tyrrhenian Sea area. J. Mar. Syst. **18**, 161–178 (1998)
Pierini, S., De Ruggiero, P., Negretti, M.E., Schiller-Weiss, I., Weiffenbach, J., Viboud, S., Valran, T., Dijkstra, H.A., Sommeria, J.: Laboratory experiments reveal intrinsic self-sustained oscillations in ocean relevant rotating fluid flows. Sci. Rep. **12**, 1375 (2022)
Pierini, S., Falco, P., Zambardino, G., McClimans, T.A., Ellingsen, I.: A laboratory study of nonlinear western boundary currents, with application to the Gulf Stream separation due to inertial overshooting. J. Phys. Oceanogr. **41**, 2063–2079 (2011)
Pierini, S., Ghil, M.: Tipping points induced by parameter drift in an excitable ocean model. Sci. Rep. **11**, 11126 (2021)
Pierini, S., Malvestuto, V., Siena, G., McClimans, T.A., Løvås, S.M.: A laboratory study of the zonal structure of western boundary currents. J. Phys. Oceanogr. **38**, 1073–1090 (2008)
Pinardi, N., Zavatarelli, M., Adani, M., Coppini, G., Fratianni, C., Oddo, P., Simoncelli, S., Tonani, M., Lyubartsev, V., Dobricic, S., Bonaduce, A.: Mediterranean Sea large-scale low-frequency ocean variability and water mass formation rates from 1987 to 2007: a retrospective analysis. Progr. Oceanogr. **132**, 318–332 (2015)
Pujol, M.-I., Faugère, Y., Taburet, G., Dupuy, S., Pelloquin, C., Ablain, M., Picot, N.: DUACS DT2014: the new multi-mission altimeter data set reprocessed over 20 years. Ocean Sci. **12**, 1067–1090 (2016)
Rahmstorf, S.: Is the Atlantic overturning circulation approaching a tipping point? Oceanography. https://doi.org/10.5670/oceanog.2024.501 (2024)

Stommel, H.: The westward intensification of wind-driven ocean currents. Trans. Am. Geophys. Union **29**, 202–206 (1948)
Stommel, H.: Thermohaline convection with two stable regimes of flow. Tellus **13**, 224–230 (1961)
Stommel, H.: The Gulf Stream, a Physical and Dynamical Description. University of California Press, Berkeley (1965)
Sverdrup, H.U.: Wind-driven currents in a baroclinic ocean; with application to the equatorial currents of the Eastern Pacific. Proc. Natl. Acad. Sci. u.s.a. **33**, 318–326 (1947)
Todd, R.E., Ren, A.S.: Warming and lateral shift of the Gulf Stream from in situ observations since 2001. Nat. Clim. Chang. **13**, 1348–1352 (2023)
Vallis, G.K.: Atmospheric and Oceanic Fluid Dynamics. Cambridge University Press, Cambridge (2006)
Van Westen, R.M., Kliphuis, M., Henk A. Dijkstra, H.A.: Physics-based early warning signal shows that AMOC is on tipping course. Sci. Adv. **10**, eadk1189 (2024)
Wieczorek, S., Ashwin, P., Luke, C.M., Cox, P.M.: Excitability in ramped systems: the compost-bomb instability. Proc. r. Soc. A **467**, 1243–1269 (2011)
Zhang, R.: Coherent surface-subsurface fingerprint of the Atlantic meridional overturning circulation. Geophys. Res. Lett. **35**, L20705 (2008)

Further Recommended Reading

Defant, A.: Physical Oceanography, vol. I. Pergamon Press, New York (1961)
Kundu, P.K., Cohen, I.M., Dowling, D.R.: Fluid Mechanics. Elsevier, Amsterdam (2012)
Marshall, J., Plumb, R.A.: Atmosphere, Ocean, and Climate Dynamics. Elsevier, Amsterdam (2008)
Monin, A.S.: Theoretical Geophysical Fluid Dynamics. Kluwer Academic Publishers, Dordrecht (1990)
Visconti, G., Ruggieri, P.: Fluid Dynamics: Fundamentals and Applications. Springer Nature Switzerland AG (2020)

Appendix A
Gauss's and Stokes' Theorems

This appendix presents the Gauss's and Stokes' theorems, which are invoked several times in the text; it is assumed that the reader has learned them in Calculus. Subsequently, some applications in the field of classical physics will be mentioned, in order to allow useful connections between the methodologies followed in the text and those adopted in other fields of physics.

- **The theorems**

Gauss's Theorem (also known as the divergence theorem):

$$\oint\!\!\!\oint_{\partial V} \mathbf{A} \cdot \mathbf{n} ds = \iiint_V \nabla \cdot \mathbf{A} dv \tag{A.1}$$

where V is an arbitrary simply connected volume, ∂V is the closed surface that delimits V, $\mathbf{A}(\mathbf{x}, t)$ is a generic vector field and $\nabla \cdot \mathbf{A}$ (also indicated as $div\mathbf{A}$) is the *divergence of* $\mathbf{A}$:

$$\nabla \cdot \mathbf{A} \equiv \left(\frac{\partial A_1}{\partial x}, \frac{\partial A_2}{\partial y}, \frac{\partial A_3}{\partial z} \right).$$

Stokes' Theorem:

$$\oint_{\partial \Gamma} \mathbf{A} \cdot d\boldsymbol{\ell} = \iint_{\Gamma} (\nabla \times \mathbf{A}) \cdot \mathbf{n} ds \tag{A.2}$$

where Γ is a simply connected surface bounded by the closed curve $\partial\Gamma$ and $\nabla \times \mathbf{A}$ (also indicated as $curl\mathbf{A}$ or $rot\mathbf{A}$) is the *curl of* $\mathbf{A}$:

$$\nabla \times \mathbf{A} \equiv \left(\frac{\partial A_3}{\partial y} - \frac{\partial A_2}{\partial z}, \frac{\partial A_1}{\partial z} - \frac{\partial A_3}{\partial x}, \frac{\partial A_2}{\partial x} - \frac{\partial A_1}{\partial y} \right).$$

S. Pierini, *Oceanic and Atmospheric Fluid Dynamics*, UNITEXT for Physics,
https://doi.org/10.1007/978-3-031-77991-6

- **Application 1: Conservative force fields**

The first application of Eq. (A.2) concerns the work done by a conservative force field. If $\mathbf{F}$ is the force field to which a point mass m is subject, then the work $L_{1,2}$ done by $\mathbf{F}$ during the displacement of the mass from point P_1 to point P_2 along a curve γ is given by the line integral:

$$L_{1,2} = \int_{\gamma} \mathbf{F} \cdot d\boldsymbol{\ell}.$$

In general, $L_{1,2}$ depends on the path γ but, if $\mathbf{F}$ is a *conservative force field* (i.e., if $\nabla \times \mathbf{F} = 0$, in which case a scalar V—the potential energy—exists such that $\mathbf{F} = -\nabla V$), then $L_{1,2}$ is independent of the path and is given by $L_{1,2} = V(P_2) - V(P_1)$. This is, for example, the case of the gravitational field and the Coulomb electrostatic field.

This energy theorem can be demonstrated thanks to Eq. (A.2). Choose any closed curve $\gamma_A \cup \gamma_B$ with γ_A going from P_1 to P_2 and γ_B going from P_2 to P_1. For Eq. (A.2) with $\mathbf{A} = \mathbf{F}$ we will have:

$$\int_{\gamma_A} \mathbf{F} \cdot d\boldsymbol{\ell} = \int_{-\gamma_B} \mathbf{F} \cdot d\boldsymbol{\ell}$$

where $-\gamma_B$ in the second integral indicates that the path is from P_1 to P_2; given the arbitrariness of $\gamma_A \cup \gamma_B$ the first part of the theorem has been proven. Moreover, since $\mathbf{F} \cdot d\boldsymbol{\ell}$ is an exact differential we will have

$$L_{1,2} = -\int_{\gamma} \nabla V \cdot d\boldsymbol{\ell} = -\int_{\gamma} dV = V(P_2) - V(P_1).$$

Thus, the second part of the theorem has also been proven.

- **Application 2: Relationship between laws of electromagnetism and Maxwell's equations**

  Maxwell's equations (which are reported for simplicity in vacuum) are

$$\nabla \cdot \mathbf{E} = \frac{\rho}{\varepsilon_0} \tag{A.3}$$

$$\nabla \cdot \mathbf{H} = 0 \tag{A.4}$$

$$\nabla \times \mathbf{E} = -\mu_0 \frac{\partial \mathbf{H}}{\partial t} \tag{A.5}$$

$$\nabla \times \mathbf{H} = \mathbf{J} + \varepsilon_0 \frac{\partial \mathbf{E}}{\partial t} \tag{A.6}$$

where $\mathbf{E}$ is the electric field, $\mathbf{H}$ the magnetic field, ρ the electric charge density, $\mathbf{J}$ the electric current density and ε_0 and μ_0 are the dielectric constant and magnetic permeability in vacuum, respectively (the force exerted on an electric point charge q in motion with velocity $\mathbf{u}$ is given by $\mathbf{F} = q(\mathbf{E} + \mu_0 \mathbf{u} \times \mathbf{H})$). These partial differential equations are obtained by applying Eqs. (A.1 and A.2) to the laws of electromagnetism in integral form. In the following, the reverse procedure is retraced.

By applying Eq. (A.1) to Eq. (A.3), Gauss's law is obtained, according to which the flux Φ of the electric field through a closed surface ∂V is proportional to the total electric charge Q contained in V:

$$\Phi(\mathbf{E}) = \frac{Q}{\varepsilon_0}.$$

By applying Eq. (A.1) to Eq. (A.4), the law is obtained according to which the flux Φ of the magnetic field through any closed surface ∂V is null:

$$\Phi(\mathbf{H}) = 0.$$

From this expression it follows that the magnetic lines of force are closed (each line of force that exits from a generic closed surface must re-enter it to ensure the nullity of the flux).

By applying Eq. (A.2) to Eq.(A.5), Faraday-Neumann's law of induction is obtained, according to which the circulation of the electric field along a closed curve γ is proportional to the opposite (Lenz's rule) of the rate of change of the magnetic flux through the surface delimited by that curve:

$$\oint_\gamma \mathbf{E} \cdot d\boldsymbol{\ell} = -\mu_0 \frac{d\Phi(\mathbf{H})}{dt}.$$

If γ coincides with an electric circuit, the circulation of $\mathbf{E}$ is called electromotive force f: this induces a current $i = -\mu_0 \dot{\Phi}(\mathbf{H})/R$ (where R is the electrical resistance of the circuit) which, for the Lenz's law, opposes the variation of the magnetic flux.

Finally, by applying Eq. (A.2) to Eq. (A.6) and neglecting the displacement current $\varepsilon_0 \dot{\mathbf{E}}$ (important for rapidly varying fields and essential for electromagnetic waves) the law is obtained according to which the circulation of the magnetic field along a closed curve γ is equal to the total electric current $I = \Phi(\mathbf{J})$ flowing through the surface delimited by that curve:

$$\oint_\gamma \mathbf{H} \cdot d\boldsymbol{\ell} = I.$$

From this formula, Biot-Savart's law is obtained, which expresses the magnetic field produced by a constant (or slowly varying) electric current.

- **Application 3: Derivation of fluid dynamics laws**

In this text, the Gauss's theorem is invoked in the derivation (i) of the pressure gradient force (Sect. 2.3), (ii) of the continuity equation (Sect. 4.2), (iii) of a property of the vortex tubes (Sect. 4.6), (iv) of Euler's and Navier–Stokes equations (Sects. 5.2 and 5.3) and (v) of the energy flux and the viscous dissipation of mechanical energy (Sect. 5.6). The same theorem is also used in Appendix B. Finally, Stokes' theorem is invoked (i) in the calculation of the vorticity contained in the boundary layer (Sect. 8.1) and (ii) in the treatment of circulation and vortex tubes (Sects. 4.6 and 8.3).

Appendix B
The Stress Tensor

In this appendix, we determine (i) the representation of stress **S** by means of the stress tensor σ, (ii) the symmetry of σ, (iii) the expression of pressure in a fluid at rest and (iv) that in a moving fluid.

- **Representation of surface forces by means of the stress tensor**

Consider an infinitesimal volume dV in the form of a tetrahedron, as in Fig. B.1.

The three orthogonal faces ds_1, ds_2, ds_3 have orthogonal unit vectors (directed outside of the volume) $-\mathbf{a}$, $-\mathbf{b}$, $-\mathbf{c}$, respectively, while $\mathbf{n}$ is the unit vector orthogonal to the fourth inclined surface ds. The resultant $d\mathcal{F}_s$ of the four surface forces applied to the tetrahedron will be given by:

$$d\mathcal{F}_s = \mathbf{S}(\mathbf{n})ds + \mathbf{S}(-\mathbf{a})ds_1 + \mathbf{S}(-\mathbf{b})ds_2 + \mathbf{S}(-\mathbf{c})ds_3 \tag{B.1}$$

(since the volume is infinitesimal, the four terms are referred to the same point $\mathbf{x}$ in the first approximation). Note that:

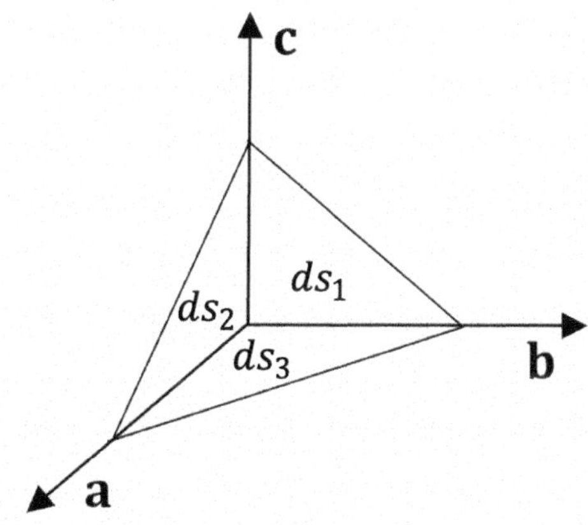

Fig. B.1 Volume element in the shape of a tetrahedron with three orthogonal faces and an inclined face

S. Pierini, *Oceanic and Atmospheric Fluid Dynamics*, UNITEXT for Physics,
https://doi.org/10.1007/978-3-031-77991-6

$$ds_1 = \mathbf{a} \cdot \mathbf{n}\, ds;\; ds_2 = \mathbf{b} \cdot \mathbf{n}\, ds;\; ds_3 = \mathbf{c} \cdot \mathbf{n}\, ds.$$

Therefore, considering Eq. (B.2), we will have:

$$(d\mathcal{F}_s)_i = \left\{ S_i(\mathbf{n}) - \sum_{j=1}^{3} \left[a_j S_i(\mathbf{a}) + b_j S_i(\mathbf{b}) + c_j S_i(\mathbf{c}) \right] n_j \right\} ds. \tag{B.2}$$

In Newton's second law, the mass is proportional to dV and, therefore, mass multiplied by acceleration is of order dV; obviously, the volume forces are also of order dV. However, the resultant of the surface forces expressed by Eq. (B.2) is of order ds. Passing to the limit in which the linear dimension of the volume tends to zero without the shape of the latter changing, due to the aforementioned disparity the only way for Newton's equation to be satisfied is that the expression in curly brackets in Eq. (B.2) is identically zero:

$$S_i(\mathbf{n}) = \sum_{j=1}^{3} \left[a_j S_i(\mathbf{a}) + b_j S_i(\mathbf{b}) + c_j S_i(\mathbf{c}) \right] n_j. \tag{B.3}$$

This implies, among other things, that in calculating the force impressed on the volume by the surface forces, one must take into account the albeit small distance between the four surfaces that delimit it and, therefore, the corresponding small difference in **S**.

Now, **S** and **n**, as vectors, do not depend on the orientation of the Cartesian coordinate system (obviously their components do), so neither can the expression in square brackets in Eq. (B.3) depend on it; in other words, Eq. (B.3) must be formally invariant under rotation of the axes. Consequently, the expression in brackets must be a second-order tensor, called the *stress tensor* σ_{ij} (whose elements are directly traceable to the terms in square brackets in Eq. (B.3)):

$$S_i(\mathbf{x}, t, \mathbf{n}) = \sum_{j=1}^{3} \sigma_{ij}(\mathbf{x}, t) n_j \tag{B.4}$$

where the dependence of **S** (and therefore $\boldsymbol{\sigma}$) on the point and time has been explicitly indicated.

- **Symmetry of the stress tensor**

We now move on to derive a fundamental property of the stress tensor: its symmetry. The moment of forces **M** (relative to a generic point **P**) exerted by the surface forces acting on the closed surface ∂V delimiting the volume V is given by

$$M_i = \oiint_{\partial V} [\mathbf{r} \times \mathbf{S}(\mathbf{n})]_i ds = \oiint_{\partial V} \sum_{j,k,\ell} \varepsilon_{ijk} r_j \sigma_{k\ell} n_\ell ds \tag{B.5}$$

where $\mathbf{r}$ is the position vector referred to $\mathbf{P}$. In the second integral, the vector product has been expressed through its matrix representation, where the tensor ε_{ijk} is called the *Levi–Civita symbol* (from the Italian mathematician Tullio Levi–Civita, 1873–1941, who, among other things, made a fundamental contribution to the development of the absolute differential calculus on a Riemannian manifold used by A. Einstein in his formulation of the theory of general relativity). The tensor is zero unless i, j and k are all different, in which case it is 1 (− 1) if i, j e k are in cyclic (non cyclic) order. The surface integral in B.5 can be transformed into a volume integral by Gauss's theorem:

$$M_i = \iiint_V \sum_{j,k,\ell} \varepsilon_{ijk} \frac{\partial \left(r_j \sigma_{k\ell}\right)}{\partial r_\ell} dv = \iiint_V \sum_{j,k,\ell} \varepsilon_{ijk} \left(\sigma_{kj} + r_j \frac{\partial \sigma_{k\ell}}{\partial r_\ell}\right) dv.$$

In the limit in which V tends to zero—together with $\mathbf{r}$—maintaining the same shape, the second term in brackets in the second integral tends to zero like $\mathbf{M}$ but faster than the first term (not multiplied by $\mathbf{r}$), that is, $\iiint \sum \varepsilon\sigma\, dv$ is of higher order compared to the other terms and must therefore be identically zero,

$$\iiint_V \sum_{j,k,\ell} \varepsilon_{ijk} \sigma_{kj} dv = 0,$$

for every choice of V. It must therefore hold

$$\sum_{j,k,\ell} \varepsilon_{ijk} \sigma_{kj} = 0$$

for every $\mathbf{x}$. It is easy to verify that this is only possible if $\boldsymbol{\sigma}$ is symmetric:

$$\sigma_{ij} = \sigma_{ji}. \tag{B.6}$$

- **Pressure in a fluid at rest**

The experimental result known as Pascal's principle, according to which in a fluid at rest ($\mathbf{u} = 0$) the stress (static pressure) is always normal to the elementary surface to which it is applied, can be theoretically supported, as shown below.

We first exploit the symmetry of $\boldsymbol{\sigma}(\mathbf{x}, t)$, which ensures its diagonalizability: this means that, at every point $\mathbf{x}$ and at every time instant t there is a Cartesian coordinate system, called the *principal axes system*, with respect to which the stress tensor is diagonal. Let $\boldsymbol{\sigma}$ be the representation of the tensor in the starting axis system; in the principal axes system we will have

$$\sigma' = \begin{pmatrix} \sigma'_{11} & 0 & 0 \\ 0 & \sigma'_{22} & 0 \\ 0 & 0 & \sigma'_{33} \end{pmatrix}. \tag{B.7}$$

Remember also that the trace of a tensor is invariant with respect to a rotation of the axes, and therefore it is a *scalar*. In the present case,

$$tr[\boldsymbol{\sigma}] = \sum_i \sigma_{ii},$$

therefore, $tr[\boldsymbol{\sigma}] = tr\left[\boldsymbol{\sigma}'\right]$. Let us decompose $\boldsymbol{\sigma}'$ as follows:

$$\sigma'_{ij} = q\delta_{ij} + d'_{ij} \tag{B.8}$$

where q is the scalar defined as

$$q = \frac{1}{3} tr[\boldsymbol{\sigma}]; \tag{B.9}$$

hence

$$tr[\mathbf{d}] = 0$$

By making explicit the matrix elements:

$$\sigma'_{ij} = \begin{pmatrix} q & 0 & 0 \\ 0 & q & 0 \\ 0 & 0 & q \end{pmatrix} + \begin{pmatrix} \sigma'_{11} - q & 0 & 0 \\ 0 & \sigma'_{22} - q & 0 \\ 0 & 0 & \sigma'_{33} - q \end{pmatrix}.$$

Consider now a spherical fluid element: the stress corresponding to the first term in Eq. (B.8) ($\mathbf{S}_q = q\mathbf{n}$) is isotropic, therefore, it will be everywhere perpendicular to the surface of the sphere, and with the same magnitude (Fig. B.2a) if the size of the volume is very small; experimental evidence suggests that the fluid element is in a state of compression, therefore $q < 0$. This uniform compression is compatible with a state of rest of the fluid.

On the other hand, the fact that $\mathbf{d}'$ has a zero trace necessarily requires that this tensor produces at least in one direction a tension (as shown, for example, in Fig. B.2b along the x'-axis, with $d'_{11} > 0$) and at least in another direction a compression (as shown in the same figure along the y'-axis, with $d'_{22} < 0$; given that in the figure $d'_{11} > \left|d'_{22}\right|$ it must be $d'_{33} = -d'_{11} - d'_{22} < 0$). This situation (which among other things would cause the deformation of the sphere into an ellipsoid) is evidently incompatible with the hypothesised absence of motion; consequently, it must be $d'_{ij} = 0$ and therefore $\sigma'_{ij} = \sigma_{ij} = q\delta_{ij}$ (all Cartesian coordinate systems are principal axes of the unit matrix δ_{ij}).

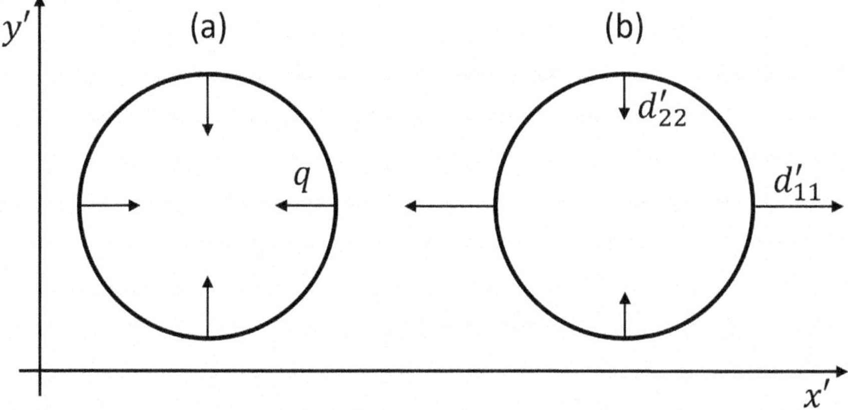

Fig. B.2 Diagram of a small sphere in a fluid at rest. The arrows indicate the local stress and x' and y' belong to the principal axes system of $\boldsymbol{\sigma}$

In summary, in a fluid at rest the stress **S** is always normal to the elementary surface to which it is applied and is defined as *static pressure*,

$$\mathbf{S}_p = -p\mathbf{n} \tag{B.10}$$

corresponding to the stress tensor

$$\sigma_{ij}^{(p)} = -p\delta_{ij} \tag{B.11}$$

where $p = -q > 0$ is a scalar.

- **Pressure in a moving fluid**

In a moving fluid it is evident that the above result does not hold: in general, the tangential stresses are not null and the normal component of the stress acting on a surface will depend on the orientation of the same. However, even in a moving fluid it is useful to have a scalar that, like the static pressure, gives a measure of the local uniform compression of the generic fluid element. This can be obtained by adopting the same definition of pressure valid in the static case (Eq. B.9):

$$p = -\frac{1}{3} tr[\boldsymbol{\sigma}]. \tag{B.12}$$

It can be shown that the pressure thus defined is equal to *minus* the average value of the normal component of the stress on the surface of a small sphere centred in **x**. Eq. (B.12) provides the classic definition of *pressure in a moving fluid*; naturally, this definition coincides with that of static pressure in the case of a fluid at rest.

It is worth noting that this purely mechanical definition of pressure does not imply in principle that this scalar coincides with that used in classical thermodynamics, which refers to states in thermodynamic equilibrium: this coincidence would strictly hold only for a uniform fluid at rest. However, there is experimental evidence that equilibrium thermodynamics is applicable with good approximation also to non-equilibrium states in non-uniform fluids typical of the most varied applications of fluid dynamics (although there are some exceptions). This authorises the use of the pressure defined by Eq. (B.12) in thermodynamic relations and in equations of state (such as, for example, in that for a perfect gas, Eq. (3.2)) considering each small fluid element as a thermodynamic system at equilibrium.

In the case of a moving fluid, the tensor **d** that appears in the decomposition

$$\sigma_{ij} = -p\delta_{ij} + d_{ij} \tag{B.13}$$

is not null, unlike the static case, and represents the *anisotropic* contribution to the stress tensor, comprising both tangential stresses and zero-sum diagonal elements (since $tr[\mathbf{d}] = 0$). In Sect. 2.4, the relationship Eq. (2.14) between **d** and the tensor **e** is introduced (see Appendix C) in the case of Newtonian fluids.

Appendix C
Relative Motion of the Fluid Near a Point

In this appendix, the relative motion of the fluid near a point is analysed. This allows us to study the deformation and rotation of a fluid element in a small time interval. In this context, the relationship linking the velocity field to the vorticity field is derived.

On various occasions, in the text, to fix ideas, an infinitesimal fluid element of cubic or spherical shape has been considered. However, it has also been emphasised that such a fluid element undergoes not only a displacement due to its velocity but, possibly, also a deformation and a rotation due to the even small differences in the velocity field at different points of the element itself. The analysis of the deformation and rotation of a fluid element presented in this appendix is relevant for two aspects discussed in the text. Firstly, a geometric interpretation of the tensor defined by Eq. (2.15) will be provided, thus clarifying the relationship between the anisotropic part of the stress tensor d_{ij} and the deformation of the fluid element in Newtonian fluids. Furthermore, the formula Eq. (4.18) linking the velocity field to the vorticity field will be demonstrated.

We want to study the geometric characteristics of the velocity field $\mathbf{u}$ in the vicinity of a point $\mathbf{x}$. Let $\mathbf{r}$ be the infinitesimal vector that identifies the position relative to $\mathbf{x}$ (Fig. C.1).

The velocity at point $\mathbf{x} + \mathbf{r}$ will be given by

$$\mathbf{u}(\mathbf{x} + \mathbf{r}, t) = \mathbf{u}(\mathbf{x}, t) + \delta \mathbf{u}$$

where the *relative velocity* of the fluid with respect to point $\mathbf{x}$ is:

$$\delta u_i = \sum_j \frac{\partial u_i}{\partial x_j} r_j = \sum_j \left(e_{ij} + \xi_{ij} \right) r_j = \delta u_i^{(s)} + \delta u_i^{(a)}. \tag{C.1}$$

The tensor $\partial u_i / \partial x_j$ is decomposed into the symmetric tensor e_{ij}, called the *rate-of-strain*, and the antisymmetric tensor ξ_{ij},

S. Pierini, *Oceanic and Atmospheric Fluid Dynamics*, UNITEXT for Physics,
https://doi.org/10.1007/978-3-031-77991-6

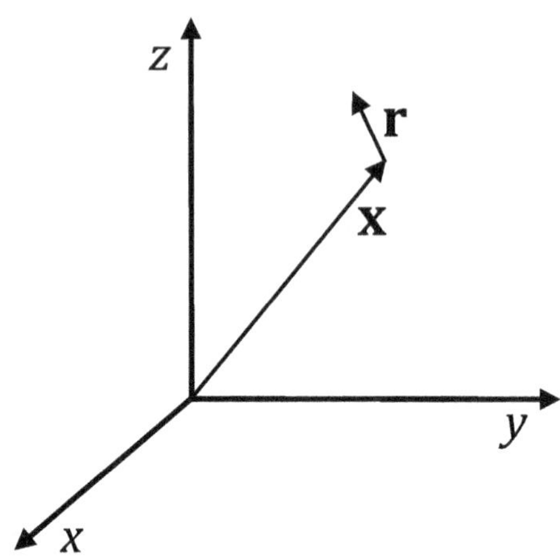

Fig. C.1 Definition of vector $\mathbf{r}$

$$e_{ij} = \frac{1}{2}\left(\frac{\partial u_i}{\partial x_j} + \frac{\partial u_j}{\partial x_i}\right); \; \xi_{ij} = \frac{1}{2}\left(\frac{\partial u_i}{\partial x_j} - \frac{\partial u_j}{\partial x_i}\right) \tag{C.2}$$

while the corresponding relative velocities are indicated respectively with $\delta u_i^{(s)}$ and $\delta u_i^{(a)}$. We now move on to consider the two terms separately.

- **Symmetric part of the relative velocity: deformation**

We start with the symmetric part which, as we will see, represents a deformation of the fluid element (that in this treatment will be without change of volume, since we will limit ourselves to considering incompressible fluids). It is possible to define a scalar (thus invariant under rotation of the coordinate axes) through the following quadratic form:

$$\Phi(\mathbf{x}, \mathbf{r}, t) = \frac{1}{2}\sum_{k,\ell} r_k r_\ell e_{k\ell}. \tag{C.3}$$

From Eq. (C.1) we have

$$\delta u_i^{(s)} = \frac{\partial \Phi}{\partial r_i}; \tag{C.4}$$

indeed:

$$\frac{\partial \Phi}{\partial r_i} = \frac{1}{2}\sum_{\ell} r_\ell e_{i\ell} + \frac{1}{2}\sum_{k} r_k e_{ki} = \sum_{j} r_j e_{ij}$$

(in the calculation, the symmetry of e_{ij} is exploited). Eq. (C.4) implies that the relative speed $\delta u_i^{(s)}$ is normal to the generic surface $\Phi = const$, with Φ a function of $\mathbf{r}$ with

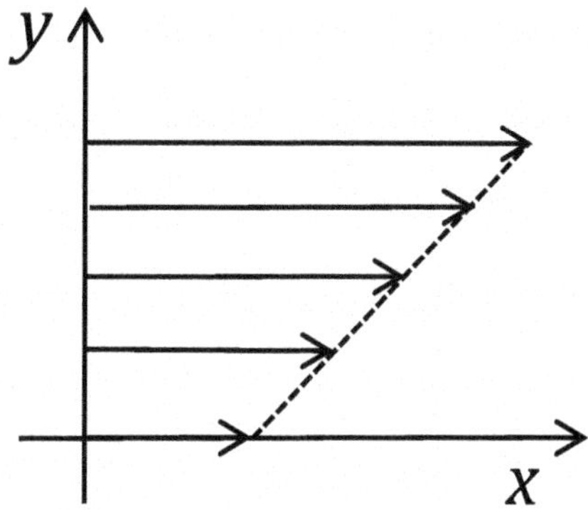

Fig. C.2 Example of shear flow

$\mathbf{x}$ fixed. If we switch to the coordinate system $\mathbf{x}'$ coinciding with the principal axes of the tensor e_{ij} (which exist because it is diagonalisable as it is symmetric), the quadratic form Eq. (C.3) is reduced to the canonical form:

$$\Phi = \frac{1}{2}\left(ar_1'^2 + br_2'^2 + cr_3'^2\right),$$

where a, b e c are the eigenvalues of e_{ij} and, therefore, the diagonal components of e'_{ij}. In that coordinate system, therefore, we will have:

$$\delta\mathbf{u}^{(s)\prime} = \left(ar_1', br_2', cr_3'\right). \tag{C.5}$$

Consequently, any material line in the vicinity of $\mathbf{x}$ which, for example, is initially parallel to the axis $\mathbf{r}_1'$, will maintain the same orientation and will change its own length at a speed equal to a (and similarly for the other two directions). Therefore, $\delta u_i^{(s)}$ will transform an initially spherical fluid element into an ellipsoid.

As an example, consider the shear flow already discussed in Sect. 4.4 (Fig. C.2):

$$\mathbf{u} = (u(y), 0, 0)$$

with $\mathbf{u}$ given, for example, by Eq. (4.20) (but now the only parameter that comes into play is the velocity shear $\partial u/\partial y$).

The symmetric tensor e_{ij} turns out to be:

$$\mathbf{e} = \begin{pmatrix} 0 & h & 0 \\ h & 0 & 0 \\ 0 & 0 & 0 \end{pmatrix}; h = \frac{1}{2}\frac{\partial u}{\partial y} > 0.$$

The relative speed associated with e_{ij} is given by Eq. (C.1):

$$\delta\mathbf{u}^{(s)} = (2r_2 h, 0, 0).$$

To reduce the relative speed in the form of Eq. (C.5), it is necessary to determine the principal axes of $\mathbf{e}$, and therefore solve the eigenvalue problem

$$\mathbf{e}\mathbf{q}^{(\ell)} = \lambda^{(\ell)}\mathbf{q}^{(\ell)}$$

with $\ell = 1, 2, 3$, where λ and $\mathbf{q}$ are the eigenvalues and eigenvectors of $\mathbf{e}$, respectively. The corresponding homogeneous system of linear algebraic equations

$$\sum_j \left(e_{ij} - \lambda^{(\ell)}\delta_{ij}\right)q_j^{(\ell)} = 0; \;\; i = 1, 2, 3 \tag{C.6}$$

is solvable if and only if

$$det(\mathbf{e} - \lambda\mathbf{I}) = 0.$$

From the corresponding third order algebraic equation

$$\begin{vmatrix} -\lambda & h & 0 \\ h & -\lambda & 0 \\ 0 & 0 & -\lambda \end{vmatrix} = -\lambda^3 + \lambda h^2 = 0$$

derive the three eigenvalues of e_{ij}:

$$\lambda^{(1)} = h = a; \; \lambda^{(2)} = -h = b; \; \lambda^{(3)} = 0 = c$$

where a, b and c are the parameters that appear in Eq. (C.5). Consequently, in the principal axes system we will have

$$e' = \begin{pmatrix} h & 0 & 0 \\ 0 & -h & 0 \\ 0 & 0 & 0 \end{pmatrix}$$

and from Eq. (C.5) we obtain:

$$\delta\mathbf{u}^{(s)\prime} = \left(hr_1', -hr_2', 0\right). \tag{C.7}$$

Note that the trace of $\mathbf{e}'$ is zero for incompressibility: $a + b + c = \sum_i e_{ii}' = \sum_i e_{ij} = \nabla \cdot \mathbf{u} = 0$.

At this point it is necessary to determine the three (normalised) eigenvectors to identify the direction of the principal axes: to do this, simply substitute the three just obtained eigenvalues into Eq. (C.6):

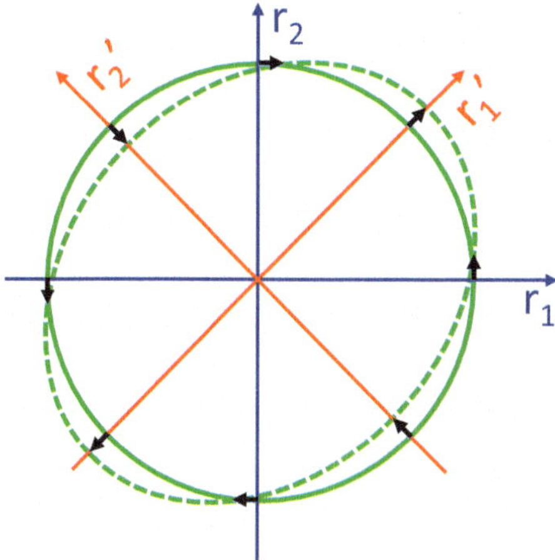

Fig. C.3 Example of pure deformation of a small fluid element with a circular section on the (x, y) plane in the shear flow here considered. The original axes are indicated in blue, the principal ones of e_{ij} in red. The arrows indicate the displacement (for a time interval dt) associated with the relative speed of pure deformation given by Eq. (C.7) at eight points on an initial circumference of radius R. The modulus of each arrow is equal to $Rhdt$. The dashed ellipsoid represents the shape that the circle would assume after the interval dt in the absence of rotation

$$\begin{cases} \ell = 1: \begin{pmatrix} -h & h & 0 \\ h & -h & 0 \\ 0 & 0 & -h \end{pmatrix} \begin{pmatrix} q_1^{(1)} \\ q_2^{(1)} \\ q_3^{(1)} \end{pmatrix} = 0 \\ \Rightarrow \mathbf{q}^{(1)} = \left(\frac{1}{\sqrt{2}}, \frac{1}{\sqrt{2}}, 0\right) \end{cases}$$

$$\begin{cases} \ell = 2: \begin{pmatrix} h & h & 0 \\ h & h & 0 \\ 0 & 0 & h \end{pmatrix} \begin{pmatrix} q_1^{(2)} \\ q_2^{(2)} \\ q_3^{(2)} \end{pmatrix} = 0 \\ \Rightarrow \mathbf{q}^{(2)} = \left(-\frac{1}{\sqrt{2}}, \frac{1}{\sqrt{2}}, 0\right) \end{cases}$$

$$\begin{cases} \ell = 3: \begin{pmatrix} 0 & h & 0 \\ h & 0 & 0 \\ 0 & 0 & 0 \end{pmatrix} \begin{pmatrix} q_1^{(3)} \\ q_2^{(3)} \\ q_3^{(3)} \end{pmatrix} = 0 \\ \Rightarrow \mathbf{q}^{(3)} = (0, 0, 1) \end{cases}$$

Therefore, the principal axes of e_{ij} form on the horizontal plane an angle of 45° with respect to the starting coordinate system. Fig. C.3 shows the deformation that a small fluid element with a circular section on the horizontal plane undergoes in a small time interval dt in the considered shear flow. This deformation rate, for Newtonian fluids, is inextricably linked to molecular viscosity, as expressed by the constitutive relationship Eq. (2.14).

- **Antisymmetric part of the relative velocity: rotation and vorticity**

The tensor ξ_{ij}, being antisymmetric, can be expressed in a completely general form as

$$\xi_{ij} = -\frac{1}{2}\sum_k \varepsilon_{ijk}\omega_k$$

where $\boldsymbol{\omega}$ is a vector to be determined based on the matrix elements of ξ (the Levi–Civita tensor ε_{ijk} was defined in Appendix B). Thanks to this expression we can rewrite $\delta u_i^{(a)}$ as follows:

$$\delta u_i^{(a)} = \sum_j \xi_{ij} r_j = -\frac{1}{2}\sum_{j,k} \varepsilon_{ijk} r_j \omega_k \equiv \frac{1}{2}(\boldsymbol{\omega} \times \mathbf{r})_i. \tag{C.8}$$

Therefore,

$$\delta \mathbf{u}^{(a)} = \frac{1}{2}\boldsymbol{\omega} \times \mathbf{r}. \tag{C.9}$$

Equation (C.9) shows that the antisymmetric part of the relative velocity represents the velocity produced in $\mathbf{x}+\mathbf{r}$ by a rigid body rotation around $\mathbf{x}$ with angular velocity $\boldsymbol{\Omega} = \boldsymbol{\omega}/2$, as is well known from rigid-body dynamics. Therefore, $\boldsymbol{\omega}$ is precisely the vorticity defined in Sect. 4.4.

At this point, we are able to determine the relationship that links the vorticity field to that of velocity. From the definition of ξ_{ij} (Eq. C.2), Eq. (C.8) provides (partial derivatives are indicated with compact notation)

$$\begin{pmatrix} 0 & u_y - v_x & u_z - w_x \\ v_x - u_y & 0 & v_z - w_y \\ w_x - u_z & w_y - v_z & 0 \end{pmatrix} = \begin{pmatrix} 0 & -\omega_3 & \omega_2 \\ \omega_3 & 0 & -\omega_1 \\ -\omega_2 & \omega_1 & 0 \end{pmatrix}$$

from which:

$$\omega_1 = w_y - v_z;\ \omega_2 = u_z - w_x;\ \omega_3 = v_x - u_y$$

or:

$$\boldsymbol{\omega} = \nabla \times \mathbf{u}. \tag{C.10}$$

With this, Eq. (4.18) anticipated in Chap. 4 has been demonstrated.

Referring to the example of the shear flow considered above, we have

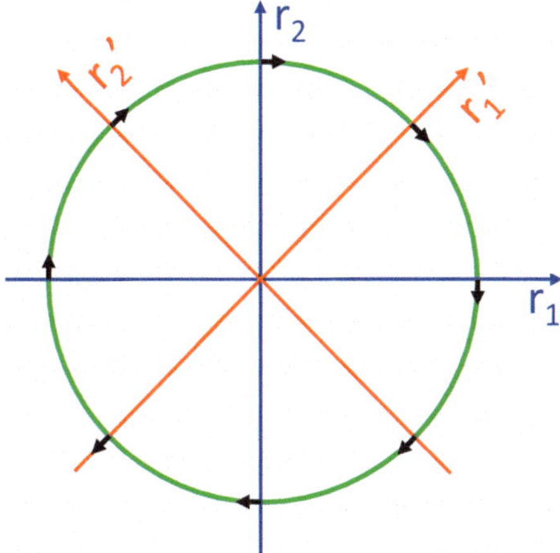

Fig. C.4 Example of pure rotation of a small fluid element with a circular cross-section on the (x, y) plane in the shear flow here considered. The original axes are indicated in blue, the principal ones of e_{ij} in red. The arrows indicate the displacement (for a time interval dt) associated with the relative velocity of pure rotation, whose magnitude is given by Eq. (C.11), at eight points on an initial circumference of radius R. The magnitude of each arrow is equal to $Rhdt$; the corresponding rotation is clockwise

$$\boldsymbol{\omega} = \left(0, 0, -\frac{\partial u}{\partial y}\right)$$

and from C.9 we obtain

$$\left|\delta \mathbf{u}^{(a)}\right| = \frac{1}{2} r |\omega| = rh. \tag{C.11}$$

Figure C.4 shows the rotation induced by the antisymmetric part of the relative velocity associated with the vorticity of a small fluid element with a circular cross-section on the horizontal plane in a small time interval dt

- **Total velocity**

In conclusion, from Eqs. (C.1, C.3, C.4 and C.8), the fluid velocity in the vicinity of point $\mathbf{x}$ is obtained:

$$u_i(\mathbf{x} + \mathbf{r}, t) = u_i(\mathbf{x}, t) + \frac{1}{2} \sum_{j,k} \left[\frac{\partial}{\partial r_i} \left(r_j r_k e_{jk} \right) - \varepsilon_{ijk} r_j \omega_k \right].$$

Figure C.5 shows the total deformation in the time interval dt of the fluid element with a circular cross-section on the horizontal plane due to the combined effect of deformation (Fig. C.3) and rotation (Fig. C.4).

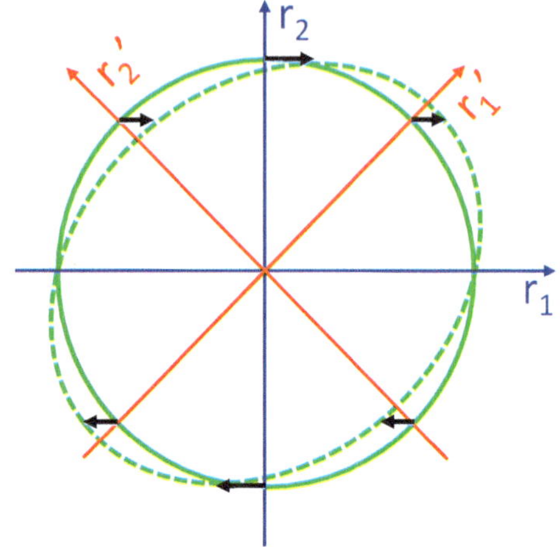

Fig. C.5 Total change, in a time interval dt, of a small fluid element with a circular cross-section on the (x, y) plane in the shear flow here considered. The arrows indicate the total displacement $\delta \boldsymbol{u} dt = \left(\delta \boldsymbol{u}^{(s)} + \delta \boldsymbol{u}^{(a)}\right) dt$

Appendix D
Kelvin's Theorem

In this appendix, two of the most powerful and elegant theorems of fluid dynamics valid for a perfect and barotropic fluid are demonstrated: Kelvin's circulation theorem and the material character of vorticity tubes and lines.

- **Kelvin's circulation theorem**

The circulation theorem is due to the British mathematician and physicist William Thompson (Lord Kelvin, 1824–1907). Let C be the circulation along a generic closed curve γ, as defined by Eq. (4.25) or equivalently by Eq. (4.26). Let us calculate the Lagrangian derivative of C, therefore considering γ as a material curve that moves with the fluid:

$$\frac{dC}{dt} = \oint_{\gamma} \frac{d}{dt}(\mathbf{u} \cdot d\boldsymbol{\ell}) = \oint_{\gamma} \frac{d\mathbf{u}}{dt} \cdot d\boldsymbol{\ell} + \oint_{\gamma} \mathbf{u} \cdot \frac{d(d\boldsymbol{\ell})}{dt}. \tag{D.1}$$

In the second line integral we have:

$$\frac{d(d\boldsymbol{\ell})}{dt} = d\mathbf{u}. \tag{D.2}$$

Indeed, as shown in Fig. D.1, we have:

$$d\boldsymbol{\ell}' = -\mathbf{u}_1 dt + d\boldsymbol{\ell} + \mathbf{u}_2 dt$$

and therefore:

$$\frac{d(d\boldsymbol{\ell})}{dt} = \frac{d\boldsymbol{\ell}' - d\boldsymbol{\ell}}{dt} = \mathbf{u}_2 - \mathbf{u}_1 = d\mathbf{u}.$$

Equation (D.2) implies

S. Pierini, *Oceanic and Atmospheric Fluid Dynamics*, UNITEXT for Physics,
https://doi.org/10.1007/978-3-031-77991-6

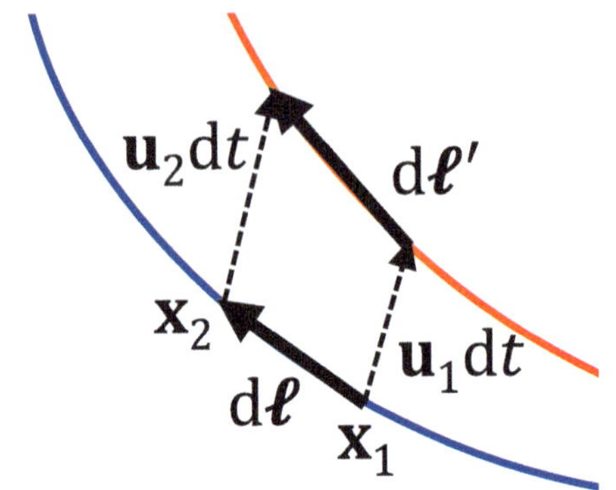

Fig. D.1 Blue line (red): portion of the curve γ at time t (t + dt). $\mathbf{u}_1$ and $\mathbf{u}_2$ are the velocities at time t related to points $\mathbf{x}_1$ and $\mathbf{x}_2 = \mathbf{x}_1 + \mathbf{d}\boldsymbol{\ell}$, respectively

$$\oint_{\gamma} \mathbf{u} \cdot \frac{d(d\boldsymbol{\ell})}{dt} = \oint_{\gamma} \mathbf{u} \cdot d\mathbf{u} = \frac{1}{2}\oint_{\gamma} d\left(|\mathbf{u}|^2\right) = 0 \tag{D.3}$$

(the last integral is null because it is applied to an exact differential along a closed curve). Therefore, thanks to Eqs. (D.3 and D.1) provides:

$$\frac{dC}{dt} = \oint_{\gamma} \frac{d\mathbf{u}}{dt} \cdot d\boldsymbol{\ell}. \tag{D.4}$$

Until now, we have referred to purely kinematic considerations; now let us consider the dynamics of a perfect fluid regulated by Euler's Eq. (5.6):

$$\frac{d\mathbf{u}}{dt} = \mathbf{g} - \frac{1}{\rho}\nabla p.$$

Substituting the acceleration into Eq. (D.4) and applying Stokes' theorem, we get ($\oint \mathbf{g} \cdot d\boldsymbol{\ell} = 0$, cf. Application 1 in Appendix A)

$$\begin{aligned}\frac{dC}{dt} &= -\oint_{\gamma} \frac{1}{\rho}\nabla p \cdot d\boldsymbol{\ell} = -\iint_{\Gamma} \left[\nabla \times \left(\frac{1}{\rho}\nabla p\right)\right] \cdot \mathbf{n}ds \\ &= -\iint_{\Gamma} \left[\frac{1}{\rho}\nabla \times \nabla p - \frac{\nabla\rho \times \nabla \mathrm{p}}{\rho^2}\right] \cdot \mathbf{n}ds = \iint_{\Gamma} \frac{\nabla\rho \times \nabla \mathrm{p}}{\rho^2} \cdot \mathbf{n}ds\end{aligned}$$

($\nabla \times \nabla p \equiv 0$) where Γ is the surface whose contour is γ ($\gamma = \partial\Gamma$). Note that the integrand function of the last integral is nothing but the baroclinic term on the right-hand side of Eq. (13.2) (see also Sect. 14.1). In conclusion, *in a perfect fluid in a state of barotropic motion* (including the particular case of a homogeneous and incompressible fluid, $\rho = const$), *Kelvin's circulation theorem* holds:

Fig. D.2 Vorticity tube

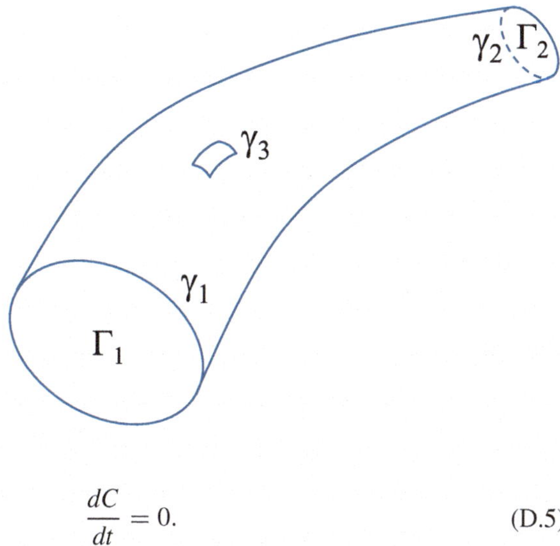

$$\frac{dC}{dt} = 0. \tag{D.5}$$

Equation (D.5) states that the circulation along a closed material curve is invariant.

- **Vorticity tubes are material volumes**

Imagine that the surface shown in Fig. D.2 at time $t = 0$ is a vorticity tube (i.e., the lateral surface is made up of vorticity lines, see also Fig. 4.10). Also imagine that, still at $t = 0$, the closed curve γ_3 lies on the vorticity tube. Since the flow through the surface enclosed by it is null, it must be $C_{\gamma_3}(t = 0) = 0$. But by Kelvin's theorem, the material evolution of γ_3 will result in $C_{\gamma_3}(t) = 0$ for every t, and this is only possible if γ_3 continues to lie on the vorticity tube. Consequently, given the arbitrariness of γ_3, the latter must itself be a material tube. In conclusion, also considering Eq. (4.29), it can be stated that, *in a perfect fluid in a state of barotropic motion, every vorticity tube is a material volume*, that maintains its intensity $\hbar$unchanged.

- **Vorticity lines are material lines**

As already discussed in Sect. 4.6, the contraction of a vorticity tube is reduced to a *vorticity line*. It is therefore obvious that, under the conditions of Kelvin's theorem, every vorticity line is a material line.

Appendix E
The Special Theory of Relativity and Relativistic Fluid Dynamics

In this appendix, we will discuss in an extremely concise manner the extension of the concepts recalled in Sect. 1.3 to motions with velocities close to that of light. This obviously goes beyond the scope of this text; however, it is considered useful and instructive to summarize these concepts to complete the Galilean principle of relativity and Newtonian dynamics examined in Chap. 1, thus also allowing a brief mention of relativistic fluid dynamics. This discussion can be a useful summary for the reader who has the basics of the theory of special relativity. For those who are not familiar with these topics (as can happen for oceanic and atmospheric science scholars), these few lines could stimulate further studies.

The *Galilean principle of relativity* (Sect. 1.3) fulfills a profound need for simplicity and elegance in the description of dynamics, as is evident not only from laboratory experiments but also from everyday life experience: the "superinertial" dynamics (Sect. 12.3) with respect to the Earth's reference frame is the same that, for example, is valid on a train that travels smoothly in rectilinear and uniform motion, whatever its speed. In Sect. 1.3 we observed how the principle of relativity applies to *inertial reference frames* (IF), each of which is linked to another by a *Galilean transformation*: if IF′ moves with respect to IF with constant velocity $\mathbf{U} = (U, 0, 0)$ (with $x' = x = 0$ at $t = 0$), then $x' = x - Ut$ and $u' = u - U$. It is important to remember that the principle of relativity manifests itself in mathematical terms in the invariance of Newton's second law under such transformations.

However, it became clear immediately after their formulation (1861–2), that Maxwell's equations in vacuum (Appendix A, Eqs. (A.3–A.6)) describing the laws of electromagnetism are *not* invariant under a Galilean transformation. Just think of the equation of electromagnetic waves: for a plane wave propagating along x this is identical to Eq. (9.44) ($A_{tt} - c^2 A_{xx} = 0$) valid for long linear and non-dispersive gravity waves; here A represents the electric field and the magnetic field and $c = (\varepsilon_0 \mu_0)^{-1/2} \cong 3 \times 10^5\,\mathrm{km\,s^{-1}}$ is the speed of light in vacuum (ε_0 and μ_0 are the dielectric constant and magnetic permeability in vacuum). It is clear that a Galilean transformation modifies the form of this equation (obviously the same is true

S. Pierini, *Oceanic and Atmospheric Fluid Dynamics*, UNITEXT for Physics,
https://doi.org/10.1007/978-3-031-77991-6

for Maxwell's equations), invalidating the Galilean principle of relativity for the laws of electromagnetism. However, this is compatible with the presence of a medium that supports the propagation of waves (such as, for example, air for sound waves). This suggested the existence of an *absolute reference frame*, privileged over all others, in which an electromagnetic signal would propagate with speed c regardless of the speed with which the signal was emitted. This hypothesis led to the introduction of the concept of *ether*, a sort of unidentified fluid at rest in the absolute reference frame, which would support the propagation of electromagnetic waves. Assuming, as seemed obvious, that Galileo's law of composition of velocities was still valid, the speed of an electromagnetic wave emitted by the Earth should have depended on the direction of propagation as a function of the speed of the Earth with respect to the ether.

In physics, an entity has meaning only if it can be observed and measured; therefore, if not the ether as a fluid, at least the absolute reference frame had to be identified experimentally: a crucial experiment was carried out for this purpose in 1887 by Albert Michelson and Edward Morley using a sophisticated interferometric system. As just observed, the speed of a light signal emitted from the Earth had to result from the composition of c and the velocity of the Earth with respect to the ether; therefore, a light signal should have propagated with different speeds in two directions perpendicular to each other. However, no difference between the two speeds was ever detected!

A series of theoretical developments conducted by physicists of the caliber of Heinrich Hertz, Henri Poincaré, Hermann Minkowski, Hendrik Lorentz, etc., culminated in the formulation of the *special* (or *restricted*) *theory* of relativity by Albert Einstein (1905a, b) which explains the (unexpected) result of that experiment, also extending the principle of relativity to electromagnetism. In short, Einstein (i) confirms the principle of relativity in IFs for all the laws of physics, therefore also for electromagnetism (thereby denying the existence of an absolute reference system, and therefore of the ether) and—on the basis of the Michelson-Morely experiment—(ii) assumes that the speed of light in a vacuum *is c in every IF*, regardless of the speed of the source or that of the observer (an axiom that is clearly absurd in classical physics).

These axioms lead to revolutionary results, confirmed with extreme precision in an infinite number of experimental verifications, restoring the laws of classical physics in the limit of velocities small with respect to c. First of all, *absolute time* (completely obvious in classical physics) is abandoned, associating a time to each IF. This means that, in the passage from IF to IF′, it is not the Galilean transformation that is satisfied but the *Lorentz transformation*, for which the time t' and the position x' in IF′ are related to t and x in IF by the relations $t' = \left(t - Ux/c^2\right)/\sqrt{1 - U^2/c^2}$ and $x' = (x - Ut)/\sqrt{1 - U^2/c^2}$; furthermore, the velocities are composed as $u' = (u - U)/\left(1 - uU/c^2\right)$. These relations imply not only the relativity of time and the simultaneity of two events, but also the relativity of length: time dilation and length contraction are absolutely real effects verified experimentally with extreme precision.

This new principle of relativity based on the Lorentz transformations (*Einstein's special principle of relativity*) is immediately verified for Maxwell's equations; in fact, these—and therefore also the wave equation—are invariant under these transformations without any modification (this was already known before the formulation of the theory of relativity). On the contrary, Newton's second equation is obviously not invariant under these new transformations, being invariant under Galileo's. To make it invariant under the Lorentz transformations, thus also guaranteeing the principle of relativity for the laws of dynamics, Einstein introduces a modified version of it that reduces to Newton's equation for velocities negligible with respect to c.

The relativistic law of dynamics has consequences of the utmost importance. Even the inertial mass is relative to the IF (while it should be noted that the electric charge is invariant); in fact, if m_0 is the *rest mass* (that is, at rest in IF), in IF′ the mass is $m = m_0/\sqrt{1 - U^2/c^2}$ (this among other things implies that *the speed of light cannot be exceeded*, since, for example, the acceleration imparted to a body by a constant force tends to zero as the speed of the body approaches that of light). Furthermore, a body with a rest mass m_0, in addition to its classical energy has also an immense additive energy expressed by the famous equation $\mathcal{E} = m_0 c^2$. This explains, among other things, the *nuclear fusion* and the stellar nucleosynthesis, as well as the *nuclear fission*, on which current nuclear power plants are based.

We conclude by mentioning the geometric formulation of the special theory of relativity. As we have seen, in classical physics, the principle of relativity requires the invariance of the equations with respect to Galilean transformations; furthermore, it is also required that the same equations are invariant with respect to the rotation of the coordinate axes (in Sect. 2.2 and Appendix B it was shown how this requirement is satisfied by the equations of fluid dynamics). Such rotation preserves the distance between two nearby points, $ds^2 = dx^2 + dy^2 + dz^2$ (this defines the Euclidean metric); on the other hand, the temporal difference between two events dt is obviously verified given the absoluteness of time. In the context of special relativity, however, ds and dt are not invariant with respect to Lorentz transformations, but the *spacetime interval* $d_R^2 = dx^2 + dy^2 + dz^2 - c^2 dt^2$ is. This suggests the introduction of a four-dimensional space, called *Minkowsky spacetime*, with a pseudo-Euclidean metric defined by d_R^2, where each point represents an event identified by the spatial coordinates $\mathbf{x} = (x, y, z)$ and a time t. In this way, *a rotation of the axes in Minkowsky spacetime represents a pure Lorentz transformation—the so-called boost—and, possibly, a rotation of the spatial axes of the IF*. Consequently, it is convenient to write the laws of physics in *covariant formalism*, that is, using four-vectors and four-tensors in such a way that the invariance under rotation in Minkowsky spacetime (*relativistic invariance*) is evident, automatically ensuring the requirements of special relativity. It is worth noting that the invariance of Maxwell's equations under Lorentz transformations requires very laborious calculations to be verified; on the other hand, if the same equations are rewritten in covariant formalism (as is possible, given what has been said so far), this requirement appears to be clearly verified without the need for any calculation.

In conclusion, by revolutionizing the concepts of time and space, Einstein extends the principle of relativity to all the laws of physics, including electromagnetism. In a subsequent, monumental generalization, Einstein (1916) formulates the *theory of general relativity* as a new theory of gravitation, which however cannot be summarized here discursively in a few lines as was done for the special theory of relativity. In general relativity, the Minkowsky -curved- spacetime (described by a pseudo-Riemannian manifold) plays a fundamental role in the formulation of the theory.

One of the main applications of the special theory of relativity concerns *high-energy physics*, where, for example, elementary particles or ions are brought—in particle accelerators—to velocities very close to that of light. In modeling the resulting collisions, relativistic versions of the equations of fluid dynamics are often used, which also take into account the electromagnetic interactions between particles (*relativistic magnetohydrodynamics*). These equations are very complex but, for the simplest version of the relativistic fluid dynamics equations (essentially the relativistic generalization of the Euler equation derived in Sect. 5.2) one can refer to Einstein (1921, cf. the paragraph "Hydrodynamical equations" and equations 52–53 therein) and to chapter XV of Landau and Lifshitz (1987). Finally, extremely interesting applications of relativistic fluid dynamics concern astrophysics, but these are the competence of general relativity.

Bibliography

Einstein, A.: Zur Elektrodynamik bewegter Körper (On the electrodynamics of moving bodies). Ann. Phys. **17**, 891–921 (1905a)
Einstein, A.: Ist die Trägheit eines Körpers von seinem Energieinhalt abhänging? (Does the inertia of a body depend upon its energy-content?). Ann. Phys. **17**, 639–641 (1905b)
Einstein, A.: Die Grundlage der allgemeinen Relativitätstheorie (The basis of the general theory of relativity). Ann. Phys. **49**, 769–822 (1916)
Einstein, A.: The Meaning of Relativity. Princeton University Press (1921)
Landau, L.D., Lifshitz, E.M.: Fluid Mechanics. Pergamon Press, Oxford (1987)

Further Recommended Readings

Batchelor, G.K.: An Introduction to Fluid Dynamics. Cambridge University Press, Cambridge (1967)
Cattaneo, C.: Introduzione alla teoria Einsteiniana della gravitazione. Veschi, Roma (1961)
Jackson, J.D.: Classical Electrodynamics. Wiley, New York (1962)
Landau, L.D., Lifshitz, E.M.: The Classical Theory of Fields. Pergamon Press, Oxford (1971)
Ougarov, V.: Théorie de la relativité restreinte. Édition Mir, Moscou (1974)
Pauli, W.: Theory of Relativity. Pergamon Press, London (1958)
Sommerfeld, A.: Electrodynamics. Academic Press, New York (1952)

S. Pierini, *Oceanic and Atmospheric Fluid Dynamics*, UNITEXT for Physics,
https://doi.org/10.1007/978-3-031-77991-6

The manufacturer's authorised representative in the EU is Springer Nature Customer Service Centre GmbH, Europaplatz 3, 69115 Heidelberg, Germany. If you have any concerns regarding our products, please contact ProductSafety@springernature.com

Printed and bound by CPI Group (UK) Ltd, Croydon, CR0 4YY
07/07/2026
02160912-0003